Synthese Library

Studies in Epistemology, Logic, Methodology, and Philosophy of Science

Volume 528

The aim of *Synthese Library* is to provide a forum for the best current work in the methodology and philosophy of science and in epistemology, all broadly understood. A wide variety of different approaches have traditionally been represented in the Library, and every effort is made to maintain this variety, not for its own sake, but because we believe that there are many fruitful and illuminating approaches to the philosophy of science and related disciplines.

Special attention is paid to methodological studies which illustrate the interplay of empirical and philosophical viewpoints and to contributions to the formal (logical, set-theoretical, mathematical, information-theoretical, decision-theoretical, etc.) methodology of empirical sciences. Likewise, the applications of logical methods to epistemology as well as philosophically and methodologically relevant studies in logic are strongly encouraged. The emphasis on logic will be tempered by interest in the psychological, historical, and sociological aspects of science. In addition to monographs *Synthese Library* publishes thematically unified anthologies and edited volumes with a well-defined topical focus inside the aim and scope of the book series. The contributions in the volumes are expected to be focused and structurally organized in accordance with the central theme(s), and should be tied together by an extensive editorial introduction or set of introductions if the volume is divided into parts. An extensive bibliography and index are mandatory.

Paula Teijeiro • Eduardo Alejandro Barrio
Editors

Metainferences in Substructural Logics

Springer

Editors
Paula Teijeiro
IIF - SADAF - CONICET
University of Buenos Aires
Buenos Aires, Argentina

Eduardo Alejandro Barrio
IIF - SADAF - CONICET
University of Buenos Aires
Buenos Aires, Argentina

ISSN 0166-6991 ISSN 2542-8292 (electronic)
Synthese Library
ISBN 978-3-032-05448-7 ISBN 978-3-032-05449-4 (eBook)
https://doi.org/10.1007/978-3-032-05449-4

This work was supported by Consejo Nacional de Investigaciones Científicas y Técnicas

This Springer imprint is published by the registered company Springer Nature Switzerland AG
The registered company address is: Gewerbestrasse 11, 6330 Cham, Switzerland

Acknowledgments

This book is the result of a collective project. Both its conception and its writing have involved a community of people. Many of the ideas, results, and approaches it contains were not only developed by the authors but also thoroughly discussed and analyzed with colleagues. Its very existence is inseparable from BA-Logic, from our Friday meetings, and from the cooperative work carried out within our research group. It is, without doubt, a reflection of shared commitment, intellectual collaboration, and care for one another.

The very formation of our research group would not have been possible without the sustained support of the Argentine State, and in particular of CONICET, whose long-term investment in scientific research has been essential to our work. The Facultad de Filosofía y Letras of Universidad de Buenos Aires has also played a fundamental role. We have learned much through the process of teaching and mentoring. Both institutions have supported our academic careers, and without them, we would not be who we are.

An essential aspect of BA-Logic is its internationalization. Over the years, we have received international funding that has enabled us to travel, strengthen international collaborations, and share our research around the world. We gratefully acknowledge the support of the British Academy (BA), the Deutscher Akademischer Austauschdienst (DAAD), the Deutsche Forschungsgemeinschaft (DFG), ECOS-SUD, and the National Endowment for the Humanities (NEH). Moreover, we want to mention that this work was supported by PLEXUS (Grant Agreement no 101086295), a Marie Sklodowska-Curie action funded by the EU under the Horizon Europe Research and Innovation Programme.

But this project has not only been made possible through institutions. It has also relied on the generosity and encouragement of individuals who have supported us from the very beginning: Romina Birman, Quentin Blomet, Pablo Cobreros, Roy Cook, Bogdan Dicher, Paul Egré, Luis Estrada, Thomas Ferguson, Saul Kripke, Volker Halbach, Hannes Leitgeb, Shay Logan. Øystein Linnebo, Carlo Nicolai, Pepe Martinez, Francesco Paoli, Elaine Pimentel, Graham Priest, Agustín Rayo, Lorenzo Rossi, and Ellie Ripley.

We also want to acknowledge every member of BA-Logic—some of whom have directly contributed to this book, and others who, even if not directly involved, are nonetheless part of this endeavor. Solidarity, friendship, and mutual support in the face of adversity have always defined our community.

We are grateful to Federico Pailos, Lucas Rosenblatt, Natalia Buacar, Ramiro Caso, Ricardo Rodriguez, Ariel Roffé, Diego Tajer, Alejandro Solares-Rojas, Aylén Bavosa Castro, Alba Cuenca, Eliana Franceschini, Juan Manuel Gagino-Di Leo, Camila Gallovich, Omar Vásquez, Daniel Grimaldi, and Ignacio Casado. We are also thankful to those who have been part of BA-Logic in the past: Noelia De Marco, Ignacio Ojea Quintana, Lavinia Picollo, and Isis Urgell.

Finally, we would like to extend our warm thanks to Otávio Bueno for his support and trust.

Building a community has always been, and will continue to be, our shared goal. We hope that this book contributes to defending the values and principles that made this project possible: solidarity, friendship, and collective commitment.

Contents

Chapter 1
Introduction

Paula Teijeiro and Eduardo Barrio

1.1 Metainferences

This is a book about old topics through newer lenses. Thus, it mischievously takes the form of a textbook. Our goal is to present classical and substructural logics, from a metainferential perspective. Bare with us.

The perspective in question is, in a nutshell, the idea that logics are made up of two types of entities. On the one hand, the standard inferences. On the other hand, metainferences, which are usually vaguely identified as "inferences between inferences." The question then is: what is it that is left out by considering only inferences, which warrants moving up a level? If metainferences are inferences between inferences, why are they not being theorized precisely by the all-encompassing theory of inferences?

The possibility of logic itself is as puzzling, if not more so, than the traditional logic puzzles such as the Liar and Curry paradoxes. In particular, it is hard to understand how it can be that logic is both useful and trivial at the same time. This triviality is usually cast as the property of being about nothing (or what might misleadingly seem more informative, about everything). We propose that the study of metainferences is the study of logic as a theory of theories, instead of logic as a theory of nothing.

For starters, one needs to have some rough understanding of what a theory is, which already can be done in a myriad ways. Theories are probably bound to comprise some sentences, truths about certain topics, but also there are several reasons to think that they could involve more than merely sentences. In particular, that they could also be made up of rules, rules that are valid in some sense specific to that particular theory:

P. Teijeiro · E. Barrio (✉)
IIF-SADAF-CONICET, University of Buenos Aires, Bulnes, Argentina

P. Teijeiro, E. A. Barrio (eds.), *Metainferences in Substructural Logics*, Synthese Library 528, https://doi.org/10.1007/978-3-032-05449-4_1

- Because they have to be complete in some sense. As Beall (2018) puts it, "(...) the task of the theorist is twofold: gather the truths about the target phenomenon, [and](...) to give as full or complete a true theory as the phenomenon allows." That is, we want to distinguish theories from other sorts of descriptions of the world, by some kind of rigor. Thus, rules tell us how to expand an initial set of sentences into an exhaustive theory.
- Because they are ever developing. On the other hand, taking theories as complete sets of sentences provides a very static and idealized point of view. Under a dynamic -more human- perspective, theories are always expanding, be it because we observe new things, or because we derive new things, or because we learn new things, etc.

There is an apparent tension above, which can be partially dissolved by saying that even though theories advance and progress through time, in each stage of their development they are idealized into some sort of partial completeness. In other words, despite there being undecided sentences, whose truth we know nothing about, we can still expect the theory to comprise—at least much of—what is entailed by the things we do know up until now. Surely logic, as a theory, provides one criterion for this completeness: if a theory comprises the sentences φ and ψ, then it has to also comprise $\varphi \wedge \psi$. Whether all other rules are in fact reducible to logical ones or not is a bit more controversial. In any case, there is certainly a wider variety of proposals on the inferentialist spectrum than on a reductionist one, and so there is a wide consensus on the importance of considering non-logical rules as well, whatever their ultimate status turns out to be.

But if a theory is made of rules besides truths, it is reasonable to expect its set of rules to also be complete in some sense, just as we ask the set of truths to be. In turn, logic, since it is a theory of theories, has to regulate not only the relations between sentences of those particular theories, but also their rules. Hence, whereas inferential rules tell us how to logically expand the class of accepted sentences, metainferences constrain the way the class of rules is extended.

Take, as Beall (2018) suggests, a theory of knowledge as an example, and let us build a toy illustration to show where metainferences come into play. It may start with some true statements, say α, β and $K\gamma$, where K is a unary knowledge operator. On top of that, it may have a rule of Factivity, which goes from $K\varphi$ to φ (instead of a Factivity axiom, which is also a possibility). The complete theory, then, will be an expansion that comprises some knowledge-authorized sentences, such as γ, and some logically-authorized ones, such as $\alpha \wedge \beta$. But maybe logic should also recommend adding the sentence $K\delta \rightarrow \delta$, independently of the fact that the theory still remains silent on whether δ or $\neg\delta$. This sentence, however, is not a consequence of the original members of the theory, but a consequence of its K-rule, by a metainference sometimes known as the Deduction Theorem: if $K\delta$ implies δ, then $K\delta \rightarrow \delta$ must be a theorem. Thus, metainferences can be seen as playing a crucial role in the expansion of a theory.

To give a detailed account of the structure of theories and their building mechanics goes beyond the scope of this book. Our hope is that the tools that can be found here can be fruitful in generating different takes on the matter. Let us say, however, that it is key for us that the following possibility remains open: the set of sentences of a theory is not necessarily closed. The standard way to understand the (even partial) completeness of theories is through closure operators. That is, if you take a set of sentences, and expand it with their consequences, and then their consequences, and so on, at the end you have applied a closure operator to the original set: you only added things (increasingness), you reached a set that cannot get bigger (idempotence) and if you had applied it to any subset, you would have obtained a subset of what you did (monotonicity). These three properties respectively correspond to the reflexivity, transitivity, and monotonicity of Tarskian consequence relations (Tarski, 1930).

But, as we will see, it might be interesting to explore non-transitive (Cobreros et al., 2014), non-monotonic and non-reflexive (French, 2022) sets of sentences, and thus the link between inferences and closure operators tumbles (Ripley, 2015 even shows that also non-contractive approaches face a similar problem). For example, there are non-transitive semantic theories, such that a paradoxical sentence λ and its also paradoxical negation $\neg\lambda$ are both part of the set of (weakly) asserted sentences. But even though the rule of Ex Falso is also part of those theories, said set of sentences is nonetheless not trivial. In other words, this set is not closed under Ex Falso, even though the rule is in fact licensed by the theory.[1] This can seem puzzling and make us wonder what are these rules good for: by making metainferences important, we have rendered inferences pointless.

Failing the alleged necessary connection between theories and closure operators is in fact one of the charges Williamson (2018) raises against that class of non-classical logics, and probably one of the more serious ones. However, there are many other ways of explaining the connection between logic and theory-building that challenge that link (see for instance Ripley, 2013; Weber, 2016; Logan, 2020; Logan, 2024).

Notice that the sort of theoretical ascension we are pointing at is anything but new. After the development of some non-classical logics which have no theorems—such as Kleene Logic (Kleene, 1952)—or have the same theorems as Classical Logic—such as the logic of Paradox (Asenjo, 1966)—it was concluded that theoremhood was insufficient to characterize a logic. This prompted an urge to characterize logics as sets of inferences, instead of merely sets of tautologies. The incorporation of metainferences into play comes from similar considerations.

[1] For a generalization of this point, see some work of BA-Logic. In particular, Barrio et al. (2020) shows a hierarchy of $\mathbb{ST}$ logics, with an infinite sequence of consequence relations holding between increasingly complex relata: formulas, inferences, metainferences, and so on. This sequence is not closed by transitivity. More details appear in the book of Pailos and Da Ré (2023). Szmuc shows non-transitive counterparts of every (non-trivial) Tarskian logic in Szmuc (2024). And Barrio, Fiore, and Pailos present logics that are different from classical logic at the level of inferences, but recover some important aspects of classical logic at every metainferential level (Barrio et al., 2024).

However, it is important to avoid the mistake of believing that we are defending a new ascent merely by inertia.

There is a crucial difference between going from theorems to inferences and going from inferences to metainferences. The concept of an inference has enough merit—thanks to its pre-theoretical importance and to its widespread scientific usage–to ground its incorporation into any theory without causing much concern. This is not the case with the concept of metainference, which has neither a folk understanding, nor such an established formulation as a technical tool. Thus, one cannot merely posit the existence of logics which coincide in the inferences they validate but differ in their metainferences. In order to have the ability to distinguish them, one has to already buy into the idea that metainferences are a theoretically meaningful concept.

However, logics without theorems—or with the same theorems as Classical logic but different valid inferences—were accepted by the community precisely because they enabled us to formulate theories that are different from those that we can formulate over Classical logic. Similarly, logics with the same inferences as Classical logic, but different metainferences allow us to formulate theories different from those that we can formulate over Classical logic. This is not to say that only non-classical logicians should be interested in this perspective. On the contrary, we believe that a metainferential formalization is worthy of consideration even for a fully classical logician, as can be illustrated by the example above.

Before moving on to characterize which metainferences are preeminent for us here, let us briefly pause to clarify a possible misunderstanding. Hunter (1971) characterizes the metatheory of logic as a theory about sentences used to express the truths of logic. This theory is, according to him, worthy of formalizing, but that formalization occurs outside of the language of the logic being theorized about. One may venture, then, that understanding logic as a set of inferences and metainferences is just a way to deal with both the logic and the metalogic in a unified system. This is certainly a possibility, and it is in fact the take that seems to be behind some works on the topic. However, we consider this take to be somewhat problematic for at least two reasons:

(a) Metatheory of logic usually concerns our means to evaluate inferences. Some metatheoretical facts are about systems of proof—such as Cut-elimination theorems—some are about models—like Compactness—and some are about their coincidence—such as Soundness and Completeness. On the contrary, metainferences are merely relations between inferences themselves, which at best can be seen as only a small part of the whole metatheory.
(b) A metatheory, inasmuch as it is a theory-of, is supposed to be completely determined by its object. Hence, the metalogic has to be completely determined by the logic it is about. One way to think metainferences in these terms is by stating that they are closure principles of consequence relations (for instance, in Cobreros et al., 2014). Nonetheless, this is not strictly our sense. If metainferences were to be understood as absolutely determined by the valid inferences, there would be not much gain in adding them to the theory, since one can simply abstract or read them off from it.

Summing up, there is a sense in which metalogic is a richer discipline, whereas metainferences are as abstract and shallow as inferences. In some other sense, metalogic is about logic, whereas we are claiming that metainferences are part of logic. Hence, metainferences are both too much and also not enough to be thought as a formalization of the metatheory of inferences.

1.2 Structural or Substructural

So, we now have that logic is a high-order theory, in the sense that it comprises two levels: the inferential logic, which accounts for the valid inferences regulating relations between sentences, and the metainferential logic, which accounts for the valid metainferences regulating relations between inferences. Thus, provided that metainferences are interesting in themselves, how do we characterize the valid ones? First of all, we need to provide technical tools to characterize Classical Logic. But also, just as non-classical logics arose from questioning the validity of certain classical inferences, the ground is laid to raise doubts about the validity of classical metainferences. Here, we are faced with two options: to question metainferences involving some specific piece of logical vocabulary or to question, more generally, the structural metainferences, which regulate the entire language. In this book we are going to consider only the second type of non-classical logics, the one we call the class of substructural logics.

Structural metainferences are closely related to what Tarski (1930) and Gentzen (1935) considered to be the tenets of logical consequence. In terms of the former, as we already mentioned, these are Reflexivity, Monotonicity, and Transitivity, which in terms of the latter, are codified by the rules of Identity, Weakening, and Cut, respectively, and are also joined by Contraction. To question these is to go against what many people consider the most basic, uncontroversial aspects of logic, mainly for the closure-related reasons we saw above. In order to remain as neutral as possible regarding which of them are actually essential—if any—we are going to provide a very minimal understanding of what an inferential logic is, requiring only closure under uniform substitution. This property guarantees that there is at least a bedrock of formality that all of these theories stand on, so they can claim the title of logic.

The study of substructural logics as those which challenge these rules long predates the metainferential perspective (see Restall, 2002; Paoli, 2013; and Negri & Von Plato, 2008, for example). However, there is a constellation of uses of that term, some of which align with ours and some of which do not. Let us then, differentiate our use from that of a substructural calculus. It is in this sense that it is sometimes said that logics being substructural is a matter of presentation (see for example Indrzejczak, 2021). But what is meant by that is that the same inferential logic can be characterized by calculi which have or lack certain structural rule. In contrast, according to our perspective, by making metainferences constitutive of the logic, substructurality ceases to be a matter of presentation: they either belong to the logic, or they do not.

This does not mean that proof systems will not play a privileged role in this book. In particular, our preferred tool—although not the only one that is going to be used– is sequent calculi. The inferential part of the logic is, as customary, determined by the provable sequents of the system. Sequent to sequent rules, on the other hand, can be regarded as codifying metainferences in the most explicit way possible.

It is worth noting, however, that there are two senses of substructurality of a calculus on the rounds: a stronger and a weaker one. According to the former, in a substructural calculus there is some structural rule which is not admissible—that is, it does not preserve provability in the calculus. According to the latter, there has to be some structural rule which is not derivable—that is, it fails to preserve provability in some extension of that calculus with possibly new, non-logical rules. We say that the second sense is weaker because it does not require that the calculus has a counterexample to the rule in question in order to count as substructural, where a counterexample in this case is an instance of the rule such that its premises are provable and its conclusion is not.

Even though it is logics and not their calculi which we primarily call substructural, we can nonetheless establish a link between the two uses of the term, and state that it is the weaker sense of substructurality of a calculus the one which has preeminence for us. The reason is that metainferences regulate the rules that might be part of any theory we want; they are not merely constraints on logical validities but on validities on a broader understanding of the term. Thus, when we take a sequent calculus to determine a logic, the metainferential level is going to be determined by its derivable rules. In some cases, the rules in question will remain admissible, whereas in others, the language will have the means for expressing a counterexample which witnesses the substructurality—either because we enrich the basic propositional vocabulary, or because such counterexamples are very easy to come by (as in the case of the non-reflexive logics discussed).

The kick-start is then Gentzen's **LK**, which can be seen as a complete formalization of Classical propositional logic. Thus, the pairing of **LK**-provable sequents with classical inferences and of **LK**-derivable inferences with classical metainferences comprises a good starting point. Much will follow just by subtracting structural rules from that calculus, although, in some cases there will be added operational rules. There are, however, some exceptions, since for some of the non-transitive and non-reflexive logics that we will consider, there are no sequent calculi developed. Luckily, there are nice characterization results for these kind of systems in terms of three-valued valuations, which are not only a very easy tool to use, but also can be seen as providing a semantics in a quite robust way. Thus, we will sometimes appeal to these valuations, and in particular to the concept of "Local Validity," in order to characterize the valid metainferences of some logics.[2]

[2] For more details about local and global validity, see Teijeiro (2021).

One very interesting aspect of thinking about substructurality in metainferential terms is that it allows us to have bi-dimensional logics, where the inferential part and the metainferential part behave quite differently. Of course, in order to make this sort of cross-level assessments, one needs to provide the tools to compare them, since each of them deals with a different kind of entity. This is why we will often resort to translations, which "flatten" metainferences into inferences, that is, functions which for instance turn an inference from φ and ψ to γ into the formula $(\varphi \wedge \psi) \to \gamma$, and then render relations between inferences as relations between formulas.

1.3 Structure and Style

This work is the result of an ensemble cast, with most authors belonging to the Buenos Aires Logic Group, and some others that are also part of PLEXUS, an international collaboration project dedicated to substructural logics. Thus, the author's voice will sound like a chorus', that we are hoping stays more or less in tune.

Despite the philosophical agenda driving it, it is hardly a book on philosophy of logic. Nevertheless, we tried to make the exposition as clear and gentle as possible, giving intuitive ground for the ideas we develop, and providing exercises to help incorporating them. Although we are going to get (very) technical, the book starts from scratch, so we are not assuming any mathematical or logical knowledge. Having said that, it is best for the reader to have at least one elementary course on logic, since the depth and complexity of at least some parts of the text are probably too much as a first approximation.

Chapters 2–5 are each dedicated to a different family of metainferential logics. This means that, even though they are, in a certain sense, self-contained, they all heavily presuppose concepts and results from chapters before them—to avoid constant repetitions. They also share a certain general two-part structure: an inferential section, and a metainferential one.

The basic framework is laid out in the first chapter, written by Bruno Da Ré, Damián Szmuc and Paula Teijeiro, which constitutes a presentation of Classical propositional Logic, from a metainferential perspective. We show Soundness and Completeness of the calculus and the semantics, both for the inferential and the metainferential level. We provide proofs of some important metatheoretical results, such as Compactness and Finiteness. The crucial result will be the Cut-elimination theorem, not only because of its valuable consequences for **LK**, but also because it illustrates how we can get the non-transitive logics in the following chapter.

The two subsequent chapters delve into the logics that started the whole metainferential outlook, and thus are given preeminent space. Chapter 3 is written by Agustina Borzi and Camillo Fiore, and deals with non-transitive logics that fall under what we will call the ST-family. Chapter 4, on the other hand, is written by Joaquín S. Toranzo Calderón and Martina Zirattu, and tackles the dual but less explored class of non-reflexive logics belonging to the TS-family. Their structure is quite parallel, and it also shares some similarities with what came before. There are also Soundness and Completeness results for several related logics, both at the inferential and the metainferential level. Finally, we can find some double-edged results that bridge these logics with well-known structural non-classical logics. As it turns out, if we translate inferences into formulas—as we mentioned above–, we can prove the equivalence between the valid metainferences of $\mathbb{ST}$ and $\mathbb{TS}$—the most prominent members of each family—and the valid inferences of $\mathbb{LP}$ and $\mathbb{K}3$, respectively. These results are both illuminating and worrisome, since they provide a better understanding of these less familiar logics, but at the same time they make us wonder whether they are worth the trouble (we believe they are).

Chapter 5 is written by Miguel Álvarez Lisboa, Yll Buzoku and Mariela Rubin, and it has the complex task of presenting logics that forfeit Weakening (and) or Contraction. This class is much more heterogeneous than the previous ones, not only because there are two structural rules involved, but also because these have greater impact on the language than the others: when either one of these rules is absent, each of the logical connectives splits in (at least) two, non-equivalent versions. However, given that these logics also have a richer, older tradition, and thus there is a lot more already said about them in other books on substructural logics, we decided to only present a specific selection, having the last chapter in mind. It is, nonetheless, considerably longer than the others.

It features some metatheoretical results in common with Chap. 2, such as a Cut-elimination theorem, but we do not provide the same thorough Soundness and Completeness characterizations. As we said, the easiest way to tackle substructural logics is proof-theoretically, and when it comes to semantics, we tried in general to privilege valuational presentations, which are the simplest, neatest way to provide a semantic interpretation for non-classical logics. However, things turn a bit muddier for logics without Weakening and Contraction, since valuations are, for the most part, not available. We decided then to introduce a section about phase semantics. They are considerably harder and much less intuitive than, for instance, Kleene semantics, but we believe the reader will benefit from this tool anyway. We show inferential Soundness and Completeness, since the metainferential part is, as far as we know, still very much unexplored.

Finally, Chap. 6 is written by Edson Bezerra and Jonathan Erenfryd, and it is the only part of this book completely devoted to philosophy of logic. On the one hand, you can find there some interpretations of the failures of the structural metainferences. On the other, we present one of the major applications that have driven research in this area, which is the formalization of theories which have self-reference and semantical predicates, and are thus prone to paradoxes. In particular, the focus will be in the famous Liar paradox, and the more recent Validity Paradox. Substructural logics, as we will see, are in an interesting position to deal with these sorts of theories, since they can provide a uniform treatment of these types of antinomies.

Bibliography

Asenjo, F. G. (1966). A calculus of antinomies. *Notre Dame Journal of Formal Logic, 7*(1), 103–105.

Barrio, E. A., Pailos, F., & Szmuc, D. (2020). A hierarchy of classical and paraconsistent logics. *Journal of Philosophical Logic, 49*, 93–120.

Barrio, E. A., Fiore, C., & Pailos, F. (2024). Meta-classical non-classical logics. *The Review of Symbolic Logic*, 1–26.

Beall, J. C. (2018). The simple argument for subclassical logic. *Philosophical Issues, 28*(1), 30–54.

Cobreros, P., Egré, P., Ripley, E., & Van Rooij, R. (2014). Reaching transparent truth. *Mind, 122*(488), 841–866.

French, R. (2022). Metasequents and tetravaluations. *Journal of Philosophical Logic, 51*(6), 1453–1476.

Gentzen, G. (1935). Untersuchungen über das logische schließen. *Mathematische Zeitschrift (english translation in Gentzen 1969), 39*(2), 176–210, 405–431.

Hunter, G. (1971). *Metalogic: An introduction to the metatheory of standard first order logic*. University of California Press.

Indrzejczak, A. (2021). *Sequents and trees*. Birkhäuser.

Kleene, S. C. (1952). *Introduction to metamathematics*. North-Holland.

Logan, S. A. (2020). Deep fried logic. *Erkenntnis, 87*(1), 257–286.

Logan, S. A. (2024). *Relevance logic*. Cambridge University Press.

Negri, S., & Von Plato, J. (2008). *Structural proof theory*. Cambridge University Press.

Pailos, F., & Da Ré, B. (2023). *Metainferential Logics*. Springer.

Paoli, F. (2013). *Substructural logics: A primer* (Vol. 13). Springer Science & Business Media.

Restall, G. (2002). *An introduction to substructural logics*. Routledge.

Ripley, E. (2013). Revising up: Strengthening classical logic in the face of paradox. *Philosophers' Imprint, 13*(5), 1–13.

Ripley, E. (2015). Contraction and closure. *Thought: A Journal of Philosophy, 4*(2), 131–138.

Szmuc, D. E. (2024). Non-transitive counterparts of every tarskian logic. *Analysis, 84*(2), 320–326.

Tarski, A. (1930). On some fundamental concepts of metamathematics. In *Logic semantics, metamathematics* (pp. 30–37). Hackett Publishing Company.

Teijeiro, P. (2021). Strength and stability. *Análisis Filosófico, 41*(2), 337–349.

Weber, Z. (2016). On closure and truth in substructural theories of truth. *Synthese, 199*(Suppl 3), 725–739.

Williamson, T. (2018). Alternative logics and applied mathmtics. *Philosophical Issues, 28*(1), 399–424.

Chapter 2
Classical Logic

Bruno Da Ré, Damián Szmuc, and Paula Teijeiro

Abstract In this chapter we will introduce Classical Logic, from both a valuational and a proof-theoretical point of view. We will start by defining some basic tools pertaining to both these frameworks, in a general fashion, so they will be useful for subsequent chapters too. Then, we will present Classical inferential Logic, through the sequent calculus **LK** and boolean bivaluations. We will then prove various metatheorical results about them, such as Soundness and Completeness, Compactness and Decidability. In the second half of the chapter, we will tackle the metainferential side, presenting different concepts of metainferential validity—both semantical and proof-theoretical. We will also show some metainferential versions of the metatheorems previously proved for the inferential part. We end the chapter by proving Cut-elimination and discussing some of its consequences.

2.1 Basic Notions

In this section, we will introduce some of the key concepts we will be dealing with throughout the book. Most of the logical phenomena we review in this book can already be observed at the propositional level. Thus, in what follows, we work with propositional systems (although this will require some rephrasing when it comes to applications, as we will see in due course).

Let us, then, detail the propositional tools. The language $\mathcal{L}$ contains a denumerable set Var of propositional variables $p, q, r, \ldots$ and a set of logical connectives $\neg, \wedge, \vee, \rightarrow$, intended to represent negation, conjunction, disjunction, and the material conditional, respectively. $FOR(\mathcal{L})$ is the set of well-formed formulas of $\mathcal{L}$, as usually construed. When talking about this language, we will employ a metalanguage enriched with lower case Greek letters $\varphi, \psi, \chi, \ldots$, which work as metavariables for formulas, upper case Greek letters $\Gamma, \Delta, \Theta, \ldots$, which work as

B. Da Ré · D. Szmuc · P. Teijeiro (✉)
UBA/IIF-SADAF-CONICET, Buenos Aires, Argentina

P. Teijeiro, E. A. Barrio (eds.), *Metainferences in Substructural Logics*, Synthese Library 528, https://doi.org/10.1007/978-3-032-05449-4_2

metavariables for multisets of formulas, as well as some set-theoretic symbols with their standard meaning—as applied to multisets.[1]

One could say that logic's basic object of study is *inference*. In particular, it is a theory about which inferences are correct, in a certain specific sense, and which ones are not. We start, then, by a formal characterization of what an inference is:

Definition 2.1.1 An *inference* (written $\Gamma \Rightarrow \Delta$) is a pair $\langle \Gamma, \Delta \rangle$, where Γ and Δ are multisets of formulas (respectively, the *premises* and *conclusions* of the inference).

Thus, our first goal is to discriminate the acceptable inferences from those that are unacceptable, and we will see how to do this both proof-theoretically and semantically. Furthermore, notice that we are allowing inferences to have not only multiple premises but also conclusions.[2] This is what Humberstone calls in (2011, p. 105) 'Multiset-Multiset framework'. Although it may seem more artificial than the Multiset-formula framework (where conclusions can only be single), it has many technical advantages. One way to hold a more intuitive grasp on multiple conclusions is to understand them disjunctively: the conjunction of the premises implies the disjunction of the conclusions.[3] Notice also that, in the current framework, premises and conclusions are allowed to be the empty set. The intuitive interpretation of a correct inference with no premises or conclusions is a matter that will be decided in slightly different ways in each logic we will review throughout this book.

A second goal stems from the initial one, by reflecting on how inferences relate to one another. This could be a purely metatheoretical interest, but one may also believe that these higher-level relations can impact which patterns of reasoning are deemed correct. For example, there may be some inferences that are correct taken in isolation, but which may lead to wrong conclusions when linked together.

---

[1] Keep in mind that a *multiset* of formulas is like a set but sensitive to the number of occurrences of each formula. For example, while $\{\varphi, \varphi\}$ is the same set as $\{\varphi\}$, they are two different multisets since the first one contains two elements, but the second one is a singleton. Although this seems an unnecessary complication here, multisets (instead of sets) will come in handy when defining some substructural logics in later chapters. For more on multisets and the usual set-theoretic operations between them, see Bimbó (2014, ch. 3).

[2] This decision could be questioned. Arguments in 'real life' seem to lead to a unique conclusion. As Rumfitt puts it in (2008, p. 79): "The rarity, to the point of extinction, of naturally occurring multiple-conclusion arguments has always been the reason why mainstream logicians have dismissed multiple-conclusion logic as little more than a curiosity." In the same direction, in (2011) Steinberger affirms that "multiple-conclusion systems fail to provide a 'natural' representation of our ordinary modes of inference". In (2005) Restall has challenged this position that multiple conclusions are not part of our inferential practice, and in (2020) Dicher has some arguments to counter those previously mentioned. Also, see Shoesmith and Smiley (1978).

[3] One of the advantages that have been widely discussed in the literature is the fact that, as pertains to Classical Logic, having multiple conclusions allows to rule out the so-called non-normal valuations—i.e. bivaluations that are not a counterexample to any classically valid argument, but that do not comply with the Boolean valuation scheme that we will discuss below in this chapter. For more on these, see Shoesmith and Smiley (1978), Oller (2014) or Oller (2016).

Hence, besides inferences, we will present a second object of study, *metainferences*, instantiating a relation between inferences:

Definition 2.1.2 A *metainference* (written $S \Rrightarrow s$) is a pair $\langle S, s\rangle$, where S is a set of inferences and s is an inference.

For readability, when exemplifying a metainference $S \Rrightarrow s$, where $S = \{\Gamma_1 \Rightarrow \Delta_1, \ldots, \Gamma_n \Rightarrow \Delta_n, \ldots\}$ and $s = \Gamma \Rightarrow \Delta$ we will graphically represent it in the following vertical fashion:

$$\frac{\Gamma_1 \Rightarrow \Delta_1 \quad \ldots \quad \Gamma_n \Rightarrow \Delta_n \quad \ldots}{\Gamma \Rightarrow \Delta}$$

Again, the goal is to discriminate which metainferences are acceptable and which are not, and this can also be done via proof-theoretic or semantic means. Moreover, given that it is much less standard to work with the Multiset-Multiset framework for metainferences than it is for inferences, when discussing the former, we will restrict ourselves to the simpler setting, only moving to the more general one if it is required.

Our definitions so far can be read as characterizing tokens, or particular instances of inferences and metainferences in the language $\mathcal{L}$. There is, however, another look at inferences and metainferences that we want to cover. One thing that most logicians would agree on is that correct logical inferences are those in which the premises justify the conclusion(s) *in virtue of their form*. The formal language in itself is an elucidation of what 'form' means. But there is a stronger, intratheoretical requirement, which also enforces the concept of logical form and is incarnated in the idea of a *schema*—that is, a class of tokens that share the same structure:

Definition 2.1.3 A *schema* is the set of all and only tokens that can be obtained from one of its members—a 'basic instance'—by uniformly substituting some propositional variable p in it by some formula φ, and possibly adding some extra premises or conclusions.

Observe that this allows for inferences such as the following to count as tokens of the inference schema known as Modus Ponens:

$$r, p \rightarrow q, p \Rightarrow q, r$$

Formulas which are not obtained by substitution on the basic instance are called the *context* of the inference, and when we represent the schema using metavariables for multisets of formulas, they can be pictured as follows:

$$\Gamma, \varphi \rightarrow \psi, \varphi \Rightarrow \psi, \Delta$$

Just like with inferences, the following example of a metainference:

$$\frac{p \Rightarrow p \wedge q}{p \Rightarrow q}$$

can be considered as a token of the following schematic metainference:

$$\frac{\varphi \Rightarrow \varphi \wedge \psi}{\varphi \Rightarrow \psi}$$

Moreover, context can be added to schematic metainferences too, as illustrated by the following:

$$\frac{\Gamma, \varphi \Rightarrow \varphi \wedge \psi, \Delta}{\Gamma, \varphi \Rightarrow \psi, \Delta}$$

As a terminological convention, when no confusion arises, we will talk of schematic metainferences including those with context.

Putting together the concepts of inference, metainference, and schema, we can venture a first characterization of what a logic is. This approximation, it should be noted, is equivalent to saying that a logic is a certain collection of inference and metainference schemas.[4]

Definition 2.1.4 A *logic* $\mathbb{L}$ is a set of inferences and metainferences, closed under uniform substitution.

In this sense, we are taking logics to be certain sets that, as such, are independent of the different (proof-theoretical, valuational) tools that can be used to characterize them.

Finally, the last basic notions that we need to get in place to properly start the investigations in this book are those of structural and substructural logic. This, however, is something that will be postponed until the end of this chapter where we will delve into several distinctions to characterize these notions as precisely as possible.

This being said, as we shall see in the remainder of this book, the question of which inferences and metainferences are acceptable in a given logic can be analyzed from both a proof-theoretic and a semantic point of view. These two angles from which to characterize or study a logical system will provide respective standards and techniques to assess inferences and metainferences in general, and structural metainferences in particular. We now turn to introducing the basics of the technical apparatuses required to understand these points of view.

[4] It is worth noting that our definition of logic may seem somewhat arbitrarily restrictive: why not consider relations between metainferences, and so on? This is perfectly possible—and it has indeed been explored, for instance, in Barrio et al. (2020) or in Pailos and Da Ré (2023). We are not going to dive too deep into such complexities, since there is enough to be said already at this level, but there will be some sections in subsequent chapters that will brush these topics.

2.1.1 Proof Theory

One way of determining what inferences hold in a given logic is to associate to that logic a certain formal calculus counting with axioms, rules, principles, and things of the sort—later verifying which inferences are deemed correct in such a calculus. Of course, calculi come in very different shapes and forms: axiomatic calculi, natural deduction calculi, tableaux calculi, and so on. Throughout this book, we will be focusing on a special kind of calculi, originating in the foundational works of Gentzen (1935), which are known as 'sequent calculi'. Let us then give you the basics of how these formalisms are constructed and how to operate with them.

A Gentzen-style sequent calculus (from now on, a *sequent calculus*) over a language $\mathcal{L}$ is an axiomatic system whose axioms are sequents, and whose rules operate over sequents. In this context, a *sequent* is an inference $\Gamma \Rightarrow \Delta$, where Γ, Δ are finite multisets of formulas. The left side of the sequent is also called *antecedent* and its right side, the *succedent* of the sequent. According to Gentzen, a natural way to understand sequents is the denotational interpretation: the sequent of the form $\gamma_1, \ldots, \gamma_n \Rightarrow \delta_1, \ldots, \delta_m$ has the same meaning as the sentence $\gamma_1 \wedge \cdots \wedge \gamma_n \rightarrow \delta_1 \vee \cdots \vee \delta_m$. This goes in line with the disjunctive reading of multiple conclusions suggested above. That being said, it shall be noted that throughout this book, the interpretation of the sequents will vary depending on the logic we are analyzing in each chapter.

While sequents play the role of axioms, some metainferences serve as the rules of the calculus. Let us now see what a rule actually is:

Definition 2.1.5 Given a sequent calculus **S**, a rule R with n premises is a schematic metainference with the following form:

$$R\,\frac{\Gamma_1 \Rightarrow \Delta_1 \quad \ldots \quad \Gamma_n \Rightarrow \Delta_n}{\Sigma \Rightarrow \Pi}$$

where $\Gamma_i \Rightarrow \Delta_i$, $\Sigma \Rightarrow \Pi$, with $1 \leq i \leq n$, are sequents. We call the sequents above the line the 'premises' of the rule R, and the sequent below the line the 'conclusion' of the rule R.

Without going into the specifics of any particular sequent calculus, it should be noted that rules can be roughly divided into two different kinds. On the one hand, we have rules that are completely general and that do not make any mention of specific logical connectives (that is, their basic instance is made entirely of atomic formulas). These rules are usually called *structural*. On the other hand, we have rules where one or more logical connectives are specifically featured (that is, they appear in their basic instance). Rules of this latter form are usually called *operational*.

Finally, a proof of a sequent $\Gamma \Rightarrow \Delta$ in a sequent calculus **S** is a finite tree, where the root is the sequent $\Gamma \Rightarrow \Delta$, each tip of the tree is an axiom and the nodes of the branches contain sequents which result from applications of the rules of the calculus to the sequents above them. This can be recursively defined as follows:

Definition 2.1.6 Given a sequent calculus **S**

- If s is an axiom of **S**, then s is a derivation D of s in **S**
- If $D_1, \ldots, D_n$ are derivations of $s_1, \ldots, s_n$ in **S**, and $s_1, \ldots s_n \Rightarrow s_m$ is an instance of an n-premise rule of **S**, then the finite tree:

$$\dfrac{\begin{array}{ccc} D_1 & & D_n \\ \vdots & & \vdots \\ s_1 & \ldots & s_n \end{array}}{s_m}$$

  is a derivation of s_m.

If a sequent has a proof in **S** we will sometimes say it is **S**-provable, or provable in **S**.

Definition 2.1.7 Given a sequent calculus **S**, $\Gamma \vdash_{\mathbf{S}} \Delta$ if and only if there exists $\Gamma' \subseteq \Gamma$ and $\Delta' \subseteq \Delta$ finite, such that $\Gamma' \Rightarrow \Delta'$ is **S**-provable.

We will sometimes employ the symbol $\vdash_{\mathbf{S}}$ without parameters on either side to refer to the class of all inferences $\Gamma \Rightarrow \Delta$ such that $\Gamma \vdash_{\mathbf{S}} \Delta$.

Having presented some basic notions regarding proof theory, we will now turn to the other way of assessing whether or not certain inferences and metainferences hold in a given logic. This alternative point of view is represented by the semantic approach.

2.1.2 *Models*

Contrary to what happens in natural language—where the semantics is given by a single, fixed interpretation—the semantics of formal languages is given by a wider class of interpretations or *models* instead. This can be done in various ways. One of these is represented by truth-valued semantics, so-called because formulas are taken to denote, in each way of interpreting the language, a truth value. Then, the meaning of complex sentences is (at least partially) determined by the meaning of their components and finally, the correctness of inferences is assessed by considering truth value assignment patterns across all interpretations.

When working with propositional languages, this is usually carried out with the help of *valuation functions*. A function of this sort is a mapping from all the formulas of the language to a given set of truth values, TV.

Definition 2.1.8 Given a propositional language $\mathcal{L}$ and a set of truth values TV, a *valuation* for $\mathcal{L}$ is a mapping v from $FOR(\mathcal{L})$ to TV.

Valuational interpretations can be regarded as semantics for the language under the assumptions that, on the one hand, the numbers 1 and 0 are taken to represent *bona fide* truth values, i.e. Truth and Falsity, and on the other hand, that meaning (in the narrow sense) of declarative sentences is provided by their truth conditions. These assumptions are contested by many people, but for now let us say that if we take them for granted, constrictions on the class of valuations can be regarded as fixing the meaning of some vocabulary of the language. In the case of a language for propositional logic, these constraints are, for example, that a formula φ and its negation $\neg\varphi$ cannot both be assigned the truth value 0 or the truth value 1. Another example could be that if a formula φ is assigned the truth value 1, then $\varphi \vee \psi$ should be assigned the truth value 1.

Sometimes, these constraints are such that they can be summarized in a truth function—i.e. a function that goes from n-tuples of truth values to some truth value. When we are in a situation like this—where every connective # has a truth function $f^{\#}$ associated with it that is represented by a corresponding truth table—it is possible to completely determine the meaning or the truth value of a complex formula by looking into the truth values of its components. In such a case, we may say that such a collection of truth functions constitutes a *valuation scheme* X and we denote the set of valuations complying with such a valuation scheme by V^{X}.

Definition 2.1.9 Given a propositional language $\mathcal{L}$ and a set of truth values TV, a *valuation v based on the valuation scheme* X for $\mathcal{L}$ is a mapping from $FOR(\mathcal{L})$ to TV such that for all n-ary connectives # of $\mathcal{L}$ and its corresponding n-ary truth function $f^{\#} \in \mathrm{X}$, it happens that $v(\#(\varphi_1, \ldots, \varphi_n)) = f^{\#}(v(\varphi_1), \ldots, v(\varphi_n))$.

Since valuations are *only* constrained with respect to the vocabulary that is considered relevant to the form of inferences, they can be used to provide definitions of validity for logics. Very generally, valid arguments can be selected as those that have no *counterexample* in a given class of valuations. What a counterexample is will vary widely from one logic to the other—as we will see in the corresponding chapters—but the intuitive idea can be framed by appealing to a situation where the premises receive a 'good' truth value and the conclusions receive a 'bad' truth value.

Definition 2.1.10 $\Gamma \vDash_V \Delta$ if and only if there is no valuation $v \in V$ which is a counterexample to $\Gamma \Rightarrow \Delta$.

As before, we will sometimes employ the symbol $\vDash_V$ without parameters on either side to mean the class of all inferences $\Gamma \Rightarrow \Delta$ such that $\Gamma \vDash_V \Delta$

In the following sections, we devote ourselves to discussing what is perhaps the most famous logical system, Classical Logic $\mathbb{CL}$. In line with our previous considerations, we will introduce this system by both proof-theoretic and semantic means, describing which are the inferences and which are the metainferences that hold within this system.

2.2 Classical Inferential Logic

2.2.1 *The Sequent Calculus LK*

There are many sequent calculi for $\mathbb{CL}$. In this book, we will mostly work with the system we call **LK**. It is important to observe that this calculus closely resembles the original one introduced by Gentzen but it's not identical to it. Nevertheless, the similarities make it justified to apply the same name to it.[5]

Definition 2.2.1 The sequent calculus **LK** is defined by the following rules, where φ^{At} means that the formula is atomic:[6]

Structural rules

$$\text{Id}\ \frac{}{\varphi^{At} \Rightarrow \varphi^{At}}$$

$$\text{LW}\ \frac{\Gamma \Rightarrow \Delta}{\Gamma, \varphi \Rightarrow \Delta} \qquad \text{RW}\ \frac{\Gamma \Rightarrow \Delta}{\Gamma \Rightarrow \varphi, \Delta}$$

$$\text{LC}\ \frac{\Gamma, \varphi, \varphi \Rightarrow \Delta}{\Gamma, \varphi \Rightarrow \Delta} \qquad \text{RC}\ \frac{\Gamma \Rightarrow \varphi, \varphi, \Delta}{\Gamma \Rightarrow \varphi, \Delta}$$

$$\text{Cut}\ \frac{\Gamma \Rightarrow \varphi, \Delta \qquad \Sigma, \varphi \Rightarrow \Pi}{\Gamma, \Sigma \Rightarrow \Delta, \Pi}$$

Operational rules

$$\text{L}\neg\ \frac{\Gamma \Rightarrow \varphi, \Delta}{\Gamma, \neg\varphi \Rightarrow \Delta} \qquad \text{R}\neg\ \frac{\Gamma, \varphi \Rightarrow \Delta}{\Gamma \Rightarrow \neg\varphi, \Delta}$$

$$\text{L}\wedge\ \frac{\Gamma, \varphi \Rightarrow \Delta}{\Gamma, \varphi \wedge \psi \Rightarrow \Delta} \qquad \text{L}\wedge\ \frac{\Gamma, \psi \Rightarrow \Delta}{\Gamma, \varphi \wedge \psi \Rightarrow \Delta}$$

$$\text{L}\vee\ \frac{\Gamma, \varphi \Rightarrow \Delta \qquad \Gamma, \psi \Rightarrow \Delta}{\Gamma, \varphi \vee \psi \Rightarrow \Delta} \qquad \text{R}\wedge\ \frac{\Gamma \Rightarrow \varphi, \Delta \qquad \Gamma \Rightarrow \psi, \Delta}{\Gamma \Rightarrow \varphi \wedge \psi, \Delta}$$

$$\text{R}\vee\ \frac{\Gamma \Rightarrow \varphi, \Delta}{\Gamma \Rightarrow \varphi \vee \psi, \Delta} \qquad \text{R}\vee\ \frac{\Gamma \Rightarrow \psi, \Delta}{\Gamma \Rightarrow \varphi \vee \psi, \Delta}$$

$$\text{L}\rightarrow\ \frac{\Gamma \Rightarrow \varphi, \Delta \qquad \Sigma, \psi \Rightarrow \Pi}{\Gamma, \Sigma, \varphi \rightarrow \psi \Rightarrow \Delta, \Pi} \qquad \text{R}\rightarrow\ \frac{\Gamma, \varphi \Rightarrow \psi, \Delta}{\Gamma \Rightarrow \varphi \rightarrow \psi, \Delta}$$

---

[5] In his original presentation in Gentzen (1935), Gentzen worked with sequences of formulas instead of multisets, as we will do. Also, his axioms were not restricted to atomic formulas. Thus, our presentation is an adaptation of Gentzen's original rules to the multiset framework.

[6] Notice that the Identity axiom is a schematic rule only in a restricted sense, because it does not encompass every substitution of the basic instance, but only the atomic ones

Before moving on, there are a couple of things to mention.

Definition 2.2.2 The formulas belonging to the contexts are called *side* formulas. The (occurrence of the) formula appearing in the conclusion which is not in the context is the *principal* formula. Finally, the (occurrences of the) formulas appearing in the premises that are not in the context are the *auxiliary* formulas.

For example, in the following rule:

$$\mathrm{R}\wedge\ \frac{\Gamma \Rightarrow \varphi, \Delta \qquad \Gamma \Rightarrow \psi, \Delta}{\Gamma \Rightarrow \varphi \wedge \psi, \Delta}$$

formulas φ, ψ are auxiliary, $\varphi \wedge \psi$ is principal, and Γ, Δ are the context, with the occurrences of the formulas in them being the side formulas.

Regarding the operational rules, we need to distinguish between two different types: the *multiplicative* ones and the *additive* ones. This distinction can be applied to rules for any n-ary operation, for $1 < n$, although it means different things in the case of single-premise and in the case of multi-premise rules. For multi-premise rules, being multiplicative means that the context in the premises can be different—and thus, are also called *context-independent*—while being additive means that contexts have to be the same—and thus, are also called *context-sharing*. The only example of a context-independent rule in **LK** is L $\rightarrow$. Alternatively, for instance, the multiplicative version of R$\wedge$ is:

$$\mathrm{R}\wedge_m\ \frac{\Gamma \Rightarrow \varphi, \Delta \qquad \Sigma \Rightarrow \psi, \Pi}{\Gamma, \Sigma \Rightarrow \varphi \wedge \psi, \Delta, \Pi}$$

For single-premise rules, on the other hand, being multiplicative means that all immediate subformulas of the principal formula are auxiliary formulas, while being additive means that is not the case. In **LK**, the only single-premise multiplicative rule is R $\rightarrow$, which features both the antecedent and the consequent of the principal conditional as auxiliary formulas in the premise of the rule. L$\wedge$ is additive because each version of the rule features only one of the conjuncts as the auxiliary formula. Again, the alternate multiplicative version of L$\wedge$ is the following rule:

$$\mathrm{L}\wedge_m\ \frac{\Gamma, \varphi, \psi \Rightarrow \Delta}{\Gamma, \varphi \wedge \psi \Rightarrow \Delta}$$

As it can easily be seen from the examples above, beyond the choices made in **LK**, any operational rule can be defined in the multiplicative or the additive fashion, and one can even combine both styles, using one of them for multi-premise and the other one for single-premise rules. In the presence of Contraction and Weakening, both formulations are equivalent in the following sense:

Definition 2.2.3 Given a sequent calculus **S**, two rules $S_1 \Rrightarrow s_1$ and $S_2 \Rrightarrow s_2$ are *interderivable* in **S** if there are two derivations: a derivation of s_1 which only uses rules in $\mathbf{S} - \{S_1 \Rrightarrow s_1\}$ or members of S_1 as axioms; and a derivation of s_2 which only uses rules in $\mathbf{S} - \{S_2 \Rrightarrow s_2\}$, or members of S_2.

The interesting case is when we consider a purely structural **S**, so we can pinpoint the equivalence of the two rules in isolation from other logical rules. However, the structural rules are not innocuous. In Chap. 5, when we present Contraction-free and Weakening-free systems, the way we formulate operational rules will affect the set of provable sequents, since additive and multiplicative connectives will no longer be interderivable.

Let's see now some examples of derivations in **LK**. For instance, the following sequent $\Rightarrow p \vee \neg p$, which is a case of the very well-known Principle of Excluded Middle (or *Tertium Non Datur*) is provable:

$$\text{RC}\,\dfrac{\text{R}\vee\,\dfrac{\text{R}\vee\,\dfrac{\text{R}\neg\,\dfrac{\text{Id}\,\dfrac{}{p \Rightarrow p}}{\Rightarrow p, \neg p}}{\Rightarrow p \vee \neg p, \neg p}}{\Rightarrow p \vee \neg p, p \vee \neg p}}{\Rightarrow p \vee \neg p}$$

Also $p \wedge \neg p \Rightarrow$, which is a case of the rule of Explosion (or *Ex Contradictione Quodlibet*):

$$\text{LC}\,\dfrac{\text{L}\wedge\,\dfrac{\text{L}\wedge\,\dfrac{\text{L}\neg\,\dfrac{\text{Id}\,\dfrac{}{p \Rightarrow p}}{p, \neg p \Rightarrow}}{p \wedge \neg p, \neg p \Rightarrow}}{p \wedge \neg p, p \wedge \neg p \Rightarrow}}{p \wedge \neg p \Rightarrow}$$

And the following two cases of very important rules of $\mathbb{CL}$: the rule of Modus Ponens:

$$\text{L}\to\,\dfrac{\text{Id}\,\dfrac{}{p \Rightarrow p} \qquad \text{Id}\,\dfrac{}{q \Rightarrow q}}{p, p \to q \Rightarrow q}$$

and Disjunctive Syllogism:

$$\text{L}\neg\,\dfrac{\text{L}\vee\,\dfrac{\text{RW}\,\dfrac{\text{Id}\,\dfrac{}{p \Rightarrow p}}{p \Rightarrow p, q} \qquad \text{RW}\,\dfrac{\text{Id}\,\dfrac{}{q \Rightarrow q}}{q \Rightarrow p, q}}{p \vee q \Rightarrow p, q}}{\neg p, p \vee q \Rightarrow q}$$

Exercises

Exercise 1 Identify which **LK** rules are exemplified by the following metainference cases. Indicate principal, auxiliary, and side formulas.

- $$\dfrac{p, \neg(q \wedge r) \Rightarrow r \vee \neg s, p \to q \qquad p, q \wedge r \Rightarrow r \vee \neg s, p \to q}{p, (q \wedge r) \vee \neg(q \wedge r) \Rightarrow r \vee \neg s, p \to q}$$

- $$\dfrac{q \wedge \neg s, p \vee (r \vee \neg s) \Rightarrow q \to r}{q \wedge \neg s, p \vee (r \vee \neg s) \Rightarrow q \to r, q \to r}$$

- $$\dfrac{\neg(p \to r) \Rightarrow p \wedge \neg q, p \wedge \neg q \qquad \neg(p \to r), \neg s \vee \neg q \Rightarrow}{\neg(p \to r), \neg(p \to r), (p \wedge \neg q) \to (\neg s \vee \neg q) \Rightarrow p \wedge \neg q}$$

Exercise 2 Formulate the additive versions of the multiplicative rules of **LK** and the multiplicative versions of the additive ones.

Exercise 3 Offer a derivation of the following sequents in **LK**:

- $\neg(p \vee q) \Rightarrow \neg p \wedge \neg q$
- $p \rightarrow q, q \rightarrow s \Rightarrow p \rightarrow s$
- $p \rightarrow (q \wedge r), q \rightarrow (q \wedge r), p \vee q \Rightarrow \neg(\neg q \vee \neg r)$
- $\Rightarrow ((p \rightarrow q) \rightarrow p) \rightarrow p$

Exercise 4 Prove that the additive and multiplicative versions of the rules for each connective are interderivable.

2.2.2 *Boolean Bivaluations*

Let us now address the issue of logical validity in $\mathbb{CL}$, semantically considered. The most common and well-known presentation of the semantics for $\mathbb{CL}$ is given in terms of a family of two-valued valuations, that is, the set of truth values is just the two-element set $\{1, 0\}$. We will call V^2 the class of all functions ranging from $FOR(\mathcal{L})$ onto that set. From those, the valuations which characterize $\mathbb{CL}$ are those complying with the Boolean valuation scheme B comprising the truth tables appearing below in Fig. 2.1. Valuations based on such a scheme are often called two-valued Boolean valuations or Boolean bivaluations because the corresponding truth functions correspond to the two-element Boolean algebra. For this reason, we will use V^{B} as a name for the set thereof and we will use B a subindex for each of the truth functions to denote that it is the Boolean truth function we are talking about.

To assess the validity of certain inferences in $\mathbb{CL}$ from a semantic point of view then, we must consider which Boolean bivaluations can be taken as counterexamples to certain inferences and which cannot be taken as such—in which case we will say that these valuations actually *satisfy* the inference in question. In the case of $\mathbb{CL}$, the most traditional semantic account of logical consequence appeals to *necessary truth-preservation* from premises to conclusions. This approach is embodied in the following definitions.

Definition 2.2.4 A bivaluation v is a *counterexample* to an inference $\Gamma \Rightarrow \Delta$ if and only if $v(\gamma) = 1$, for all $\gamma \in \Gamma$, and $v(\delta) = 0$, for all $\delta \in \Delta$. Otherwise, the valuation v *satisfies* the inference.

Fig. 2.1 The two-valued Boolean truth tables

	$f_{\text{B}}^{\neg}$
1	0
0	1

$f_{\text{B}}^{\wedge}$	1	0
1	1	0
0	0	0

$f_{\text{B}}^{\vee}$	1	0
1	1	1
0	1	0

$f_{\text{B}}^{\rightarrow}$	1	0
1	1	0
0	1	1

Definition 2.2.5 An inference $\Gamma \Rightarrow \Delta$ is V^{B}-*valid* (written $\Gamma \vDash_{V^{\mathrm{B}}} \Delta$) if and only if there is no valuation $v \in V^{\mathrm{B}}$ which is a counterexample to it. Equivalently, if and only if it is satisfied by all valuations $v \in V^{\mathrm{B}}$.

The value 1 is called 'designated', because it is the one that is to be preserved in valid arguments. As we will see later in this book, many non-classical logics are obtained by expanding the set of designated or the set of undesignated values, whereas some other logics—in particular, some of the substructural ones considered in following chapters—get rid of the idea of designated value and value preservation altogether.

With the help of these tools, let's now see some examples of inferences that are valid according to Boolean bivaluations. We will review the examples discussed in the previous section, showing that they are also deemed correct according to the semantic point of view.

First, notice that the inference $\Rightarrow p \vee \neg p$ is valid. On the one hand, if $v(p) = 1$, then $v(p \vee \neg p) = 1$ and, on the other, if $v(p) = 0$ then $v(\neg p) = 1$, whence $v(p \vee \neg p) = 1$. Thus, $\vDash_{V^{\mathrm{B}}} p \vee \neg p$. Formulas that receive a designated value in every valuation (like $p \vee \neg p$) are called *tautologies*. Also, the inference $p \wedge \neg p \Rightarrow$ is valid. On the one hand, if $v(p) = 0$, then $v(p \wedge \neg p) = 0$ and, on the other, if $v(p) = 1$ then $v(\neg p) = 0$, whence $v(p \wedge \neg p) = 0$. Thus, $p \wedge \neg p \vDash_{V^{\mathrm{B}}}$. Formulas that do not receive a designated value in any valuation (like $p \wedge \neg p$) are called *contradictions*.

Notice that the division between contradictions and tautologies is not exhaustive, i.e. some formulas are not tautologies nor contradictions. These formulas are called *contingencies*. It's easy to check that if φ is a contingency then both $\Rightarrow \varphi$ and $\varphi \Rightarrow$ are V^{B}-invalid.

Let's see now the two examples above of inferences with non-empty premises and non-empty conclusion. First, the inference $p, p \rightarrow q \Rightarrow q$ is valid because if there were a counterexample, i.e. a Boolean bivaluation v such that $v(p) = v(p \rightarrow q) = 1$ and $v(q) = 0$, this would imply that $v(p \rightarrow q) = 0$, a contradiction. Thus, there is no such counterexample, and therefore $p, p \rightarrow q \vDash_{V^{\mathrm{B}}} q$. Finally, the inference $\neg p, p \vee q \Rightarrow q$ is valid because if there were a Boolean bivaluation v such that $v(\neg p) = v(p \vee q) = 1$ and $v(q) = 0$, this would imply that $v(p) = 1$ and that $v(\neg p) = 0$, a contradiction. Thus, there is no such counterexample and therefore $\neg p, p \vee q \vDash_{V^{\mathrm{B}}} q$.

Now, given that these inferences were also derivable in the sequent calculus **LK**, we may wonder whether there is a deeper match between these two approaches. In the following section, we discuss this.

Exercises

Exercise 5 Determine whether the following sequents are V^{B}-valid. If they are not, provide a counterexample:

- $\neg(p \rightarrow r) \Rightarrow p \wedge \neg q, p \wedge \neg q$
- $\neg s \vee \neg q \Rightarrow$
- $\neg(p \rightarrow r), (p \vee \neg q) \rightarrow (\neg s \vee \neg q) \Rightarrow \neg s \vee \neg q$

Exercise 6 Prove the Monotonicity of $\vDash_{V^{\mathrm{B}}}$: if $\Gamma' \vDash_{V^{\mathrm{B}}} \Delta'$ where Γ', Δ' are multisets of formulas, then $\Gamma \vDash_{V^{\mathrm{B}}} \Delta$, for every Γ and Δ such that $\Gamma' \subseteq \Gamma$ and $\Delta' \subseteq \Delta$.

Exercise 7 Let φ be such that $\nvDash_{V^{\mathrm{B}}} \varphi$. Prove that it has an instance that is a contradiction. Additionally, prove that, if φ is contingent, then it also has an instance that is a tautology.

Exercise 8 A set of formulas Γ is satisfiable if there is a valuation v such that $v(\gamma) = 1$, for all $\gamma \in \Gamma$. A maximally satisfiable set Γ_u is a satisfiable set such that $\Gamma_u \cup \varphi$ is not satisfiable, for any $\varphi \notin \Gamma_u$. Prove that if $\Gamma_u \vDash_{V^{\mathrm{B}}} \psi$, then $\psi \in \Gamma_u$.

2.2.3 *Inferential Soundness and Completeness*

Proof systems and semantics are alternative ways of characterizing a logic, and thus, one wants to match a given proof system to a given semantics, and vice versa. The reason is that the congruence of these tools speaks in favor of them: the reliability of proof systems is enforced by a compelling class of semantic interpretations, and the other way around. In this regard, this section aims to prove the coincidence of the set of acceptable inferences determined according to Boolean bivaluations and the set of acceptable inferences determined by **LK**. The result establishing that every inference acceptable according to the former is also according to the latter is called 'Soundness', whereas the result establishing that every inference acceptable according to the latter is also according to the former is called 'Completeness'.

The Soundness theorem is proved by induction on the height of the proof, which intuitively means the number of nodes of its longest branch.

Definition 2.2.6 Let D be a derivation. The *height* of D, $h(D)$ is defined as follows:

- $h(D) = 0$ if D is the proof of an axiom
- $h(D) = h(D') + 1$ if D is a derivation which extends D' by a one-premise rule.
- $h(D) = max\{h(D') + h(D'')\} + 1$ if D is a derivation which extends D' and D'' by a two-premise rule.

Theorem 2.2.7 (Soundness) *If* $\Gamma \vdash_{\mathbf{LK}} \Delta$, *then* $\Gamma \vDash_{V^{\mathrm{B}}} \Delta$.

Proof If $\Gamma \vdash_{\mathbf{LK}} \Delta$ there are finite Γ' and Δ', such that $\Gamma' \subseteq \Gamma$ and $\Delta' \subseteq \Delta$ and $\Gamma' \Rightarrow \Delta'$ is **LK**-provable. Now, we need to prove that $\Gamma' \vDash_{V^{\mathrm{B}}} \Delta'$. For the base case, if $\Gamma' \Rightarrow \Delta'$ is an axiom, then it is V^{B}-valid because all axioms are V^{B}-valid as it is easy to verify. For the inductive step, if $\Gamma' \Rightarrow \Delta'$ has a proof of height n, then the sequents immediately above it have proofs shorter than n, and by the inductive hypothesis, they are V^{B}-valid. Given that all rules in **LK** preserve V^{B}-validity (Exercise 9), it follows that $\Gamma' \vDash_{V^{\mathrm{B}}} \Delta'$. Thus, by Exercise 6, $\Gamma \vDash_{V^{\mathrm{B}}} \Delta$. □

Soundness is, as usual, the easy half of the pair. To prove Completeness, on the other hand, one needs to figure out a way to construct countermodels for proof-theoretically invalid inferences. When proof theory is framed in terms of sequents, there is a property that the rules of a calculus might have, which can lend a hand in finding such a valuation:

Definition 2.2.8 Given a sequent calculus **S**, a rule R:

$$R\ \frac{\Gamma_1 \Rightarrow \Delta_1 \quad \dots \quad \Gamma_n \Rightarrow \Delta_n}{\Sigma \Rightarrow \Pi}$$

is *invertible*, if whenever the conclusion of R is **S**-provable, then each premise is **S**-provable.

For example, the rule R$\wedge$ is invertible, and in fact, we can show something a bit stronger, which is that it is *height preserving invertible*, which means that not only there is a proof of each premise, but also that there is a proof of height at most as great as the proof of the conclusion:

Fact 2.2.9 R$\wedge$ is height-preserving invertible.

Proof By induction on the height of the derivation of $\Gamma \Rightarrow \Delta, \varphi \wedge \psi$. For the base case, assume there is a proof D of height 1 of $\Gamma \Rightarrow \Delta, \varphi \wedge \psi$—since it features a complex formula, it cannot be an axiom and thus, there is no proof of height 0. Then, $\varphi \wedge \psi$ has to be principal, because it is not an atomic formula. Thus, it is either the consequence of R$\wedge$, or of RW. If it is a consequence of R$\wedge$, then $\Gamma \Rightarrow \Delta, \varphi$ and $\Gamma \Rightarrow \Delta, \psi$ are $\varphi \Rightarrow \varphi$ and $\psi \Rightarrow \psi$, where φ and ψ are atomic formulas. Hence, they are axioms and thus, provable with height less or equal to 1. If $\Gamma \Rightarrow \Delta, \varphi \wedge \psi$ is a consequence of RW, then $\Gamma \Rightarrow \Delta$ is an axiom, and then $\Gamma \Rightarrow \Delta, \varphi$ and $\Gamma \Rightarrow \Delta, \psi$ follow by RW with a proof of height less or equal than 1.

For the inductive step, assume $\Gamma \Rightarrow \Delta, \varphi \wedge \psi$ has a proof D of height $k+1$. If $\varphi \wedge \psi$ is principal and a consequence of R$\wedge$ or RW, then the reasoning is analogous to the base step. If it is a consequence of RC, then there is a proof of $\Gamma \Rightarrow \Delta, \varphi \wedge \psi, \varphi \wedge \psi$ of height k. By induction hypothesis, there are proofs of $\Gamma \Rightarrow \Delta, \varphi \wedge \psi, \varphi$ and $\Gamma \Rightarrow \Delta, \varphi \wedge \psi, \psi$ with height less or equal than k. Again by induction hypothesis, there are proofs of $\Gamma \Rightarrow \Delta, \varphi, \varphi$ and $\Gamma \Rightarrow \Delta, \psi, \psi$ with height less or equal than k, and hence, by RC, we can prove $\Gamma \Rightarrow \Delta, \varphi$ and $\Gamma \Rightarrow \Delta, \psi$ with height less or equal than $k+1$. If $\varphi \wedge \psi$ is not principal, then the sequent comes from either a single-premise or a two-premise rule. Assume the latter (the former is analogous), so the premises are $\Gamma' \Rightarrow \Delta', \varphi \wedge \psi$ and $\Gamma'' \Rightarrow \Delta'', \varphi \wedge \psi$. By inductive hypothesis, there are proofs of $\Gamma' \Rightarrow \Delta', \varphi$ and $\Gamma' \Rightarrow \Delta', \psi$ and of $\Gamma'' \Rightarrow \Delta'', \varphi$ and $\Gamma'' \Rightarrow \Delta'', \psi$ with height less or equal than k. Hence, we can construct a proof of height less or equal than $k+1$ of $\Gamma \Rightarrow \Delta, \varphi$ and $\Gamma \Rightarrow \Delta, \psi$ by applying the last rule of the original proof to the first and the third sequents, and to the second and the fourth, respectively. □

Invertible rules are really useful because they provide a root-first proof-search: put succinctly, one starts from the sequent one wants to prove and tries to construct a proof from the bottom, applying the rules upside-down, until reaching the axioms. If no axiom is reached, then there is no proof for the sequent.

For multi-premise rules, it is the shared-context ones which are invertible. For mixed-contexts rules, invertibility fails because given a certain provable sequent, there is more than one possible set of premises that could justify it. However, the variation does not come from the fact that there are different possible auxiliary formulas, split into different rules, as in the single-premise case. Instead, the problem arises because contexts are allowed to be different, and thus all combinations of formulas in the contexts of the conclusion are possible contexts for the premises. Hence, even though there are not many possible rules, there are still many possible instances of the same rule which could have led to the conclusion. **LK** does have a majority of shared-context rules but still counts L $\rightarrow$ as an exception. As an illustration, given the provable sequent

$$p \rightarrow q, p, \neg q \Rightarrow$$

all the possible instances of L $\rightarrow$ with it as a conclusion are:

$$\frac{\Rightarrow p \qquad q, \neg q, p \Rightarrow}{p \rightarrow q, p, \neg q \Rightarrow} \qquad \frac{\neg q, p \Rightarrow p \qquad q \Rightarrow}{p \rightarrow q, p, \neg q \Rightarrow}$$

$$\frac{\neg q \Rightarrow p \qquad q, p \Rightarrow}{p \rightarrow q, p, \neg q \Rightarrow} \qquad \frac{p \Rightarrow p \qquad q, \neg q \Rightarrow}{p \rightarrow q, p, \neg q \Rightarrow}$$

But not all of these metainferences are invertible since not all of them have provable premises (indeed, only the last one does). And this is why L $\rightarrow$ is not invertible.

So, summarizing, we said that having invertible rules was useful to prove Completeness because one could develop a proof-search procedure that starts from the root, which, if it fails to find a proof, can also provide a recipe for a counterexample.[7] But then, it turns out that **LK** is not fully invertible. These obstacles, however, are not insurmountable, if we take the previous considerations into account. The sort of process we present in what follows is called a *reduction*. The general idea is that, given a sequent, we consider, for each of its formulas, what sequents would be needed to prove them. By successively reasoning in this way, we generate a tree where in each node the formulas get simpler as we go along. Eventually, when there are no formulas left to reduce, we either arrive at a tree all whose leaves have some shared formula on the left and right of the sequent (closed leaves), or else the tree has at least one leaf which does not have such a formula (an open leaf). In the first case, the tree can be used to build either a proof (in our case,

---

[7] There are sequent calculi for Classical Logic containing only invertible rules, such as the calculus **G3c** developed by Ketonen (1944). See e.g. Indrzejczak (2021) and Negri and Von Plato (2008), for details.

the tree itself will be the proof, although this is not always so) and in the second case, the open leaf serves as a guide to building a counterexample of the root sequent.

The construction of the tree operates in stages and steps. Each step reduces a formula and extends the tree by adding sequents above a given leaf. Each stage is a series of steps—i.e. a subtree—where all formulas which were generated in the previous stage are reduced. We will now present this in full detail.

Definition 2.2.10 A *reduction* of a sequent $\Gamma \Rightarrow \Delta$ is a rooted tree constructed as follows:

- **Stage 0**: Write $\Gamma \Rightarrow \Delta$ as the root.
- **Stage** $n+1$: If there are no formulas left to reduce in any leaf, then if there is an open leaf, halt the process. If all leaves are closed, then extend each branch by successively removing every atom except one which is repeated in both sides of the sequent. If, on the contrary, there are still formulas left to reduce, then take an open leaf and a formula φ (following some enumeration of formulas) which was generated in stage n and reduce it. Steps can be of three kinds, depending on the main connective of the formula φ to be reduced and its position on the sequent:

  Case (a). Negations, conjunctions on the right, disjunctions on the left, and conditionals on the right. Extend the tree with the sequent or sequents which are the result of applying the inverse of the rule corresponding to φ's main connective and position.

  Case (b). Conjunctions on the left and disjunctions on the right. Extend the tree by repeating the sequent on the leaf and adding an extra copy of φ. Then extend it twice with the sequent which is the result of applying each inverse of the rule corresponding to φ's main connective and position (once for each copy of φ).

  Case (c). Conditionals on the left. Extend the tree by successively adding one extra copy of each formula ψ in the original sequent other than φ, until the sequent $\Gamma, \Gamma, \varphi \Rightarrow \Delta, \Delta$ is reached. Then, extend it with the sequents which are the result of applying to the leaf the inverse of L $\rightarrow$, using one copy of Γ and one copy of Δ as context for each premise.

  When all formulas from stage n have been reduced, move on to the next stage.

Case (a) is the easiest. For example, if the formula being reduced is a conjunction on the right, the result of the reduction is just an instance of the rule R$\wedge$:

$$\dfrac{\vdots}{\dfrac{\Gamma \Rightarrow \varphi \wedge \psi, \Delta}{\vdots}} \quad \text{is transformed into} \quad \dfrac{\dfrac{\dfrac{\vdots}{\Gamma \Rightarrow \varphi, \Delta} \qquad \dfrac{\vdots}{\Gamma \Rightarrow \psi, \Delta}}{\Gamma \Rightarrow \varphi \wedge \psi, \Delta}}{\vdots}$$

Case (b) is a bit more complex. In the example below we have to reduce a disjunction on the right, and in the result of the reduction we can read, top-down, two successive applications of R$\vee$, followed by an application of RC:

$$\dfrac{\vdots}{\Gamma \Rightarrow \varphi \vee \psi, \Delta} \quad \text{is transformed into} \quad \dfrac{\dfrac{\dfrac{\dfrac{\vdots}{\Gamma \Rightarrow \varphi, \psi, \Delta}}{\Gamma \Rightarrow \varphi, \varphi \vee \psi, \Delta}}{\Gamma \Rightarrow \varphi \vee \psi, \varphi \vee \psi, \Delta}}{\Gamma \Rightarrow \varphi \vee \psi, \Delta}$$

Finally, for the more complex case (c), if we read the reduction top-down, we can see an application of L $\rightarrow$ followed by successive applications of Contraction on both sides of the sequents.

$$\dfrac{\vdots}{\Gamma, \varphi \rightarrow \psi \Rightarrow \Delta} \quad \text{is transformed into} \quad \dfrac{\dfrac{\dfrac{\vdots}{\Gamma \Rightarrow \varphi, \Delta} \quad \dfrac{\vdots}{\Gamma, \psi \Rightarrow \Delta}}{\Gamma, \Gamma, \varphi \rightarrow \psi \Rightarrow \Delta, \Delta}}{\vdots \atop \Gamma, \varphi \rightarrow \psi \Rightarrow \Delta}$$

As a final example, we show the complete reduction of the sequent $(p \wedge q) \rightarrow \neg r, p \wedge q \Rightarrow \neg r$, which read top-down, is a somewhat convoluted proof of said sequent:

$$\dfrac{\dfrac{\dfrac{\dfrac{\dfrac{\dfrac{\dfrac{\dfrac{\dfrac{\dfrac{\dfrac{\dfrac{\dfrac{}{p \Rightarrow p}\,\text{Id}}{p, q \Rightarrow p}\,\text{LW}}{p, q, r \Rightarrow p}\,\text{LW} \quad \dfrac{\dfrac{\dfrac{}{q \Rightarrow q}\,\text{Id}}{p, q \Rightarrow q}\,\text{LW}}{p, q, r \Rightarrow q}\,\text{LW}}{p, q, r \Rightarrow p \wedge q}\,\text{R}\wedge \quad \dfrac{\dfrac{\dfrac{}{r \Rightarrow r}}{p, r \Rightarrow r}\,\text{LW}}{p, q, r \Rightarrow r}\,\text{LW} \atop p, q, r, \neg r \Rightarrow \quad \text{L}\neg}{(p \wedge q) \rightarrow \neg r, p, q, r, p, q, r \Rightarrow}\,\text{L}\rightarrow}{(p \wedge q) \rightarrow \neg r, p, q, r, p, q \Rightarrow}\,\text{LC}}{(p \wedge q) \rightarrow \neg r, p, q, r, p \Rightarrow}\,\text{LC}}{(p \wedge q) \rightarrow \neg r, p, q, r \Rightarrow}\,\text{LC}}{(p \wedge q) \rightarrow \neg r, p, q \Rightarrow \neg r}\,\text{R}\neg}{(p \wedge q) \rightarrow \neg r, p, p \wedge q \Rightarrow \neg r}\,\text{L}\wedge}{(p \wedge q) \rightarrow \neg r, p \wedge q, p \wedge q \Rightarrow \neg r}\,\text{L}\wedge}{(p \wedge q) \rightarrow \neg r, p \wedge q \Rightarrow \neg r}\,\text{LC}$$

Of course, the fact that we obtained a proof is not accidental, and it can be generalized by stating that any closed reduction tree will be a proof of its root, as the following lemma shows:

Lemma 2.2.11 *If the reduction tree for* $\Gamma \Rightarrow \Delta$ *is closed, then* $\Gamma \vdash_{\mathbf{LK}} \Delta$

Proof On the one hand, any tree is always finite. This is because, first, the number of formulas in Γ and Δ is finite; second, each step generates a finite number of new nodes and a finite number of new formulas; third, each stage has a finite number of steps; fourth, at the end of each stage, the formulas on the leaves have less complexity[8] than the leaves of the previous stages.

[8] The complexity of a formula is the number of connectives appearing it.

On the other hand, the leaves of a closed tree are all axioms of **LK**.

Finally, each node is the result of the application of an **LK**-rule. The initial part consists of applications of Weakening. Sequents that were generated as in case (a) are premises of R$\neg$, L$\neg$, R$\wedge$, L$\vee$, or R $\rightarrow$. Those that were generated as in case (b) are premises of Contraction, L$\wedge$ or R$\vee$. Lastly, those that were generated as in case (c) are premises of Contraction or L $\rightarrow$. □

The above lemma guarantees that, when a sequent corresponds to a proof-theoretically invalid inference, the reduction process will output an open tree, and we can then exploit that open tree to construct a countermodel for the inference. In other words, we can show that the system is complete:

Theorem 2.2.12 (Finite Completeness) *For any finite* Γ, Δ, *if* $\Gamma \models_{V^{\mathrm{B}}} \Delta$, *then* $\Gamma \vdash_{\mathbf{LK}} \Delta$.

Proof If $\Gamma \nvdash_{\mathbf{LK}} \Delta$, then because of Lemma 2.2.11, the reduction tree of $\Gamma \Rightarrow \Delta$ has open branches. For any of these, let $\Gamma_u \Rightarrow \Delta_u$ be the sequent which results from the sequents $\Gamma_i \Rightarrow \Delta_i$ in some open branch, where Γ_u is the union of all Γ_i in the branch, and Δ_u is the union of all Δ_i in the branch. We then prove by induction on the complexity of the formulas that $\Gamma_u \Rightarrow \Delta_u$ has a V^{B}-counterexample. For the base case, take any valuation v which assigns 1 to every atomic formula in Γ_u and 0 to every atomic formula in Δ_u. Since the leaf is open, there are no propositional variables in common between antecedent and consequent, and by the construction of the tree, this implies there are no propositional variables in common below. Thus, there will be such an assignment. Assume then that v is such that $v(\gamma) = 1$ and $v(\delta) = 0$ for every $\gamma \in \Gamma_u$ and $\delta \in \Delta_u$ of complexity less than k, and let φ be a formula of complexity k. Let's see some cases:

Case (a). $\varphi = (\psi \wedge \xi)$ and $\varphi \in \Gamma_u$. By the reduction process, both ψ and ξ must be in some Γ_i, and thus, in Γ_u. Since their complexity is less than k, by induction hypothesis $v(\psi) = 1$ and $v(\xi) = 1$, and thus, $v(\psi \wedge \xi) = 1$

Case (b). $\varphi = (\psi \rightarrow \xi)$ and $\varphi \in \Gamma_u$. By the reduction process, either ψ is in some Δ_i or ξ is in some Γ_i, and thus, either $\psi \in \Gamma_u$ or $\xi \in \Delta_u$. Since their complexity is less than k, by induction hypothesis either $v(\psi) = 0$ or $v(\xi) = 1$, and thus, $v(\psi \rightarrow \xi) = 1$.

The rest of the cases are similar. Since $\Gamma' \subseteq \Gamma_u$ and $\Delta' \subseteq \Delta_u$, we know v witnesses the fact that $\Gamma' \not\models_{V^{\mathrm{B}}} \Delta'$. □

Now, the above theorem only states that the proof theory is sufficient with regard to finite inferences, since what is being reduced are sequents, and sequents are, by definition, finite. However, there is another result that can be used to strengthen this claim, one that states that finite inferences are all one needs to consider. This follows from a property known as 'Compactness', which is a property of certain kinds of sets:

Definition 2.2.13 A set of formulas Γ is *finitely satisfiable* iff for every finite set $\Gamma' \subseteq \Gamma$, there is a v such that $v(\gamma) = 1$, for every $\gamma \in \Gamma'$. Γ is *maximally finitely* *satisfiable* iff it is finitely satisfiable and for every formula φ, either $\varphi \in \Gamma$ or $\varphi \cup \Gamma$ is not finitely satisfiable

Compactness states that finitely satisfiable sets are satisfiable *tout court*. To prove that, we will employ a kind of strategy usually called proof *à la* Henkin, which consists of showing that all finitely satisfiable sets can be extended to maximal sets and that it is these maximal sets that are satisfiable. Thus, the first step towards Compactness is the following lemma:

Lemma 2.2.14 *Every finitely satisfiable set is a subset of a maximally finitely satisfiable set.*

Proof Let Γ be a finitely satisfiable set. Following some enumeration of formulas, extend Γ in stages, by adding at stage n the formula φ_n if $\Gamma_{n-1} \cup \{\varphi_n\}$ remains finitely satisfiable. Take Γ_u to be the union of all Γ_n. We show that Γ_u is maximally finitely satisfiable. First, it is finitely satisfiable because if it were not, there would be a finite unsatisfiable $\Gamma' \subseteq \Gamma_u$, but given Γ' is finite, that would imply there is some stage m such that $\Gamma' \subseteq \Gamma_m$ and thus, Γ_m is not finitely satisfiable, which is impossible by construction. Second, it is maximal because if a formula φ_m is not in Γ_u, it is because it cannot be added in stage m of the construction without losing finite satisfiability. □

Then, we show Compactness by exploiting the special properties that maximal sets have, properties that ease the task of finding a valuation that makes all the formulas in them true.

Theorem 2.2.15 (Compactness) *Every finitely satisfiable set is satisfiable*

Proof Let Γ be a finitely satisfiable set of formulas. Extend it to a maximally finitely satisfiable set Γ_u. Let v be a valuation function that assigns value 1 to every atomic formula in Γ_u, and only them. We show that $v(\varphi) = 1$ iff $\varphi \in \Gamma_u$. First, we need a lemma, which states that, given its maximality, Γ_u contains exactly one of $\varphi, \neg\varphi$ (see Exercise 11). The proof that v behaves as expected is by induction on the complexity of φ. The base case is by definition of v. For the induction step, we show the following two cases:

Case (a). $\varphi = \neg\psi$. If $\varphi \in \Gamma_u$, then $\psi \notin \Gamma_u$. By inductive hypothesis, $v(\psi) = 0$, and thus, $v(\varphi) = 1$. The converse is analogous.

Case (b). $\varphi = \psi \wedge \xi$. If $v(\varphi) = 1$, then $v(\psi) = 1$ and $v(\xi) = 1$. By inductive hypothesis, they are both in Γ_u. But $\{\psi, \xi, \neg(\psi \wedge \xi)\}$ is not satisfiable, and thus, $\neg(\psi \wedge \xi) \notin \Gamma_u$. By the lemma, then, $\psi \wedge \xi \in \Gamma_u$. For the converse, if $v(\varphi) = 0$, then either $v(\psi) = 0$ or $v(\xi) = 0$. By the inductive hypothesis, at least one of them, say, ψ, is not in Γ_u, and thus, by the lemma, $\neg\psi$ is. But $\{\neg\psi, \psi \wedge \xi\}$ is not satisfiable, and thus, $\varphi \notin \Gamma_u$.

□

The crucial consequence of Compactness that we were after is then that finite arguments constitute the core of Classical Logic:

Corollary 2.2.16 (Finiteness) *If* $\Gamma \vDash_{V^{\mathrm{B}}} \Delta$, *then there are finite* $\Gamma' \subseteq \Gamma$ *and* $\Delta' \subseteq \Delta$ *such that* $\Gamma' \vDash_{V^{\mathrm{B}}} \Delta'$.

Proof Assume, for the contrapositive, that there is no finite $\Gamma' \subseteq \Gamma$ and $\Delta' \subseteq \Delta$ such that $\Gamma' \vDash_{V^{\mathrm{B}}} \Delta'$. Thus, there is some valuation $v \in V^{\mathrm{B}}$ such that $v(\gamma) = 1$ for all $\gamma \in \Gamma'$ but $v(\delta) = 0$ for all $\delta \in \Delta'$. Thus, all finite sets $\{\Gamma', \neg\Delta'\}$ are satisfiable. Thus, by Compactness (2.2.15), the set $\{\Gamma, \neg\Delta\}$ is satisfiable. Thus, $\Gamma' \nvDash_{V^{\mathrm{B}}} \Delta'$

□

The full form of Completeness follows immediately from the finite one in Theorem 2.2.12 and Finiteness above:

Corollary 2.2.17 (Completeness) *If* $\Gamma \vDash_{V^{\mathrm{B}}} \Delta$, *then* $\Gamma \vdash_{\mathbf{LK}} \Delta$.

Before moving on to the consequences of Completeness, let us briefly point a methodological commentary. Here, we opted to present a constructive proof of (finite) Completeness and a Henkin-style proof of Compactness, then obtaining full Completeness as a corollary. The proof is constructive in the sense that, for any given sequent, it does not merely tell you that there is *some* proof or *some* countermodel for it, but provides an algorithm to find such a derivation or countermodel.

It is worth noticing, however, that Henkin (1949) originally went the other way around: he devised his strategy to prove the strong form of Completeness, from which Compactness follows as a corollary (See Exercise 12). Proving Completeness à la Henkin means showing that a proof-theoretically invalid argument from Γ to Δ has a countermodel by extending the set $\{\Gamma, \neg\Delta\}$ to a maximal set, and then showing that the maximal set is satisfiable (as we did with Compactness, the difference being that in this case, the set is maximally consistent instead of maximally finitely satisfiable).

We now turn our attention to two very important results, which follow very easily from Completeness. First, notice that it complements Lemma 2.2.11 in guaranteeing that the reduction procedure serves as an algorithm for proof search: if the resulting tree is closed, the sequent is provable, and if it is open, then it is not. Then, this implies that:

Corollary 2.2.18 (Decidability) *The provability relation of* **LK** *is decidable.*

Note that the property of being a $\mathbb{CL}$-valid argument can be decided directly via the particular reduction trees we defined, but that they are not necessary, in the sense that truth tables also serve as a simpler decision algorithm.

The second notable thing following from Completeness stems from the fact that the proofs generated by reduction do not employ Cut at any step.

Definition 2.2.19 Given a sequent calculus **S**, a primitive rule R is *eliminable* from it if and only if the calculus $\mathbf{S}^*$ which results from removing R from **S** is such that $\vdash_{\mathbf{S}} = \vdash_{\mathbf{S}^*}$.

And since Completeness guarantees that these trees are all that one needs to take into account to determine provability, we can conclude that the rule of Cut is not indispensable to prove any sequent:

Corollary 2.2.20 (Cut-elimination) *Cut is eliminable from* **LK**.

Notice that the Cut-elimination Theorem itself does not make reference or concern the class of Boolean bivaluations V^{B} in any essential way. It is, in fact, a purely proof-theoretic property: every **LK**-provable sequent has a Cut-free **LK**-proof. Thus, even though it can easily be presented as a corollary to Completeness, as we did here, one may wonder how may one prove it without this roundabout method. In Sect. 2.4, we will indeed provide an alternative demonstration that does not appeal to valuations of any kind.

Exercises

Exercise 9 Prove that all rules in **LK** preserve V^{B}-validity.

Exercise 10 Are all of the rules of **LK** invertible? Prove which ones are invertible and which ones are not.

Exercise 11 Let Γ_u be a maximally finitely satisfiable set. Show that it contains exactly one of φ, $\neg\varphi$.

Exercise 12 Show that Compactness is a corollary of the strong form of Completeness.

2.3 Classical Metainferential Logic

So far, we have dealt with tools for determining which inferences are valid and which are not, and that covers the first level of what we understand here as 'Classical Logic'. And even though, as we already stated, having a set of valid inferences may not be enough to fully select a set of valid metainferences, those same tools (i.e. sequent calculi and valuations) that were used in previous sections can be now exploited to characterize metavalidity. The present section is devoted then to that enterprise, reviewing, in turn, the proof-theoretic approach, the valuational one, and finally, their coincidence, as it pertains to $\mathbb{CL}$.

2.3.1 Derivability and Admissibility

Sequent calculi are especially well suited for present purposes because—although they were primarily designed with inferences in mind—they are in fact already framed in metainferential terms, in the sense that their explicit rules are metainferences. Thus, one can extract a concept of metainferential validity from those rules. Before delving into the details, we want to point that, in the same way as what happened in the inferential case, it is natural to define basic proof-theoretic notions of validity which are finite, since rules of a calculus are finite schematic metainferences made out of sequents. But we are interested in a general account of metainferences, and hence, we need to somehow show that these are enough to encompass all validities, whether finite or not. For that, we will define two general relations, and then, at the end of the chapter, discuss some Finiteness results, and their limitations.

To get things started, we present two different alternative relations between sequents, which are the possible candidates to characterize metavalidity: the notion of *admissibility*, and the notion of *derivability*.

Definition 2.3.1 Given a sequent calculus **S**, a finite metainference token $S \Rrightarrow s$ is *admissible* in **S** if whenever the premises are **S**-provable, the conclusion is **S**-provable also. A rule is admissible if all of its instances are admissible.

As an illustration, the following metainference token is admissible in **LK**, because neither premise is **LK**-provable:

$$\text{Meta Disjunctive Syllogism}\ \dfrac{\Rightarrow p \vee q \qquad \Rightarrow \neg p}{\Rightarrow q}$$

Definition 2.3.2 Given a sequent calculus **S**, a finite metainference token $S \Rrightarrow s$ is *derivable* in **S** if s is provable in the calculus which results from adding to **S** the elements in S as axioms. A rule is derivable if all of its instances are derivable.

The same token of Meta Disjunctive Syllogism is derivable as well, as it can be shown by the following derivation, in which $\{\Rightarrow p \vee q, \Rightarrow \neg p\}$ are added to **LK** as axioms, and then the conclusion is proven in the extended calculus:

$$\text{Cut}\ \dfrac{\text{Cut}\ \dfrac{\text{Axiom}\ \dfrac{}{\Rightarrow p \vee q} \qquad \text{L}\vee\ \dfrac{\text{RW}\ \dfrac{\text{Id}\ \dfrac{}{p \Rightarrow p}}{p \Rightarrow p, q} \qquad \text{RW}\ \dfrac{\text{Id}\ \dfrac{}{q \Rightarrow q}}{q \Rightarrow p, q}}{p \vee q \Rightarrow p, q}}{\Rightarrow p, q} \qquad \text{Cut}\ \dfrac{\text{Axiom}\ \dfrac{}{\Rightarrow \neg p} \qquad \text{L}\neg\ \dfrac{\text{Id}\ \dfrac{}{p \Rightarrow p}}{p, \neg p \Rightarrow}}{p \Rightarrow}}{\Rightarrow q}$$

Thus, we have two, at least *prima facie* different candidates to provide a general proof-theoretically-based concept of metainferential validity:

Definition 2.3.3 Given a sequent calculus **S**, a set of inferences S and an inference $\Gamma \Rightarrow \Delta$, we say that $S \Vdash^{A}_{\mathbf{S}} \Gamma \Rightarrow \Delta$ if and only if either $\Gamma \vdash_{\mathbf{S}} \Delta$, or there exists a finite $S_f \subseteq S$, with an element $\Gamma_n \Rightarrow \Delta_n \in S_f$ such that $\Gamma_n \nvdash_{\mathbf{S}} \Delta_n$.

Definition 2.3.4 Given a sequent calculus **S**, a set of sequents S and a sequent $\Gamma \Rightarrow \Delta$, we say that $S \Vdash^{D}_{\mathbf{S}} \Gamma \Rightarrow \Delta$ if and only if there exists a finite $S_f \subseteq S$, such that $\Gamma \vdash_{\mathbf{S}+} \Delta$, where **S+** is the result of adding S_f to **S** as axioms.

We need first to understand how different they are. In some cases, the two concepts align perfectly, like for instance when considering **LK** and schemas. This property of calculi is known as 'structural Completeness', as described by Pogorzelski (1971) and Pogorzelski and Wojtylak (2008):

Definition 2.3.5 (Structural Completeness) A calculus **S** is *structurally complete* iff derivable rules in **S** are the same as admissible rules in **S**.

To prove that **LK** is structurally complete, we first show how to turn sequents into formulas, in a way that transforms admissible rules into provable sequents:

Definition 2.3.6 (Horizontalization) Given a sequent $\Gamma \Rightarrow \Delta$, the corresponding formula that *horizontalizes* it is:

- $\bigwedge \Gamma \to \bigvee \Delta$, if $\Gamma \neq \emptyset \neq \Delta$
- $\bigvee \Delta$, if $\Gamma = \emptyset$ and $\Delta \neq \emptyset$
- $\neg \bigwedge \Gamma$, if $\Gamma \neq \emptyset$ and $\Delta = \emptyset$
- $q \vee \neg q \Rightarrow p \wedge \neg p$, if $\Gamma = \emptyset$ and $\Delta = \emptyset$

where $\bigwedge \Gamma = \gamma_1 \wedge \cdots \wedge \gamma_n$ and $\bigvee \Delta = \delta_1 \vee \cdots \vee \delta_m$, if $\Gamma = \{\gamma_1, \ldots, \gamma_n\}$ and $\Delta = \{\delta_1, \ldots, \delta_m\}$. We refer to all these as $\bigwedge \Gamma \to \bigvee \Delta$ letting the context disambiguate in each case.

Lemma 2.3.7 *Let the following rule*

$$\frac{\Gamma_1 \Rightarrow \Delta_1 \ldots \Gamma_n \Rightarrow \Delta_n}{\Sigma \Rightarrow \Pi}$$

be **LK***-admissible. Then, the following inference schema*

$$\bigwedge \Gamma_1 \to \bigvee \Delta_1, \ldots, \bigwedge \Gamma_n \to \bigvee \Delta_n \Rightarrow \bigwedge \Sigma \to \bigvee \Pi$$

is **LK***-provable.*

Proof Assume the latter is not **LK**-provable. Hence, there is an instance of it, say $\bigwedge \Gamma_1' \to \bigvee \Delta_1', \ldots, \bigwedge \Gamma_n' \to \bigvee \Delta_n' \Rightarrow \bigwedge \Sigma' \to \bigvee \Pi'$, which is not **LK**-provable. By Completeness, this means that this instance is V^{B}-invalid. In other words, there is a valuation $v \in V^{\mathrm{B}}$ that satisfies each of the premises of the inference but dissatisfies its conclusion. Now, we build another instance based on the valuation v, replacing in $\bigwedge \Gamma_1' \to \bigvee \Delta_1', \ldots, \bigwedge \Gamma_n' \to \bigvee \Delta_n' \Rightarrow \bigwedge \Sigma' \to \bigvee \Pi'$ each propositional letter p_i occurring in it by $p \wedge \neg p$ if $v(p_i) = 0$ and by $p \vee \neg p$ if $v(p_i) = 1$. Let this be $\bigwedge \Gamma_1'' \to \bigvee \Delta_1'', \ldots, \bigwedge \Gamma_n'' \to \bigvee \Delta_n'' \Rightarrow \bigwedge \Sigma'' \to \bigvee \Pi''$. Not only this new instance is an instance of the inference schema we started with, but also every valuation $v \in V^{\mathrm{B}}$ satisfies the premises of the new instance and does not satisfy its conclusion. Also, since it is invalid, by Soundness we know that $\bigwedge \Gamma_1'' \to \bigvee \Delta_1'', \ldots, \bigwedge \Gamma_n'' \to \bigvee \Delta_n'' \Rightarrow \bigwedge \Sigma'' \to \bigvee \Pi''$ is not **LK**-provable.

Having proved that, suppose the following rule:

$$\frac{\Gamma_1 \Rightarrow \Delta_1 \ldots \Gamma_n \Rightarrow \Delta_n}{\Sigma \Rightarrow \Pi}$$

is **LK**-admissible. So, every instance of it should be **LK**-admissible. In particular, the following metainference instance:

$$\frac{\Gamma_1'' \Rightarrow \Delta_1'' \ldots \Gamma_n'' \Rightarrow \Delta_n''}{\Sigma'' \Rightarrow \Pi''}$$

should be **LK**-admissible. By definition, this means that either its conclusion is **LK**-provable or some premise is not. If we had a proof in **LK** of $\Sigma'' \Rightarrow \Pi''$, then, by using L$\wedge$, R$\vee$, R $\to$ and LW we would have a proof of $\bigwedge \Gamma_1'' \to \bigvee \Delta_1'', \ldots, \bigwedge \Gamma_n'' \to \bigvee \Delta_n'' \Rightarrow \bigwedge \Sigma'' \to \bigvee \Pi''$, which cannot be the case. Thus, it must be the case that some of the premises of the metainference instance above is not **LK**-provable, say $\Gamma_i'' \Rightarrow \Delta_i''$. However, if this inference is not **LK**-provable then by Completeness $\Gamma_i'' \Rightarrow \Delta_i''$ must be V^{B}-invalid, which implies that there is a valuation $v \in V^{\mathrm{B}}$ such that satisfies all of the premises but dissatisfies all the conclusions. However, this valuation would not satisfy $\bigwedge \Gamma_i'' \to \bigvee \Delta_i''$ which cannot be the case. Either way, we arrive at a contradiction and, thus, the referred rule isn't **LK**-admissible. This completes the proof. □

Fact 2.3.8 **LK** is structurally complete.

Proof Assume the following rule:

$$\frac{\Gamma_1 \Rightarrow \Delta_1 \ldots \Gamma_n \Rightarrow \Delta_n}{\Sigma \Rightarrow \Pi}$$

is **LK**-admissible. Thanks to Lemma 2.3.7, this implies that the sequent $\bigwedge \Gamma_1 \to \bigvee \Delta_1, \ldots, \bigwedge \Gamma_n \to \bigvee \Delta_n \Rightarrow \bigwedge \Sigma \to \bigvee \Pi$ is **LK**-provable. On the other hand, in the system which results from adding $\Gamma_1 \Rightarrow \Delta_1, \ldots, \Gamma_n \Rightarrow \Delta_n$ to **LK**, the sequents $\Rightarrow \bigwedge \Gamma_1 \to \bigvee \Delta_1$ through $\Rightarrow \bigwedge \Gamma_n \to \bigvee \Delta_n$ are provable. Hence, by n-applications of Cut, one can derive $\Rightarrow \bigwedge \Sigma \to \bigvee \Pi$, which, in turn, allows one to derive $\Sigma \Rightarrow \Pi$ through a number of self-evident steps—left as Exercise 15. □

Interestingly, though, while admissibility and derivability coincide concerning rules in **LK**, this is not always the case. There are other presentations of Classical inferential Logic (and other logics) where some rules are left as merely admissible—that is, they are admissible but not derivable.

Furthermore, these notions do not even come together for all metainferences in **LK**. The coincidence detailed above concerns schemas, whereas something different happens in the case of tokens. Moreover, this is true not only for **LK** but for any reasonable proof system (see Exercise 16): as the example below shows, even if a schema is both inadmissible and underivable, some of its underivable instances will be admissible.

Fact 2.3.9 Not all **LK**-admissible metainference tokens are **LK**-derivable.

Proof The metainference token:

$$\frac{p \Rightarrow q}{r \Rightarrow s}$$

is **LK**-admissible but not **LK**-derivable. □

Finally, it is worth mentioning that the converse direction is true for every metainference (not only tokens, but also schemas), and every sequent calculi whatsoever.

Fact 2.3.10 Given a sequent calculus **S**, all derivable rules are admissible.

Proof See Exercise 14. □

Now, given then that there are two concepts that do not always coincide, it is fair to wonder whether one of them is preferable to the other. There is no clear answer to this question, given that both candidates have their relative advantages. On the one hand, if one's aim is to have a minimal presentation of the logic, in the sense of having as few explicit rules as possible—for instance, to ease the proofs of metatheoretical results for it—then leaving some rules as merely admissible can be convenient.

On the other hand, derivable rules are stable, in that they continue to be derivable even when adding additional axioms. Admissible rules, on the contrary, are more 'contingent', in the sense that even if they come out valid from the calculus as it stands, they may cease to hold if we extend it with new explicit rules (we will see a particular example of this in the next chapter). Hence, having all desired rules as derivable instead of merely admissible might be preferable—for example, because it guarantees that the logical grounds are settled before one starts to build other kinds of theories on top of that calculus.

Thus, while both accounts of validity are legitimate and useful for their own purposes, we believe that when the goal is to settle what the metainferences that ought to be considered part of a logic are, it seems preferable to take the derivable rules of a suitable calculus. This decision does not matter that much in the present chapter, but, as we will see, is not innocuous in the chapters that follow.

Exercises

Exercise 13 Determine whether the following are true or false: (i) All derivable rules are eliminable, (ii) All admissible rules are eliminable.

Exercise 14 Prove that for all sequent calculi, all derivable metainferences are admissible.

Exercise 15 Complete the proof of Fact 2.3.8.

Exercise 16 Explain why 'for any reasonable proof system', there must be underivable but admissible metainference tokens.

2.3.2 Global, Local, and Absolutely Global validity

Having scrutinized the phenomenon of metainferential validity from the proof-theoretical perspective, it is now time to face the task from the point of view of valuations. We will consider three ways to do so. The two initial accounts of validity are usually referred to as 'global' and 'local' conceptions of metainferential validity. Finally, there is a third concept—initially developed by Garson (1990) and also studied by Humberstone (1996)—dubbed 'global validity'. Here, because we are already employing this denomination for a different concept, we will refer to it as 'absolute global validity'.

Let us start by bringing the attention to some general issues, that will in fact truly manifest their importance in the next section. Contrary to what happens with proof theory, we do not need to define valuational validity over a finite relation; each metainference is instead evaluated on its own, and not based on some elementary rules. However, when showing the connection between proof-theoretical and valuational metavalidities, the corresponding Soundness and Completeness theorems will be first established only for finite metainferences. We will finish by discussing when can these in turn be extended to full Soundness and Completeness via the Compactness property.

However, this independence that *prima facie* valuational concepts have over rules does not imply that rules cannot be exploited to generate a class of intended interpretations. Like in the case of inferential validity, the sets of valuations used to define logical consequence can be obtained in two ways. On the one hand, they may have some intrinsic semantic interest, for instance Boolean bivaluations can be obtained by reflecting on the meaning of certain classical connectives. But also, they can be extracted from a set of metainferences, such as the rules of a sequent calculus, so that they can be seen as capturing the semantic content codified by those metainferences. Thus, in what follows the reader will find the corresponding definitions of metavalidity, along with a corresponding concept of range of a set of rules. We will show that, in the case of $\mathbb{CL}$, either choosing the set of intended Boolean valuations or taking some range from **LK** will output equivalent results. This, however, will not necessarily hold in the logics considered in subsequent chapters.

Having cleared that up, let us then start first by examining the global approach. Just like classical inferential validity is understood as truth-preservation, metainferential validity is, according to these standards, also the preservation of some property from the premises to the conclusion of the metainference. In the case of global validity, it straightforwardly corresponds to the preservation of inferential validity, from premises to conclusion.

Definition 2.3.11 A metainference $S \Rrightarrow s$ is *globally* V-valid (written $S \Vdash^{G}_{V} s$) if and only if whenever the premises in S are V-valid, the conclusion s is V-valid.

Intuitively, the classical concept of global metainferential validity arises when we take V to be the set of Boolean bivaluations V^{B}. This is indeed the result we obtain when we define global validity by appealing to the global range of the classical calculus **LK**.

Definition 2.3.12 The *global range in* V of a set of metainferences $\mathbf{S}$ is the maximum set of valuations $V_i \subseteq V$ such that all members of $\mathbf{S}$ are globally V_i-valid.

Fact 2.3.13 The global range in V^2 of the rules of **LK** is V^{B}.

Proof For the left-to-right direction, let v be a non-Boolean valuation. There are four possible cases, corresponding to whether the valuation violates the conditions for each connective:

Case (a). v is not $\neg$-Boolean. Then, there is a φ such that $v(\varphi) = v(\neg\varphi) = 1$ or $v(\varphi) = v(\neg\varphi) = 0$. Assume the former possibility holds, and take an instance of L$\neg$ of the following form:

$$\frac{\varphi \Rightarrow \varphi}{\varphi, \neg\varphi \Rightarrow}$$

Given that all valuations satisfy the premises, but v is a counterexample to the conclusion, v cannot be in the global range in V^2 of **LK**. If the latter possibility holds instead, then again there is a dual instance, this time of R$\neg$, such that all valuations satisfy the premises but v does not satisfy the conclusion. Hence, v is not in the global range in V^2 of the rules of **LK**.

Case (b). v is not $\vee$-Boolean. Then, there is a φ and a ψ such that either one of them is true, but $v(\varphi \vee \psi) = 0$ or else $v(\varphi) = v(\psi) = 0$ but $v(\varphi \vee \psi) = 1$. Assume the latter possibility holds, and take an instance of L$\vee$ of the following form:

$$\frac{\varphi \Rightarrow \varphi \qquad \psi \Rightarrow \psi}{\varphi \vee \psi \Rightarrow \varphi, \psi}$$

As before, given that all valuations satisfy the premises, but v is a counterexample to the conclusion, v cannot be in the global range in V^2 of **LK**.

If the former possibility holds instead, then analogously, v invalidates the dual instance of R$\vee$. Hence, v is not in the global range of the rules of **LK**.

The rest of the cases are left as an exercise. The right-to-left direction is dual.

□

Second, comes the local approach, which relies on the concept of satisfaction of an inference by a valuation, introduced in Definition 2.2.4. Thus, local validity corresponds to the preservation of satisfaction by a valuation, from premises to conclusions.

Definition 2.3.14 A metainference $S \Rrightarrow s$ is *locally* V-valid (written $S \Vdash^{L}_{V} s$) if and only if all V-valuations which satisfy the premises S, satisfy the conclusion s.

Again, it is natural to define the set of classically valid metainferences as those which are locally V^{B}-valid. As it happens with global validity, the same set is characterized by taking the local range of **LK**.

Definition 2.3.15 The *local range in* V of a set of metainferences **S** is the maximum set of valuations $V_i \subseteq V$ such that all members of **S** are locally V_i-valid.

It can be easily seen that local validity implies global validity, not only in the classical case but in general, concerning any set of valuations V.

Fact 2.3.16 All locally V-valid metainference are globally V-valid.

Proof See Exercise 17. □

This inclusion of local validity in global validity is proper for metainference cases:

Fact 2.3.17 Not all globally V-valid metainferences are locally V-valid.

Proof The counterexample is the same as the one in Fact 2.3.9. □

However, the result above does not mean that the local range is properly included in the global range. They both coincide in the case of **LK**.

Fact 2.3.18 The local range in V^2 of the rules of **LK** is V^{B}.

Proof Analogous to Fact 2.3.13. □

In the classical case, in a way analogous to what happened in proof theory, the mismatch between global and local validity disappears when we consider schemas. As an illustration, the schema which goes from $\varphi \Rightarrow \psi$ to $\gamma \Rightarrow \delta$—which has instances that are globally but not locally valid—is neither locally nor globally V^{B}-valid in general, given that it has the following globally V^{B}-invalid instance:

$$\frac{p \Rightarrow p}{p \Rightarrow q}$$

We now state and prove this classical coincidence of metavalidities.

Fact 2.3.19 All globally V^{B}-valid metainference schemas are locally V^{B}-valid, and vice versa.

Proof Let $S \Rrightarrow s$ be a locally V^{B}-invalid metainference schema. Thus, it has a locally V^{B}-invalid instance $S_i \Rrightarrow s_i$. Let v be a valuation that does not satisfy $S_i \Rrightarrow s_i$, and replace every propositional variable such that $v(p) = 1$ by the formula $p \vee \neg p$ and every propositional variable such that $v(p) = 0$ by the formula $p \wedge \neg p$. The result is a globally invalid instance of $S \Rrightarrow s$. □

The proof above can be generalized to many other sets of valuations, whenever the language is powerful enough to have formulas representing the values in their codomain, as it was done in the classical case.

So far, then, we presented the first two concepts of metavalidity, and established some conditions for their coincidence. We now consider the third one, that we call 'absolute global validity', which was introduced by Garson (1990), and thoroughly examined by Humberstone (1996). The reason for the name we give it here comes from the fact that determining a metainference to be absolutely globally valid requires assessing the global validity of the metainference in question concerning not one set of valuations V, but a multiplicity of sets of valuations contained in a family $\mathcal{V}$.

Definition 2.3.20 A metainference $S \Rrightarrow s$ is *absolutely globally* $\mathcal{V}$-valid (written $S \Vdash^{AG}_{\mathcal{V}} s$) if and only if it is globally V-valid, for every $V \in \mathcal{V}$.

In both previous cases, to obtain the classical definition we just took the set V to be V^{B}. In this case, however, we need a set of sets of valuations, and the choice may be less obvious. One natural candidate is the powerset of V^{B}, but it seems safer to start by taking, once again, a range from **LK**.

Definition 2.3.21 The *absolute global range in* V of a set of metainferences $\mathbf{S}$ is the maximum set $\mathcal{V}$ of sets of valuations $V_i \subseteq V$ such that all members of $\mathbf{S}$ are absolutely globally $\mathcal{V}$-valid.

Now, we can be reassured that the initial choice was appropriate, given that the following is true:

Fact 2.3.22 The absolute global range in V^2 of the rules of **LK** (written $\mathcal{V}^{\mathbf{LK}}$) is the powerset of their local range in V^2.

Proof For the left-to-right direction, let $V_1 \subseteq V^2$ be a set that is not in the powerset of V^{B}. This means that there is a $v_1 \in V_1$ which is not Boolean. We show that it V_1-globally invalidates some rule of **LK**, say $S \Rrightarrow s$.

Case (a). v_1 is not $\neg$-Boolean. Assume it is such that $v(\varphi) = v(\neg\varphi) = 1$ (the other possibility is analogous). Then, take the instance:

$$\frac{\varphi \Rightarrow \varphi}{\neg\varphi, \varphi \Rightarrow}$$

Since all valuations in V^2 satisfy the premises, the metainference above must be V_1-invalid. Thus, $V_1 \notin \mathcal{V}^{\mathbf{LK}}$.

Case (b). v_1 is not $\vee$-Boolean. Assume it is such that $v(\varphi) = v(\psi) = 0$ but $v(\varphi \vee \psi) = 1$ (the other possibilities are analogous). Then, take the instance:

$$\frac{\varphi \Rightarrow \varphi \qquad \psi \Rightarrow \psi}{\varphi \vee \psi \Rightarrow \varphi, \psi}$$

Since all valuations in V^2 satisfy the premises, the metainference above must be V_1-invalid. Thus, $V_1 \notin \mathcal{V}^{\mathbf{LK}}$.

The other cases are left as an exercise. For the right-to-left direction, let $V_1 \subseteq V^2$ be a set that is not in $\mathcal{V}^{\mathbf{LK}}$. Then, there is a rule of **LK** which has an instance $S \Rrightarrow s$ that is invalidated by V_1. This means there is a $v \in V_1$ such that it is a local counterexample to $S \Rrightarrow s$. Then, v cannot be Boolean, and thus, V_1 is not a member of the powerset of V^{B}. □

Given these concepts, the fact that absolute global validity is, in some sense, the stronger of the three concepts can be stated in the following way:

Fact 2.3.23 If V_i is the local range in V of a set of metainferences **S**, then for every $v_i \in V_i$, the singleton $\{v_i\}$ belongs to the absolutely global range in V of **S**.

Proof Assume v_i is in the local range in V of $S \Rrightarrow s$. Then, for each member of S, it either satisfies its conclusion or does not satisfy some of its premises. Hence, all members of **S** preserve $\{v_i\}$-validity. Thus, $\{v_i\}$ is in the absolutely global range in V of $S \Rrightarrow s$. □

This means that, if a metainference is locally invalid according to **S** (i.e. there is a counterexample to it in the local range in V of **S**), it will be absolutely globally invalid (i.e. there is a corresponding counterexample in the absolutely global range in V of **S**). The converse, once again, will not hold in general. Consider a calculus with all structural rules of **LK**, plus the following operational rules, which we will call $\mathbf{S_1}$:

$$\frac{\Rightarrow \varphi(\psi)}{\Rightarrow \varphi \vee \psi} \qquad \frac{\Rightarrow \varphi \vee \psi \qquad \varphi(\psi) \Rightarrow}{\Rightarrow \psi(\varphi)}$$

This characterization of disjunction is pretty similar to the one in **LK**. It is straightforward to check that the valuations belonging to the local range in V^2 of $\mathbf{S_1}$ are all those satisfying the following conditions:

- If $v(\varphi) = 1$ or $v(\psi) = 1$, then $v(\varphi \vee \psi) = 1$
- If $v(\varphi) = v(\psi) = 0$, then $v(\varphi \vee \psi) = 0$

That is, their local range is exactly the set of classical valuations for disjunction. This in turn means that the rules validate every classical inference for a language with disjunction. That is, inferentially, they seem to be enough. Metainferentially, however, they are in some sense weaker than those of **LK**. In particular, even though they render the following schema locally valid:

$$P_1 \frac{\varphi \Rightarrow \qquad \psi \Rightarrow}{\varphi \vee \psi \Rightarrow}$$

there is in the absolute global range in V^2 of $\mathbf{S_1}$ an element which is the set $\{v_1, v_2\}$, where v_1 and v_2 behave as follows:

$$v_1(\varphi) = v_1(\psi) = 0 \text{ and } v_1(\varphi \vee \psi) = 1 \qquad v_2(\varphi) = v_2(\psi) = v_2(\varphi \vee \psi) = 0$$

Given that the rule P_1 is not globally valid in $\{v_1, v_2\}$, this thus shows that local validity can *strictly* include absolute global validity.

But then again, this does not occur in the case of **LK**, where absolute $\mathcal{V}^{\mathbf{LK}}$-global validity and local V^{B}-validity do coincide.

2.3.3 Metainferential Soundness and Completeness

So far we presented two proof-theoretic concepts of metainferential validity (derivability and admissibility) and three valuational ones (global, local, and absolute global validity). One may wonder, then, whether in the case of $\mathbb{CL}$ as presented by **LK** and Boolean bivaluations, these align in some interesting way—just like it happens at the inferential level with the **LK**-provable sequents and the V^{B}-valid inferences. The answer is positive. We will refer to the results establishing these connections as metainferential Soundness and Completeness, for they respectively represent the fact that every metainference deemed valid from a proof-theoretic stance will also be considered so from a semantic point of view, and vice versa. As we already advanced, since the relations of derivability and admissibility are finite, these coincidences will be restricted to rules and then extended whenever possible.

First, the notions of admissibility and global validity can be quite easily seen to coincide given some reasonable assumptions, since the former is the preservation of provability over a calculus, while the latter is the preservation of validity. Thus, when some very natural conditions are met, these two are matched:

Theorem 2.3.24 *Given a set of valuations V and a sequent calculus* **S** *that is sound and complete concerning the set of V-valid inferences, a rule is admissible in* **S** *if and only if it is globally V-valid.*

Proof From left to right, assume a rule is admissible in **S** but is not globally V-valid. Then, there is an instance of the schema such that every premise of it is V-valid, but its conclusion is V-invalid. By inferential Soundness, since the conclusion is V-invalid it is not **S**-provable, and by inferential Completeness given that every premise is V-valid, each premise must be **S**-provable. But this is impossible since the rule is admissible in **S**.

From right to left, assume a rule is globally V-valid but it is not admissible in **S**. Then, there must be an instance of it such that each premise is **S**-provable but its conclusion is not. However, by inferential Soundness, if the premises are **S**-provable then they are V-valid. So, since the schema is globally V-valid, the conclusion must be V-valid also, and by inferential Completeness it should be provable. □

From this general result and the inferential Soundness and Completeness results for **LK** discussed earlier, the following fact ensues:

Fact 2.3.25 (Metainferential Soundness and Completeness 1) A rule is admissible in **LK** if and only if it is globally V^{B}-valid.

Regarding the concept of derivability, it can also be paired with a semantic counterpart by appealing to the notion of absolute global validity defined above. The next result was originally proved in (1996) by Humberstone:

Theorem 2.3.26 *Given a set of sets of valuations $\mathcal{V} \subseteq \wp(V^2)$ that is the absolute global range in V^2 of a set of rules* **S** *that includes all the structural rules, a rule is derivable in* **S** *if and only if it is absolutely globally $\mathcal{V}$-valid.*

Using the results proved before, about the presentation of $\mathbb{CL}$ that we have scrutinized so far, this has the following implication:

Fact 2.3.27 (Metainferential Soundness and Completeness 2) A rule is derivable in **LK** if and only if it is absolutely globally $\wp(V^{\mathrm{B}})$-valid.

We now move on to discussing what we may call a series of *collapse results*. These will pertain to necessary or sufficient conditions for some of the notions of metainferential validity to coincide—besides the coincidences mentioned in the theorems above.

This being said we start presenting one of the sufficient conditions for *global and absolutely global validity to collapse*:

Theorem 2.3.28 *Let* V *be a set of valuations;* **S** *be a structural and structurally complete sequent calculus, which is also sound and complete for the set of* V*-valid inferences; and* $\mathcal{V} \subseteq \wp(V^2)$ *be the absolute global range in* V^2 *of* **S**. *Then, a rule is derivable in* **S** *if and only if it is globally* V*-valid, and similarly it is admissible in* **S** *if and only if it is absolutely globally* $\mathcal{V}$*-valid.*

Proof Immediate from the results above. □

In turn, this has the next immediate consequence regarding our favored presentation of $\mathbb{CL}$ and the coincidence of admissibility, derivability, global and absolutely global validity concerning the corresponding ranges:

Fact 2.3.29 A rule is derivable in **LK** if and only if it is globally V^{B}-valid, and similarly it is admissible in **LK** if and only if it is absolutely globally $\wp(V^{\mathrm{B}})$-valid.

These results may or may not leave the concept of local validity without a natural proof-theoretical counterpart. As we shall see next, sometimes (when it collapses with absolute global validity) it coincides with derivability, and some other times (when it collapses with global validity) it coincides with admissibility, and there is no third intermediate candidate—not that we know of, for that matter. This, however, does not mean it should be discarded. On the contrary, its use is quite widespread given, for instance, the fact that it is stable like derivability, but more tractable than absolute global validity.[9] Moreover, there are some metainferential logics which have a valuational characterization in terms of local validity, but for which there is no known calculus. As it happened with the proof-theoretic notions, these are matters which will gain weight in subsequent chapters. There, local validity will be taken as the default standard for characterizing metainferences.

Next, we show several circumstances under which local validity collapses into the other two notions of metainferential validity. Let us, on the one hand, observe what may be sufficient for *local validity to collapse with global validity*. For this result, we need some definitions, first.

[9] The discussion about what is the correct notion of metainferential validity (or even if it makes sense at all to have the discussion) is very prolific and in the last years many authors, including us, have actively participated in it. For instance, see Teijeiro (2021) or Pailos and Da Ré (2023, ch.10).

Definition 2.3.30 A set of valuations V is *truth-constant complete* if and only if for every element a in the set of truth-values, there is a formula φ_a such that for all valuations $v \in V$, $v(\varphi_a) = a$.

Theorem 2.3.31 *Given a set of valuations V based on a valuation scheme X that is truth-constant complete, and a sequent calculus* **S** *that is sound and complete for the set of V-valid inferences, a rule is admissible in* **S** *if and only if it is locally V-valid.*

Proof From left to right, assume a rule is not locally V-valid. Then, it has an instance $S \Rrightarrow s$ that is locally V-invalid, such that there is a valuation $v \in V$ which satisfies its premises but not its conclusion. Take the substitution instance $S' \Rrightarrow s'$ which substitutes every propositional variable p such that $v(p) = a$, for the corresponding truth-constant φ_a such that for all $v \in V$, $v(\varphi_a) = a$. Every valuation satisfies the premises of and none satisfies the conclusion of this instance $S' \Rrightarrow s'$, which makes it globally V-invalid. Hence, the rule in question is not globally V-valid. Therefore, given the results above, it is not admissible in **S**. From right to left, assume a rule is locally V-valid. Then, it is globally V-valid and given the results above it is admissible in **S**. □

This has the effect, regarding $\mathbb{CL}$ as presented via **LK** and Boolean bivaluations, of implying the following, which we already foreshadowed a few pages before when mentioning that global and local V^{B}-validity coincided for rules:

Fact 2.3.32 A rule is admissible in **LK** if and only if it is locally V^{B}-valid.

Notice, on the other hand, that the next result can also be proved given some reasonable assumptions, leading to a scenario where *local and absolutely global validity collapse*:

Theorem 2.3.33 *Given a set of valuations V and a sequent calculus* **S** *that includes all the structural rules such that $\wp(V)$ is the absolute global range in V^2 of* **S**, *a rule is derivable in* **S** *if and only if it is locally V-valid.*

Proof From left to right, assume a rule is locally V-invalid. Thus, there is a valuation $v \in V$ satisfying the premises but not the conclusion. Now, take the set $\{v\}$. This set belongs to the absolute global range of **S**, given that this is no other than $\wp(V)$. Thus, since the metainference in question is not globally $\{v\}$-valid, it is not absolutely globally $\wp(V)$-valid. Whence, it is not derivable in **S**. From right to left, assume a rule is not derivable in **S**. Thus, it is absolutely globally $\wp(V)$-invalid. Hence, there is a $V_i \subseteq V$ such that the premises of the metainference in question are V_i-valid but the conclusion isn't. Hence, there is a $v \in V_i$ such that it is a counterexample to the conclusion but not the premises of said metainference. But $v \in V$ and, thus, the metainference isn't locally V-valid either. □

As regards $\mathbb{CL}$, then, the statement appearing down here becomes true:

Fact 2.3.34 A rule is derivable in **LK** if and only if it is locally V^{B}-valid.

2.3.4 Metainferential Finiteness

To finish this section, we address the question of Finiteness. Recall that, in the case of inferences, the core of the problem was to show Compactness (2.2.15), that is, that finite satisfiability of a set of formulas was enough for satisfiability. However, this is not true for sets of inferences in general. The fact that inferences themselves may have infinitely many premises or conclusions allows for sets of inferences which are finitely satisfiable but not satisfiable *tout court*.

Fact 2.3.35 Sets of inferences are not compact with respect to V^B.

Proof Take the set $S = \{p \Rightarrow : p \in Var\} \cup \{\Rightarrow Var\}$, where Var is the set of all propositional variables. Any finite subset of S is satisfiable, given that for any $\Gamma \subseteq Var$ which is finite, there is a valuation assigning 0 to members of Γ and 1 to some element not in it. However, S is not satisfiable, given that there is no valuation assigning 0 to all propositional variables and 1 to some. □

From this, it follows that there are locally valid metainferences which are essentially infinite:

Fact 2.3.36 There are metainference tokens such that $S \Vdash^{L}_{VB} s$ but there is no finite subset $S_f \subseteq S$ such that $S_f \Vdash^{L}_{VB} s$

Proof Because of Fact 2.3.35, the following metainference is locally valid, but no finite submetainference is:

$$\frac{p_1 \Rightarrow \quad \ldots \quad p_i \Rightarrow \quad \ldots \quad \Rightarrow Var}{\Rightarrow}$$

□

Let us call cases such as the one above 'infinitary':

Definition 2.3.37 Let $\Vdash$ be a validity relation for metainferences. A metainference token $S \Rrightarrow s$ is *infinitary* (with respect to $\Vdash$) if $S \Vdash s$, but there is no finite subset $S_f \subseteq S$ such that $S_f \Vdash s$. If $S \Rrightarrow s$ is valid, but some $S_f \Rrightarrow s$ is also valid, $S \Rrightarrow s$ is *finitary*. A metainference schema is *finitary* iff all of its instances are finitary. A validity relation $\Vdash$ is *infinitary* iff there is an infinitary metainference $S \Rrightarrow s$ such that $S \Vdash s$

The same example displayed in Fact 2.3.36 witnesses that absolute global validity is infinitary too.

Fact 2.3.38 There are metainference tokens such that $S \Vdash^{AG}_{VB} s$ but there is no finite subset $S_f \subseteq S$ such that $S_f \Vdash^{AG}_{VB} s$

You can construct the corresponding counterexamples to the finite cases by taking the singleton $\{v\}$, where v is the local counterexample. The absolute global validity of the infinite metainference follows from the fact that no subset of $\wp(V_B)$ satisfies all premises at once.

Notice that the metainferential Completeness theorems that we stated (Facts 2.3.25 and 2.3.27, and the collapsing results) are what we called 'Finite Completeness' (Theorem 2.2.12), that is, they are relativized to rules, which have a finite number of premises. Thus lies the impossibility to reason from Completeness to Finiteness, which, as we already said, was Henkin's own path. Nevertheless, the negative results above are not all there is to say on this matter, and we can—as we did in the inferential case—reason from some finitary relations to full Completeness.

The first finitary relation is the global one:

Fact 2.3.39 If $S \Vdash^{G}_{V_B} s$, then $S \Rrightarrow s$ is finitary.

Proof Assume for all finite $S_f \subseteq S$, $S_f \nVdash^{G}_{V} s$. This means that s is invalid, while every element of every finite subset of S is valid. But that is just to say that all elements of S are valid, while the conclusion is not, and thus, $S \nVdash^{G}_{V} s$ □

Since locally valid metainferences are globally valid, all infinitary locally valid metainferences will be globally valid too. But because global validity is finitary, and local *invalidity* does not imply global *invalidity* for metainference tokens, these globally valid metainferences will have a globally valid (although locally invalid) finite submetainference. Indeed, in the example above, given that all premises are invalid, *all* finite submetainferences are in fact globally valid. Hence, the fact that local validity implies global validity does not preclude that local infinitariness does not imply global infinitariness.

One may be suspicious of this mismatch, if one recalls that local invalidity does imply global invalidity under the right circumstances—that is, if one has a language with formulas expressing every truth value. But that result holds only for schemas. Indeed, if we take the token above to be the basic instance of a metainferential schema, then that schema has an instance which is not infinitary (i.e., if p is replaced by a tautology, then $p \vee \neg p \Rightarrow \Vdash^{L}_{V} \Rightarrow$).

In fact, all locally valid schemas have a finitary instance:

Fact 2.3.40 If $S \Rrightarrow s$ is a schematic metainference and $S \Vdash^{L}_{V_B} s$, then it has a finitary instance

Proof Let $S \Rrightarrow s$ be a schematic metainference. Either all members of S are V_B-valid schematic inferences, or some is not. If they are all V_B-valid, then the conclusion must be a V_B-valid schematic inference too. Then, for any finite subset of the premises S_f, $S_f \Vdash^{L}_{V_B} s$. If some element $s_n \in S$ is not V_B-valid, then it has an instance s^i_n which is counterexemplified by every valuation[10] (Exercise 19). Thus, the submetainference which goes from s^i_n to the corresponding instance of the conclusion s^i is locally valid. Hence, $S \Rrightarrow s$ has a finitary instance. □

The second finitary relation is the local one, but this time relativized to metainferences made out of sequents. That is, the inferential premises of the rule

[10] That is, the instance is an *anti-inference*, in terms of Chap. 4, see Definition 4.3.5.

may be infinite, but they are such that each of them has a finite number of elements as premises and conclusions.

Fact 2.3.41 For all metainferences $S \Rrightarrow s$ where S is a set of sequents and s is a sequent, and such that $S \Vdash\!\vDash^{L}_{VB} s$, there is a finite subset $S_f \subseteq S$ such that $S_f \Vdash\!\vDash^{L}_{VB} s$

Proof Assume $S \Vdash\!\vDash^{L}_{VB} s$. Since all members of S and s are sequents, their horizontalizations (2.3.6) are well defined. Call them Γ_S and φ_s, respectively. Given that $S \Rrightarrow s$ is locally valid, the inference $\Gamma_S \Rightarrow \varphi_s$ is valid, by Exercise 18. Thus, by inferential Finiteness (2.2.16), there is a finite subset Γ_f of Γ_S such that $\Gamma_f \vDash_{VB} \varphi_s$. Thus, the metainference that goes from the sequents whose horizontalizations are in Γ_f to s is locally valid, by Exercise 18. □

Given all results above, there are many general forms of Soundness and Completeness that follow. We present one of them, as an illustration:

Theorem 2.3.42 $S \Vdash\!\vDash^{G}_{VB} s$ *if and only if* $S \Vdash\!\vdash^{A}_{\mathbf{LK}} s$

Proof Let s be $\Gamma \Rightarrow \Delta$. From left to right, assume $S \Vdash\!\vDash^{G}_{VB} s$. Hence, by Finiteness (2.3.39) there is a finite $S_f \subseteq S$ such that $S_f \Vdash\!\vDash^{G}_{VB} s$. Hence, either some $s_n \in S_f$ is not V_B-valid or s is. Suppose that the latter holds. Thus, by Full Completeness (2.2.17), $\Gamma \vdash_{\mathbf{LK}} \Delta$ and hence, $S \Vdash\!\vdash^{A}_{\mathbf{LK}} s$. Now suppose instead that some s_n, say $\Gamma_n \Rightarrow \Delta_n$ is not V_B-valid. It follows by Monotonicity that no subinference is valid either. By Soundness, none of the finite ones are provable, and thus, $\Gamma_n \nvdash_{\mathbf{LK}} \Delta_n$. Therefore, $S \Vdash\!\vdash^{A}_{\mathbf{LK}} s$.

From right to left, assume $S \Vdash\!\vdash^{A}_{\mathbf{LK}} s$. Hence, by definition there is a finite S_f such either $\Gamma_n \nvdash_{\mathbf{LK}} \Delta_n$, for some $\Gamma_n \Rightarrow \Delta_n \in S$, or $\Gamma \vdash_{\mathbf{LK}} \Delta$. If the latter holds, by Soundness, $\Gamma \vDash_{VB} \Delta$. If the former holds, by Full Completeness $\Gamma_n \nvDash_{VB} \Delta_n$. Therefore, $S \Vdash\!\vDash^{G}_{VB} s$. □

Exercises

Exercise 17 Prove that if a metainference is locally V-valid, it is also globally V-valid.

Exercise 18 Prove that a rule is locally V_B-valid iff its horizontalization is V_B-valid.

Exercise 19 Prove that if a schematic inference is not V_B-valid, then it has an instance such that for every $v \in V_B$, v does not satisfy that instance.

2.4 Cut-elimination

In this section, we will prove that the rule Cut is eliminable in the sequent calculus **LK**. It was Gentzen in (1935) who first proved this result, referring to it as the *Hauptsatz*. The intuitive meaning behind the proof of the elimination of Cut is that every proof in **LK** of a given sequent $\Gamma \Rightarrow \Delta$ can be transformed into another proof of the same sequent in **LK** such that the rule Cut is not applied. In this vein, proving Cut-elimination equals to showing that the rule of Cut is redundant and we could get rid of it without affecting the set of provable sequents.[11]

In order to simplify the proof, in this section, by LW and RW we will refer to the following pair of rules:

$$\text{LW}\ \frac{\Gamma \Rightarrow \Delta}{\Gamma, \Gamma' \Rightarrow \Delta} \quad \text{RW}\ \frac{\Gamma \Rightarrow \Delta}{\Gamma \Rightarrow \Delta', \Delta}$$

with Γ', Δ' nonempty. This modification does not make any substantial difference and allows us to shorten the proofs. In fact, we leave to the reader (Exercise 20) to prove that these rules are interderivable with those presented in Definition 2.2.1.

Following Gentzen, instead of directly proving that Cut is eliminable in **LK**, we will show that an equivalent rule is eliminable and then we will use that result to prove Cut-elimination. Let's start with the definition of such a rule, and we will delay to the end of this section the reason why we cannot prove the elimination of Cut directly:

Definition 2.4.1 Let Mix be the following rule:

$$\text{Mix}\ \frac{\Gamma \Rightarrow \Delta, \overbrace{\varphi, \dots, \varphi}^{n\text{-occurrences}} \qquad \overbrace{\varphi, \dots, \varphi}^{m\text{-occurrences}}, \Sigma \Rightarrow \Pi}{\Gamma, \Sigma \Rightarrow \Delta, \Pi}$$

with $\varphi \notin \Delta, \Sigma$.

So, basically the rule Mix allows to cut in one step all of the occurrences of a given formula, while the rule Cut only allows to cut one occurrence at each step. However, both rules are equivalent in presence of Weakening and Contraction. Let's show first that Cut is derivable from Mix and Weakening:

$$\text{RW}\ \frac{\text{LW}\ \frac{\text{Mix}\ \frac{\Gamma \Rightarrow \Delta, \varphi \qquad \varphi, \Sigma \Rightarrow \Pi}{\Gamma, \Sigma' \Rightarrow \Delta', \Pi}}{\Gamma, \Sigma \Rightarrow \Delta', \Pi}}{\Gamma, \Sigma \Rightarrow \Delta, \Pi}$$

where the Weakening rules were used because φ can occur in Δ, Σ (producing in the above derivation Δ', Σ') and Mix erases all of these occurrences, so we need to add them by Weakening.

[11] For a comprehensive presentation of different proofs of Cut-elimination in **LK** and related systems for $\mathbb{CL}$, see Indrzejczak (2021).

Now, we need to show that Mix is derivable from Cut and Contraction:

$$\dfrac{\text{RC } n-1 \text{ times}\ \dfrac{\Gamma \Rightarrow \Delta, \overbrace{\varphi, \ldots, \varphi}^{n\text{-occurrences}}}{\Gamma \Rightarrow \Delta, \varphi} \qquad \text{LC } m-1 \text{ times}\ \dfrac{\overbrace{\varphi, \ldots, \varphi}^{m\text{-occurrences}}, \Sigma \Rightarrow \Pi}{\Sigma, \varphi \Rightarrow \Pi}}{\Gamma, \Sigma \Rightarrow \Delta, \Pi}\ \text{Cut}$$

So, now that we know that both rules are interderivable, the strategy in what follows is to replace Cut by Mix in **LK** and show that Mix is eliminable in the resulting proof system. As a corollary, we will conclude that Cut is eliminable in **LK**.

Definition 2.4.2 Let **LK*** be the sequent calculus which results from replacing the rule Cut by the rule Mix in **LK**.

Now, we can state the main result of this section:

Theorem 2.4.3 *The rule Mix is eliminable in* **LK***.

Before providing this proof, let's outline our strategy. First, it's worth mentioning—since there are some different strategies to prove this very well-known result—that we will base our proof on Paoli's work in (2013). What we are going to do is to transform each derivation in **LK*** containing one application of Mix into a derivation in **LK*** of the same sequent but not containing any application of the rule Mix. The following definition states formally how we will denote a derivation with only one application of the Mix rule:

Definition 2.4.4 A proof D of the sequent $\Gamma, \Sigma \Rightarrow \Delta, \Pi$ is a $Mix(\varphi)$-proof if it has the following form:

$$\dfrac{\dfrac{\begin{matrix} D' \\ \vdots \end{matrix}}{\Gamma \Rightarrow \Delta, \underbrace{\varphi, \ldots, \varphi}_{n\text{-occurrences}}} \qquad \dfrac{\begin{matrix} D'' \\ \vdots \end{matrix}}{\underbrace{\varphi, \ldots, \varphi}_{m\text{-occurrences}}, \Sigma \Rightarrow \Pi}}{\Gamma, \Sigma \Rightarrow \Delta, \Pi}\ \text{Mix}$$

such that D', D'' don't contain any applications of Mix.

Now, the proof will be done via a double induction, so first, we need to present the measures of the induction:

Definition 2.4.5 The *rank* of a sequent s in a $Mix(\varphi)$-proof D (written $r(s)$) is defined as follows:

- If s belongs to the left branch (i.e. the one that ends with $\Gamma \Rightarrow \Delta, \overbrace{\varphi, \ldots, \varphi}^{n\text{-occurrences}}$), then $r(s)$ is the height of the longest upward branch of the sequents containing φ in the succedent, minus one.

- If s belongs to the right branch (i.e. the one that ends with $\overbrace{\varphi, \ldots, \varphi}^{m\text{-occurrences}}, \Sigma \Rightarrow \Pi$), $r(s)$ is the height of the longest upward branch of the sequents containing φ in the antecedent, minus one.
- $r(\Gamma, \Sigma \Rightarrow \Delta, \Pi) = r(\Gamma \Rightarrow \Delta, \overbrace{\varphi, \ldots, \varphi}^{n\text{-occurrences}}) + r(\overbrace{\varphi, \ldots, \varphi}^{m\text{-occurrences}}, \Sigma \Rightarrow \Pi)$.

The rank of a subderivation D' of a $Mix(\varphi)$-proof D of the sequent s is denoted as $r(D')$ and it is $r(s)$.

Definition 2.4.6 The *grade* of a subderivation D' of a $Mix(\varphi)$-proof D of a sequent s is equal to the complexity of the formula φ, and denoted as $g(D')$.

Definition 2.4.7 The *index* of a subderivation D' of a $Mix(\varphi)$-proof D of a sequent s ($i(D')$) is the ordered pair $i(D') = \langle g(D'), r(D') \rangle$.

In this definition, $i(D') < i(D)$ if and only if $g(D') < g(D)$ or $g(D') = g(D)$ and $r(D') < r(D)$.

Before proving the main result of this section, let's see an example of these measures. Consider the following $Mix(p)$-proof and call it D:

$$
\begin{prooftree}
\AxiomC{$p \Rightarrow p$}
\LeftLabel{LW}
\UnaryInfC{$q, p \Rightarrow p$}
\LeftLabel{RW}
\UnaryInfC{$q, p \Rightarrow p, p$}
\AxiomC{$q \Rightarrow q$}
\LeftLabel{LW}
\UnaryInfC{$q, p \Rightarrow q$}
\LeftLabel{RW}
\UnaryInfC{$q, p \Rightarrow q, p$}
\LeftLabel{R$\wedge$}
\BinaryInfC{$p, q \Rightarrow p \wedge q, p$}
\LeftLabel{RW}
\UnaryInfC{$p, q \Rightarrow p \wedge q, p, p$}
\LeftLabel{RW}
\UnaryInfC{$p, q \Rightarrow p \wedge q, p, p, p$}
\AxiomC{$p \Rightarrow p$}
\LeftLabel{L$\neg$}
\UnaryInfC{$p, \neg p \Rightarrow$}
\LeftLabel{LW}
\UnaryInfC{$p, p, \neg p \Rightarrow$}
\LeftLabel{L$\wedge$}
\UnaryInfC{$p \wedge q, p, \neg p \Rightarrow$}
\LeftLabel{LW}
\UnaryInfC{$p \wedge q, p, p, \neg p \Rightarrow$}
\LeftLabel{Mix}
\BinaryInfC{$p, q, p \wedge q, \neg p \Rightarrow p \wedge q$}
\end{prooftree}
$$

Let's call D_1 the subproof ending in $p, q \Rightarrow p \wedge q, p, p, p$ (i.e. the left subproof), and D_2 the subproof ending in $p \wedge q, p, p, \neg p \Rightarrow$ (i.e. the right subproof). Their respective ranks are $r(D_1) = r(p, q \Rightarrow p \wedge q, p, p, p) = 5$ and $r(D_2) = r(p \wedge q, p, p, \neg p \Rightarrow) = 4$. Thus, $r(D) = r(p, q, p \wedge q, \neg p \Rightarrow p \wedge q) = 9$. Regarding the grade, $g(D_1) = g(D_2) = g(D) = 0$. So the index in each case is the following: $i(D_1) = \langle 0, 5 \rangle$, $i(D_2) = \langle 0, 4 \rangle$ and $i(D) = \langle 0, 9 \rangle$.

Having all of this been presented, we are in a position to prove Theorem 2.4.3:

Proof of Theorem 2.4.3. As we mentioned, we will prove that Mix is eliminable in **LK***.To do so, we will show that every $Mix(\varphi)$-proof can be transformed into a proof not containing any application of Mix. The induction will be carried over the index of a $Mix(\varphi)$-proof D, $i(D)$, where D has the following form:

$$
\begin{prooftree}
\AxiomC{D'}
\noLine
\UnaryInfC{$\vdots$}
\UnaryInfC{$\Gamma \Rightarrow \Delta, \overbrace{\varphi, \ldots, \varphi}^{n\text{-occurrences}}$}
\AxiomC{D''}
\noLine
\UnaryInfC{$\vdots$}
\UnaryInfC{$\overbrace{\varphi, \ldots, \varphi}^{m\text{-occurrences}}, \Sigma \Rightarrow \Pi$}
\LeftLabel{Mix}
\BinaryInfC{$\Gamma, \Sigma \Rightarrow \Delta, \Pi$}
\end{prooftree}
$$

with D', D'' not containing any application of the Mix rule. The strategy is to reduce the rank of the $Mix(\varphi)$-proof where φ is not principal, by permutations on the order of the application of the rules in the derivation. In the case that both premises have φ as principal we will reduce the grade of the $Mix(\varphi)$-proof, applying the Mix rule over less complex formulas than φ (the technique of permuting the order of the application of the formulas doesn't work in this case, since these permutations could augment the rank). Let's see all of this in detail.

Case (a). Assume on the first place that $i(D) = \langle 0, 0\rangle$. In this case, since $g(D) = 0$, φ must be a propositional variable. Now, since $r(D) = 0$, we have two options: either $\Gamma \Rightarrow \Delta, \varphi, \ldots, \varphi$ is an axiom of the form $\varphi \Rightarrow \varphi$ or φ was introduced in the last step by RW, and the same for $\varphi, \ldots, \varphi, \Sigma \Rightarrow \Pi$.[12] So, we need to consider four cases:

(Case i)

$$\text{Mix}\ \dfrac{\varphi \Rightarrow \varphi \qquad \varphi \Rightarrow \varphi}{\varphi \Rightarrow \varphi}$$

(Case ii)

$$\text{Mix}\ \dfrac{\varphi \Rightarrow \varphi \qquad \text{LW}\ \dfrac{\begin{matrix}\vdots \\ \Sigma \Rightarrow \Pi\end{matrix}}{\varphi, \ldots, \varphi, \Sigma \Rightarrow \Pi}}{\varphi, \Sigma \Rightarrow \Pi}$$

(Case iii)

$$\text{Mix}\ \dfrac{\text{RW}\ \dfrac{\begin{matrix}\vdots \\ \Gamma \Rightarrow \Delta\end{matrix}}{\Gamma \Rightarrow \Delta, \varphi, \ldots, \varphi} \qquad \varphi \Rightarrow \varphi}{\Gamma \Rightarrow \varphi, \Delta}$$

(Case iv)

$$\text{Mix}\ \dfrac{\text{RW}\ \dfrac{\begin{matrix}\vdots \\ \Gamma \Rightarrow \Delta\end{matrix}}{\Gamma \Rightarrow \Delta, \varphi, \ldots, \varphi} \qquad \text{LW}\ \dfrac{\begin{matrix}\vdots \\ \Sigma \Rightarrow \Pi\end{matrix}}{\varphi, \ldots, \varphi, \Sigma \Rightarrow \Pi}}{\Gamma, \Sigma \Rightarrow \Pi, \Delta}$$

In *Case i* the application of Mix is redundant. In *Case ii* we can transform them into a Mix-free derivation as follows:

(Transformation of Case ii)

$$\text{LC } m\text{-times}\ \dfrac{\text{LW}\ \dfrac{\begin{matrix}\vdots \\ \Sigma \Rightarrow \Pi\end{matrix}}{\varphi, \ldots, \varphi, \Sigma \Rightarrow \Pi}}{\varphi, \Sigma \Rightarrow \Pi}$$

Notice that in this case, we could also have done it using only Weakening in $\Sigma \Rightarrow \Pi$, which suggests that the transformations are not unique. And using a dual strategy we can transform *Case iii* into a Mix-free derivation. In *Case iv*, we have the following transformation:

(Transformation of Case iv)

$$\text{RW}\ \dfrac{\text{LW}\ \dfrac{\begin{matrix}\vdots \\ \Sigma \Rightarrow \Pi\end{matrix}}{\Gamma, \Sigma \Rightarrow \Pi}}{\Gamma, \Sigma \Rightarrow \Delta, \Pi}$$

[12] Unless necessary, for readability reasons, we omit here and throughout the proof the superscripts m and n indicating the number of occurrences of the formula φ.

Case (b). Let's assume now, on the second place, that $i(D) = \langle 0, k \rangle$, with $1 \leq k$. This means that the mix formula is a propositional variable, but the rank of the derivation is greater than 0. Here, we need to consider two main cases: either $r(\Gamma \Rightarrow \Delta, \varphi, \ldots, \varphi) \geq 1$ or $r(\varphi, \ldots, \varphi, \Sigma \Rightarrow \Pi) \geq 1$. Assume that $r(\Gamma \Rightarrow \Delta, \varphi, \ldots, \varphi) \geq 1$ (the other case will be dual). Then, φ is side, auxiliary or principal in $\Gamma \Rightarrow \Delta, \varphi, \ldots, \varphi$. So, we need to distinguish these three cases:

(Subcase 1) Suppose first that φ is a side formula. In this case, the idea is to transform the derivation into another one where the rank of the derivation decreases. The way to do it is by pushing the application of the Mix rule upwards.

For example, assume $\Gamma = \Gamma^*, \chi \wedge \psi$. So we have the following derivation where the last rule applied on the left is L$\wedge$ and φ is side:

$$\dfrac{\mathrm{L}\wedge\,\dfrac{\begin{matrix}\vdots\\ \psi, \Gamma^* \Rightarrow^{l} \Delta, \varphi, \ldots, \varphi\end{matrix}}{\chi \wedge \psi, \Gamma^* \Rightarrow^{l+1} \Delta, \varphi, \ldots, \varphi} \qquad \begin{matrix}\vdots\\ \varphi, \ldots, \varphi, \Sigma \Rightarrow^{r} \Pi\end{matrix}}{\chi \wedge \psi, \Gamma^*, \Sigma \Rightarrow^{l+1+r} \Delta, \Pi}\,\mathrm{Mix}$$

where the superscripts indicate the rank, i.e. $l + 1 + r = k$. By pushing the Mix upwards we can transform it into the following derivation:

$$\dfrac{\dfrac{\begin{matrix}\vdots\\ \psi, \Gamma^* \Rightarrow^{l} \Delta, \varphi, \ldots, \varphi\end{matrix} \qquad \begin{matrix}\vdots\\ \varphi, \ldots, \varphi, \Sigma \Rightarrow^{r} \Pi\end{matrix}}{\psi, \Gamma^*, \Sigma \Rightarrow^{l+r} \Delta, \Pi}\,\mathrm{Mix}}{\chi \wedge \psi, \Gamma^*, \Sigma \Rightarrow \Delta, \Pi}\,\mathrm{L}\wedge$$

which has a Mix-proof subderivation with lower rank ($l + r = k - 1$). And so, by the inductive hypothesis, this last derivation can be transformed into a Mix-free proof. Let's see another example of this method, where $\Delta = \Delta^*, \chi \wedge \psi$. In this case, the last rule applied on the left premise is R$\wedge$, and again φ is a side formula:

$$\dfrac{\dfrac{\begin{matrix}\vdots\\ \Gamma \Rightarrow^{l} \Delta^*, \psi, \varphi, \ldots, \varphi\end{matrix} \qquad \begin{matrix}\vdots\\ \Gamma \Rightarrow^{s} \Delta^*, \chi, \varphi, \ldots, \varphi\end{matrix}}{\Gamma \Rightarrow^{max\{l,s\}+1} \chi \wedge \psi, \Delta^*, \varphi, \ldots, \varphi}\,\mathrm{R}\wedge \qquad \begin{matrix}\vdots\\ \varphi, \ldots, \varphi, \Sigma \Rightarrow^{r} \Pi\end{matrix}}{\Sigma, \Gamma \Rightarrow^{max\{l,s\}+1+r} \Pi, \Delta^*, \chi \wedge \psi}\,\mathrm{Mix}$$

where $k = max\{l, s\} + 1 + r$. Again, by pushing the application of Mix upwards we can transform this derivation into:

$$\dfrac{\dfrac{\begin{matrix}\vdots\\ \Gamma \Rightarrow^{l} \Delta^*, \psi, \varphi, \ldots, \varphi\end{matrix} \quad \begin{matrix}\vdots\\ \varphi, \ldots, \varphi, \Sigma \Rightarrow^{r} \Pi\end{matrix}}{\Sigma, \Gamma \Rightarrow^{l+r} \Pi, \Delta^*, \psi}\,\mathrm{Mix} \qquad \dfrac{\begin{matrix}\vdots\\ \Gamma \Rightarrow^{s} \Delta^*, \chi, \varphi, \ldots, \varphi\end{matrix} \quad \begin{matrix}\vdots\\ \varphi, \ldots, \varphi, \Sigma \Rightarrow^{r} \Pi\end{matrix}}{\Sigma, \Gamma \Rightarrow^{s+r} \Pi, \Delta^*, \chi}\,\mathrm{Mix}}{\Gamma, \Sigma \Rightarrow \chi \wedge \psi, \Delta^*, \Pi}\,\mathrm{R}\wedge$$

where the applications of Mix have rank $s + r$ and $l + r$, and it's straightforward to check that both are lesser than $max\{l, s\} + 1 + r$. Thus, we have reduced the rank of the application of Mix, and by the inductive hypothesis, there is a Mix-free derivation of both sequents $\Sigma, \Gamma \Rightarrow \Pi, \Delta^*, \psi$ and $\Sigma, \Gamma \Rightarrow \Pi, \Delta^*, \chi$. Therefore, there is a Mix-free derivation of $\Gamma, \Sigma \Rightarrow \chi \wedge \psi, \Delta^*, \Pi$.

The same strategy applies to the other rules and are left as an exercise to the reader (for this, check the case of the conditional which is slightly more complex because the contexts are not the same).

(Subcase 2) Assume now that φ is an auxiliary formula. In this case, the strategy is the same as in the previous case, i.e. to push the application of the Mix rule upwards and apply it with a lower rank. However, in some cases, we also need to apply some Weakening or Contraction. For instance, let's consider the case that the last rule applied is L $\rightarrow$:

$$\text{Mix}\ \dfrac{\text{L}\!\rightarrow\ \dfrac{\begin{matrix}\vdots\\ \Gamma_1 \Rightarrow^{l_1} \Delta_1, \varphi, \varphi, \dots, \varphi\end{matrix} \qquad \begin{matrix}\vdots\\ \psi, \Gamma_2 \Rightarrow^{l_2} \Delta_2, \varphi, \dots, \varphi\end{matrix}}{\Gamma_1, \Gamma_2, \varphi \rightarrow \psi \Rightarrow^{max\{l_1, l_2\}+1} \Delta_1, \Delta_2, \varphi, \dots, \varphi} \qquad \begin{matrix}\vdots\\ \varphi, \dots, \varphi, \Sigma \Rightarrow^{r} \Pi\end{matrix}}{\Sigma, \Gamma_1, \Gamma_2, \varphi \rightarrow \psi \Rightarrow^{max\{l_1, l_2\}+1+r} \Pi, \Delta_1, \Delta_2}$$

where the rank $k = max\{l_1, l_2\} + 1 + r$. Now, we push the Mix rule upwards, but to obtain the same sequent we need to apply some structural rules:

$$\text{RC several times}\ \dfrac{\text{LC several times}\ \dfrac{\text{L}\!\rightarrow\ \dfrac{\text{RW}\ \dfrac{\text{Mix}\ \dfrac{\begin{matrix}\vdots\\ \Gamma_1 \Rightarrow^{l_1} \Delta_1, \varphi, \varphi, \dots, \varphi\end{matrix} \qquad \begin{matrix}\vdots\\ \varphi, \dots, \varphi, \Sigma \Rightarrow^{r} \Pi\end{matrix}}{\Sigma, \Gamma_1 \Rightarrow^{l_1+r} \Pi, \Delta_1}}{\Sigma, \Gamma_1 \Rightarrow \Pi, \Delta_1, \varphi} \qquad \text{Mix}\ \dfrac{\begin{matrix}\vdots\\ \psi, \Gamma_2 \Rightarrow^{l_2} \Delta_2, \varphi, \dots, \varphi\end{matrix} \qquad \begin{matrix}\vdots\\ \varphi, \dots, \varphi, \Sigma \Rightarrow^{r} \Pi\end{matrix}}{\psi, \Sigma, \Gamma_2 \Rightarrow^{l_2+r} \Pi, \Delta_2}}{\Gamma_1, \Gamma_2, \Sigma, \Sigma, \varphi \rightarrow \psi \Rightarrow \Delta_1, \Delta_2, \Pi, \Pi}}{\Gamma_1, \Gamma_2, \Sigma, \varphi \rightarrow \psi \Rightarrow \Delta_1, \Delta_2, \Pi, \Pi}}{\Gamma_1, \Gamma_2, \Sigma, \varphi \rightarrow \psi \Rightarrow \Delta_1, \Delta_2, \Pi}$$

Here again we have reduced the rank of the Mix-proof and thus by inductive hypothesis we have a Mix-free proof of the sequent. The other cases are left to the reader as exercises.

(Subcase 3) Finally, assume that φ is a principal formula. Since here we are assuming that $g(D) = 0$ (i.e. the formula is a propositional variable), we only need to consider two cases. Either the occurrence of φ was obtained by Weakening or by Contraction:

(Case i)

$$\text{Mix}\ \dfrac{\text{RC}\ \dfrac{\begin{matrix}\vdots\\ \Gamma \Rightarrow^{l} \Delta, \varphi, \dots, \varphi, \varphi\end{matrix}}{\Gamma \Rightarrow^{l+1} \Delta, \varphi, \dots, \varphi} \qquad \begin{matrix}\vdots\\ \varphi, \dots, \varphi, \Sigma \Rightarrow^{r} \Pi\end{matrix}}{\Gamma, \Sigma \Rightarrow^{l+1+r} \Pi, \Delta}$$

(Case ii)

$$\text{Mix}\ \dfrac{\text{RW}\ \dfrac{\begin{matrix}\vdots\\ \Gamma \Rightarrow^{l} \Delta, \varphi, \dots, \varphi\end{matrix}}{\Gamma \Rightarrow^{l+1} \Delta, \varphi, \dots, \varphi, \varphi} \qquad \begin{matrix}\vdots\\ \varphi, \dots, \varphi, \Sigma \Rightarrow^{r} \Pi\end{matrix}}{\Gamma, \Sigma \Rightarrow^{l+1+r} \Pi, \Delta}$$

In both cases, we only reduce the rank of the derivation by moving the application of Mix upwards:

(Transformation of Case i)

$$\dfrac{\begin{matrix}\vdots & & \vdots \\ \Gamma \Rightarrow^{l} \Delta, \varphi, \ldots, \varphi, \varphi & & \varphi, \ldots, \varphi, \Sigma \Rightarrow^{r} \Pi\end{matrix}}{\Gamma, \Sigma \Rightarrow^{l+r} \Pi, \Delta}\ \text{Mix}$$

(Transformation of Case ii)

$$\dfrac{\begin{matrix}\vdots & & \vdots \\ \Gamma \Rightarrow^{l} \Delta, \varphi, \ldots, \varphi & & \varphi, \ldots, \varphi, \Sigma \Rightarrow^{r} \Pi\end{matrix}}{\Gamma, \Sigma \Rightarrow^{l+r} \Pi, \Delta}\ \text{Mix}$$

Since we reduce the rank, we can apply the inductive hypothesis and conclude that both derivation can be transformed into Mix-free proofs.

Case (c). Now, suppose that $i(D) = \langle k, 0\rangle$, with $1 \leq k$. In this case, Since $r(D) = 0$, the formula φ occurs for the first time in the succedent in $\Gamma \Rightarrow \Delta, \varphi, \ldots, \varphi$ and in the antecedent in $\Sigma, \varphi, \ldots, \varphi \Rightarrow \Pi$. In other words, φ must be principal in both sequents and cannot be obtained by Contraction. In the case that one of the sequents is obtained by Weakening we have the following situation:

$$\text{Mix}\ \dfrac{\text{RW}\ \dfrac{\begin{matrix}\vdots \\ \Gamma \Rightarrow \Delta\end{matrix}}{\Gamma \Rightarrow \Delta, \varphi, \ldots, \varphi} \qquad \begin{matrix}\vdots \\ \Sigma, \varphi, \ldots, \varphi \Rightarrow \Pi\end{matrix}}{\Gamma, \Sigma \Rightarrow \Pi, \Delta}$$

that can be transformed into a Mix-free proof directly as follows:

$$\text{LW}\ \dfrac{\text{RW}\ \dfrac{\begin{matrix}\vdots \\ \Gamma \Rightarrow \Delta\end{matrix}}{\Gamma \Rightarrow \Delta, \Pi}}{\Gamma, \Sigma \Rightarrow \Pi, \Delta}$$

In the case that both sequents are obtained by operational rules the strategy is to transform each Mix-proof into a derivation where the rule Mix is applied over a less complex formula, reducing the grade of the derivation, and thus the index. So, let's see some cases (the other cases are left to the reader as exercises):

(Case i)

$$\text{Mix}\ \dfrac{\text{R}\vee\ \dfrac{\begin{matrix}\vdots \\ \Gamma \Rightarrow \Delta, \psi\end{matrix}}{\Gamma \Rightarrow \Delta, \psi \vee \chi} \qquad \text{L}\vee\ \dfrac{\begin{matrix}\vdots & & \vdots \\ \Sigma, \psi \Rightarrow \Pi & & \Sigma, \chi \Rightarrow \Pi\end{matrix}}{\Sigma, \psi \vee \chi \Rightarrow \Pi}}{\Gamma, \Sigma \Rightarrow \Pi, \Delta}$$

(Case ii)

$$\text{Mix}\ \dfrac{\text{R}\rightarrow\ \dfrac{\begin{matrix}\vdots \\ \Gamma, \psi \Rightarrow \Delta, \chi\end{matrix}}{\Gamma \Rightarrow \Delta, \psi \rightarrow \chi} \qquad \text{L}\rightarrow\ \dfrac{\begin{matrix}\vdots & & \vdots \\ \Sigma_1 \Rightarrow \Pi_1, \psi & & \Sigma_2, \chi \Rightarrow \Pi_2\end{matrix}}{\Sigma_1, \Sigma_2, \psi \rightarrow \chi \Rightarrow \Pi_1, \Pi_2}}{\Gamma, \Sigma_1, \Sigma_2 \Rightarrow \Pi_1, \Pi_2, \Delta}$$

And each of this derivations can be transformed into the following ones (reducing the complexity of the Mix-proof):

(Transformation of Case i)

$$\text{RW}\ \dfrac{\text{LW}\ \dfrac{\text{Mix}\ \dfrac{\begin{matrix}\vdots & & \vdots \\ \Gamma \Rightarrow \Delta, \psi & & \Sigma, \psi \Rightarrow \Pi\end{matrix}}{\Gamma, \Sigma^{*} \Rightarrow \Pi, \Delta^{*}}}{\Gamma, \Sigma \Rightarrow \Pi, \Delta^{*}}}{\Gamma, \Sigma \Rightarrow \Pi, \Delta}$$

(Transformation of Case ii)

$$\dfrac{\dfrac{\begin{matrix}\vdots\\ \Gamma, \psi \Rightarrow \Delta, \chi\end{matrix} \qquad \begin{matrix}\vdots\\ \Sigma_1 \Rightarrow \Pi_1, \psi\end{matrix}}{\Gamma^*, \Sigma_1 \Rightarrow \Pi_1^*, \Delta, \chi}\,\text{Mix} \qquad \begin{matrix}\vdots\\ \Sigma_2, \chi \Rightarrow \Pi_2\end{matrix}}{\dfrac{\dfrac{\Gamma^*, \Sigma_1, \Sigma_2^* \Rightarrow \Pi_1^*, \Delta^*}{\Gamma, \Sigma_1, \Sigma_2 \Rightarrow \Pi_1^*, \Delta^*}\,\text{LW}}{\Gamma, \Sigma_1, \Sigma_2 \Rightarrow \Pi_1, \Delta}\,\text{RW}}\,\text{Mix}$$

where in the transformations the last applications of Weakening recover the occurrences that could be erased by the applications of Mix (in the case that there are no such occurrences, these Weakenings can be omitted). Note that in all of these transformations we are applying the rule Mix in less complex formulas, and therefore by inductive hypothesis we have a Mix-free derivation of all of these sequents.

Case (d). Finally, assume that $i(D) = \langle k, j\rangle$, with $1 \leq k, j$. In this last case, we need to consider three cases: either φ is auxiliary, it is side or it is principal. In the first two cases, it's straightforward to check that the strategy of pushing the application of Mix upwards does the job, as was shown before. Let's just consider the case that φ is principal, and $r(\Gamma \Rightarrow \Delta, \varphi, \ldots, \varphi) \geq 1$. Assume also that $\varphi \notin \Gamma, \Pi$ (otherwise by Weakening and Contraction we can obtain a Mix-free proof of $\Gamma, \Sigma \Rightarrow \Delta, \Pi$ (left to the reader)). Then, we need to consider how φ is obtained by checking all of the possible last rules. Let's consider just the case that $\varphi = \psi \wedge \chi$ (the other cases are left to the reader). Then, we have the following derivation:

$$\dfrac{\dfrac{\begin{matrix}\vdots\\ \Gamma \Rightarrow^i \Delta, \psi\wedge\chi, \ldots, \psi\end{matrix} \qquad \begin{matrix}\vdots\\ \Gamma \Rightarrow^l \Delta, \psi\wedge\chi, \ldots, \chi\end{matrix}}{\Gamma \Rightarrow^{max\{i,l\}+1} \Delta, \psi\wedge\chi, \ldots, \psi\wedge\chi}\,\text{R}\wedge \qquad \begin{matrix}\vdots\\ \Sigma, \psi\wedge\chi, \ldots, \psi\wedge\chi \Rightarrow^r \Pi\end{matrix}}{\Gamma, \Sigma \Rightarrow^{max\{i,l\}+1+r} \Delta, \Pi}\,\text{Mix}$$

Now we proceed to transform this proof into a derivation where Mix is applied with lower ranks as follows:

$$\dfrac{\dfrac{\dfrac{\begin{matrix}\vdots\\ \Gamma \Rightarrow^i \Delta, \psi\wedge\chi, \ldots, \psi\end{matrix} \quad \begin{matrix}\vdots\\ \Sigma, \psi\wedge\chi, \ldots, \psi\wedge\chi \Rightarrow^r \Pi\end{matrix}}{\Gamma, \Sigma \Rightarrow^{i+r} \Delta, \Pi, \psi}\,\text{Mix} \quad \dfrac{\begin{matrix}\vdots\\ \Gamma \Rightarrow^l \Delta, \psi\wedge\chi, \ldots, \chi\end{matrix} \quad \begin{matrix}\vdots\\ \Sigma, \psi\wedge\chi, \ldots, \psi\wedge\chi \Rightarrow^r \Pi\end{matrix}}{\Gamma, \Sigma \Rightarrow^{l+r} \Delta, \Pi, \chi}\,\text{Mix}}{\Gamma, \Sigma \Rightarrow^0 \Delta, \Pi, \psi\wedge\chi}\,\text{R}\wedge \quad \begin{matrix}\vdots\\ \Sigma, \psi\wedge\chi, \ldots, \psi\wedge\chi \Rightarrow^r \Pi\end{matrix}}{\dfrac{\dfrac{\Gamma, \Sigma, \Sigma \Rightarrow^r \Delta, \Pi, \Pi}{\Gamma, \Sigma \Rightarrow \Delta, \Pi, \Pi}\,\text{LC several times}}{\Gamma, \Sigma \Rightarrow \Delta, \Pi}\,\text{RC several times}}\,\text{Mix}$$

Notice that in this derivation all of the applications of Mix have a lower rank than $max\{i, l\} + 1 + r$, so we can transform each subproof into a Mix-free proof and finally, we can obtain a Mix-free proof of $\Gamma, \Sigma \Rightarrow \Delta, \Pi$. With this, we conclude the proof.

□

With this, we finish the proof of the elimination of Mix in **LK***. Because Mix and Cut are interderivable in the presence of Weakening and Contraction, we have the following corollary:

Corollary 2.4.8 *The rule Cut is eliminable in* **LK**.

Now, we can explain why we need to prove the elimination of Mix and we cannot use a similar strategy to prove directly the elimination of Cut in **LK**. The problem noticed by Gentzen is the following. Assume we have the following derivation:

$$\text{Cut}\ \dfrac{\text{RC}\ \dfrac{\Gamma \Rightarrow \Delta, \varphi, \varphi}{\Gamma \Rightarrow \Delta, \varphi} \qquad \varphi, \Sigma \Rightarrow \Pi}{\Gamma, \Sigma \Rightarrow \Delta, \Pi}$$

When transforming this derivation by using the technique of pushing the Cuts upwards and permuting the order of the application of the rules, we have the following situation:

$$\text{RC}\ \dfrac{\text{LC}\ \dfrac{\text{Cut}\ \dfrac{\text{Cut}\ \dfrac{\Gamma \Rightarrow \Delta, \varphi, \varphi \qquad \varphi, \Sigma \Rightarrow \Pi}{\Gamma, \Sigma \Rightarrow \Delta, \Pi, \varphi} \qquad \varphi, \Sigma \Rightarrow \Pi}{\Gamma, \Sigma, \Sigma \Rightarrow \Delta, \Pi, \Pi}}{\Gamma, \Sigma \Rightarrow \Delta, \Pi, \Pi}}{\Gamma, \Sigma \Rightarrow \Delta, \Pi}$$

where it was intended to reduce the rank of the application of Cut in the derivation. However, as it is easy to notice in the second application of Cut the rank of φ can increase. The solution to this problem is eliminating an equivalent rule (the rule Mix), which in some sense allows us to contract as many formulas as possible in one step, decreasing the height of the longest branch of the derivation containing that formula.

Exercises

Exercise 20 Show that the following pair of rules:

$$\text{LW}\ \dfrac{\Gamma \Rightarrow \Delta}{\Gamma, \Gamma' \Rightarrow \Delta} \qquad \text{RW}\ \dfrac{\Gamma \Rightarrow \Delta}{\Gamma \Rightarrow \Delta', \Delta}$$

with Γ', Δ' nonempty is interderivable with the pair LW and RW presented in Definition 2.2.1.

Exercise 21 Complete the proof of Theorem 2.4.3, by considering the rest of the connectives in Cases (b), (c), and (d).

2.4.1 Consequences of Cut-elimination

Before moving on, we would like to highlight some consequences of the Corollary 2.2.20, i.e. Cut-elimination. For instance, Cut-free systems with only introduction rules (e.g. **LK** without Cut) enjoy the *subformula property*. Intuitively speaking, since the only rule that allows to eliminate formulas in a derivation is Cut, once Cut is absent, all the derivations consist in successions of applications of introduction rules, and therefore all of the formulas that appear in the derivation must occur as subformulas of the conclusion. More formally:

Definition 2.4.9 Let D be a proof in a sequent calculus **S**:

$$\frac{\begin{matrix} D \\ \vdots \end{matrix}}{\Gamma \Rightarrow \Delta}$$

D has the *subformula property* if and only if all of the formulas occurring in the sequents in D are subformulas of Γ or Δ.

Definition 2.4.10 A sequent calculus **S** has the *subformula property* if and only if every proof D in **S** has the subformula property.

Now, as a consequence of the Cut-elimination Theorem, we can state the following:

Fact 2.4.11 For every proof D of a sequent in **LK**, there exists a proof D' of the same sequent enjoying the subformula property.

Proof It's a direct consequence of Cut-elimination. □

And, as a consequence:

Corollary 2.4.12 **LK** *without the rule Cut has the subformula property.*

Proof It's straightforward and we left the proof to the reader. □

As we mentioned, this is not surprising at all since all of the rules of **LK** are introduction rules except Cut. So, although it is very common to find a direct association between Cut-free systems and subformula property, this relation depends on the absence of other elimination rules. For instance, if we remove Cut from **LK** but we add the following rule:

$$\text{Meta Modus Ponens}\ \frac{\Gamma \Rightarrow \varphi \rightarrow \psi, \Delta \qquad \Sigma \Rightarrow \varphi, \Pi}{\Gamma, \Sigma \Rightarrow \psi, \Delta, \Pi}$$

then, in this system, the following proof is legitimate:

$$
\begin{prooftree}
\AxiomC{$q \Rightarrow q$}
\LeftLabel{R$\neg$}
\UnaryInfC{$\Rightarrow q, \neg q$}
\LeftLabel{R$\vee$}
\UnaryInfC{$\Rightarrow q \vee \neg q$}
\LeftLabel{LW}
\UnaryInfC{$p \vee \neg p \Rightarrow q \vee \neg q$}
\LeftLabel{R$\rightarrow$}
\UnaryInfC{$\Rightarrow (p \vee \neg p) \rightarrow (q \vee \neg q)$}
\AxiomC{$p \Rightarrow p$}
\LeftLabel{R$\neg$}
\UnaryInfC{$\Rightarrow p, \neg p$}
\LeftLabel{R$\vee$}
\UnaryInfC{$\Rightarrow p \vee \neg p$}
\LeftLabel{Meta Modus Ponens}
\BinaryInfC{$\Rightarrow q \vee \neg q$}
\end{prooftree}
$$

and in the conclusion there is no occurrence of $p, \neg p, p \vee \neg p$. Of course, notice that this is not an optimal derivation. Actually, $q \vee \neg q$ has a simpler proof (which suggests that Meta Modus Ponens is also eliminable). However, what we are showing is that in the Cut-free proof systems where some elimination rules are present, you can produce legitimate proofs where the subformula property doesn't hold.

Another very important consequence of Cut-elimination (or strictly speaking, from the subformula property) is conservativity.

Definition 2.4.13 A sequent calculus $\mathbf{S}'$ defined over a language $\mathcal{L}'$ *conservatively extends* a sequent calculus $\mathbf{S}$ defined over a language $\mathcal{L} \subseteq \mathcal{L}'$, if and only if every $\mathbf{S}'$-provable sequent whose vocabulary is in $\mathcal{L}$ has a proof in $\mathbf{S}$.

Putting things differently, a conservative extension of a system doesn't prove new sequents containing only the old vocabulary. In other words, restricted to the common vocabulary, they are the same system. The following fact is very important for reasons that will become clear in the next chapters:

Fact 2.4.14 The extension of any Cut-free system (containing only introduction rules) with introduction rules results in a conservative extension.

Proof Immediate from the subformula property. □

This implies that if we add to **LK** without Cut a pair of introduction rules regimenting the behavior of new non-logical vocabulary this new system will be a conservative extension over the logical vocabulary of **LK**. This will be particularly relevant when we discuss theories of truth and we define the truth predicate via introduction rules in Cut-free systems.

A somewhat related phenomenon is the following. As we mentioned, any extension of a Cut-free system with introduction rules is a conservative extension. Not only in the extended system it is not possible to prove any new sequent containing only the old vocabulary, but also the empty sequent cannot be proved if in the base system it couldn't. Let's put these things more formally.

Definition 2.4.15 A sequent calculus $\mathbf{S}$ is trivial if and only if every sequent $\Gamma \Rightarrow \Delta$ is $\mathbf{S}$-provable.

Notice that in any sequent calculus that contains the rules LW and RW, deriving the empty sequent is equivalent to being trivial (see Exercise 22). So, in particular, we have the following:

Fact 2.4.16 Let $\mathbf{S}'$ be a sequent calculus defined over a language $\mathcal{L}'$ by adding some introduction rules to another sequent calculus **S** defined only by introduction rules over a language $\mathcal{L} \subseteq \mathcal{L}'$. If **S** is not trivial and it is Cut-free then $\mathbf{S}'$ is not trivial.

Proof Immediate from the subformula property. □

Exercises

Exercise 22 Prove that in any sequent calculus that contains the rules LW and RW, deriving the empty sequent is equivalent to being trivial.

2.5 Substructural Logics

In this short and last section of this chapter, we introduce the definition of substructural logic that we will assume throughout this book. As we already mentioned, in Definition 2.1.4 a logic is a set of inferences and metainferences, closed under uniform substitution. Thus, in particular, to define Classical Logic we need to determine a particular set of metainference schemas, call them *classical metainferences* (that's what we did in Sect. 2.3). Among them, we can distinguish between structural and operational ones, the latter involving some logical vocabulary explicitly in their formulations, while the former being completely independent of it. This being said, we can present our definition:

Definition 2.5.1 A logic $\mathbb{L}$ is substructural if it does not contain some classical structural metainference.

The main structural metainferences that will be considered in this book are those corresponding to the structural rules of **LK**: Cut, Identity, Contraction, and Weakening. Thus, in the next chapters, different logics lacking some of these four classically structural metainferences will be thoroughly studied.

Bibliography

Barrio, E. A., Pailos, F., & Szmuc, D. (2020). A hierarchy of classical and paraconsistent logics. *Journal of Philosophical Logic, 49*, 93–120.

Bimbó, K. (2014). *Proof theory: Sequent calculi and related formalisms*. CRC Press.

Dicher, B. (2020). Hopeful monsters: A note on multiple conclusions. *Erkenntnis, 85*(1), 77–98.

Garson, J. W. (1990). Categorical semantics. In J. M. Dunn, & A. Gupta (Eds.), *Truth or consequences: Essays in honor of nuel belnap* (pp. 155–175). Springer.

Gentzen, G. (1935). Untersuchungen über das logische schließen. *Mathematische Zeitschrift (english translation in Gentzen 1969), 39*(2), 176–210, 405–431.

Henkin, L. (1949). The completeness of the first-order functional calculus. *Journal of Symbolic Logic, 14*(3), 159–166.

Humberstone, L. (1996). Valuational semantics of rule derivability. *Journal of philosophical logic, 25*(5), 451–461.

Humberstone, L. (2011). *The connectives*. MIT Press.

Indrzejczak, A. (2021). *Sequents and trees*. Birkhäuser.

Ketonen, O. (1944). *Untersuchungen zum Pradikatenkalkul. Annales Academiae Scientiarum Fennicae Series A. I., 32*.

Negri, S., & Von Plato, J. (2008) . *Structural proof theory*. Cambridge University Press.

Oller, C. (2014). Is classical negation a contradictory-forming operator? *Notae Philosophicae Scientiae Formalis, 3*(1), 1–7.

Oller, C. (2016). Contradictoriness, paraconsistent negation and non-intended models of classical logic. In H. Andreas, & P. Verdée (Eds.), *Logical studies of paraconsistent reasoning in science and mathematics* (pp. 103–109). Springer.

Pailos, F., & Da Ré, B. (2023). *Metainferential logics*. Springer.

Paoli, F. (2013). *Substructural logics: A primer* (Vol. 13). Springer Science & Business Media.

Pogorzelski, W. A. (1971). Structural completeness of the propositional calculus. *Bulletin de l'Académie Polonaise des Sciences, 19*(5), 349–351.

Pogorzelski, W. A., & Wojtylak, P. (2008). *Completeness theory for propositional logics*. Birkhäuser.

Restall, G. (2005). Multiple conclusions. In *Logic, methodology and philosophy of science: Proceedings of the twelfth international congress* (pp. 189–205). Citeseer.

Rumfitt, I. (2008). Knowledge by deduction. *Grazer Philosophische Studien, 77*(1), 61–84.

Shoesmith, D., & Smiley, T. (1978). *Multiple-conclusion logic*. Cambridge University Press.

Steinberger, F. (2011). Why conclusions should remain single. *Journal of Philosophical Logic, 40*(3), 333–355.

Teijeiro, P. (2021). Strength and stability. *Análisis Filosófico, 41*(2), 337–349.

Chapter 3
Non-Transitive Logics

Agustina Borzi and Camillo Fiore

Abstract This chapter studies in depth, both from a model- and a proof-theoretic perspective, various logics in the so-called "strict-tolerant" family. We provide results of Inferential and Metainferential Soundness and Completeness. We analyze the local, global and absolute global *st*-ranges of the relevant calculi, relative to the set of all trivaluations for the language. We explain the well-known relationship between the metainferences and inferences valid in *st*- and *tt*-consequence relations, respectively. Finally, we show that the strict-tolerant phenomenon generalizes to every non-trivial Tarskian logic. Most of the contents of this chapter are known to the literature. Some of them, however, are either new or correct some (small) mistakes in other articles.

3.1 Introduction

In this chapter we study a number of non-transitive logics that fall under what is known as the strict-tolerant family; the most prominent among them is logic $\mathbb{ST}$, which, as we will see later (Chap. 6), has been extensively applied to deal with paradoxes of various kinds.

The logics in the strict-tolerant family can be introduced both from a proof-theoretic and a model-theoretic standpoint. Very roughly, from a proof-theoretic perspective, they result from taking a calculus where the rule of Cut is eliminable and dropping this rule—while perhaps also adding some elimination rules in the process. In the resulting systems Cut is still admissible, meaning that the provable sequents are closed under it; typically, however, the rule is no longer derivable.[1]

[1] The qualification "typically" is because the rule *may* remain derivable after all. Intuitively, this would happen if the calculus had independent resources to derive the rule even after we dropped it. But none of the cases we will consider will be of this nature.

A. Borzi · C. Fiore (✉)
UBA/IIF-SADAF-CONICET, Bulnes, Argentina

P. Teijeiro, E. A. Barrio (eds.), *Metainferences in Substructural Logics*, Synthese Library 528, https://doi.org/10.1007/978-3-032-05449-4_3

Meanwhile, from a model-theoretic standpoint, the logics in the family result from adding to one's semantics a new, non-classical value, and adopting a notion of consequence according to which this value is designated for conclusions, but not for premises. The intended result is that the new system has the same valid inferences as the original one, and hence Cut is globally valid in it;[2] however, the rule is no longer valid in the local sense.

The structure of the chapter is somehow analogous to that of Chap. 2. The bulk of the chapter is contained in Sects. 3.2 and 3.3, where we study the strict-tolerant counterparts of Classical Logic—finding our main character, $\mathbb{ST}$, among them. Section 3.2 focuses on the inferential level, while Sect. 3.3 focuses on the metainferential one; both sections start with the proof theory of our logics, then tackle their model theory, and then end with the correspondence results between them. Section 3.4 studies a phenomenon that caught quite a bit of attention in the recent literature, namely, the close relationship that exists between the logics in the strict-tolerant family and certain paraconsistent logics. Finally, Sect. 3.5 studies the strict-tolerant counterparts of logics that are potentially non-classical; specifically, it presents a general result according to which, for every non-trivial Tarskian logic, there is another logic which belongs to the strict-tolerant family and serves as its non-transitive counterpart, in the sense that it shares the same valid inferences as the former but locally invalidates Cut.

3.2 Inferential Strict-Tolerant Logic

This section delves into strict-tolerant logics from the inferential standpoint. The first part focuses on proof theory, and deals with two sequent calculi: $\mathbf{LK}^{-c}$ and $\mathbf{LK}^{-c}_{+\mathrm{E}}$. The former can be found in Ripley (2013a,b); the latter in Dicher and Paoli (2019). The second part focuses on model theory, and it mainly deals with strict-tolerant consequence defined over two sets of valuations: the strong Kleene valuations, and the Schütte valuations, respectively introduced by Kleene (1952) and Girard (1987). Last, but not least, the third part of the section relates the proof- and the model-theoretic approaches by means of Inferential Soundness and Completeness results.

3.2.1 *Proof Theory*

The first of the calculi that we will study is obtained by just dropping Cut:

[2] This 'intended result' might fail if the language is rich enough; see Exercise 27.

Definition 3.2.1 The sequent calculus $\mathbf{LK}^{-c}$ results from **LK** by removing the rule of Cut.

One way of reading the Cut-elimination result for **LK** (see Corollary 2.4.8) is as saying that Cut is admissible in $\mathbf{LK}^{-c}$. In fact, one can in general say:

Fact 3.2.2 An explicit rule R of a sequent calculus **S** is eliminable if and only if R is admissible in the calculus $\mathbf{S}^{-R}$ obtained by subtracting R from **S**.

Proof Exercise 23. □

Since Cut is admissible in $\mathbf{LK}^{-c}$, we can safely state that the calculus proves exactly the same sequents as **LK**.

Nevertheless, Cut is not derivable in $\mathbf{LK}^{-c}$, which also has an impact in the derivability of other rules (as we will discuss in Sect. 3.3.1). To recover some of the derivable metainferences of **LK**, one option is to add *elimination rules* to $\mathbf{LK}^{-c}$:

Definition 3.2.3 The sequent calculus $\mathbf{LK}^{-c}_{+\mathrm{E}}$ results from $\mathbf{LK}^{-c}$ by adding the following rules:

Elimination rules

$$\mathrm{L}\neg\mathrm{E}\,\frac{\Gamma, \neg\varphi \Rightarrow \Delta}{\Gamma \Rightarrow \varphi, \Delta} \qquad \mathrm{R}\neg\mathrm{E}\,\frac{\Gamma \Rightarrow \neg\varphi, \Delta}{\Gamma, \varphi \Rightarrow \Delta}$$

$$\mathrm{L}\wedge\mathrm{E}\,\frac{\Gamma, \varphi \wedge \psi \Rightarrow \Delta}{\Gamma, \varphi, \psi \Rightarrow \Delta} \qquad \mathrm{R}\wedge\mathrm{E}\,\frac{\Gamma \Rightarrow \varphi \wedge \psi, \Delta}{\Gamma \Rightarrow \varphi, \Delta} \qquad \mathrm{R}\wedge\mathrm{E}\,\frac{\Gamma \Rightarrow \varphi \wedge \psi, \Delta}{\Gamma \Rightarrow \psi, \Delta}$$

$$\mathrm{L}\vee\mathrm{E}\,\frac{\Gamma, \varphi \vee \psi \Rightarrow \Delta}{\Gamma, \varphi \Rightarrow \Delta} \qquad \mathrm{L}\vee\mathrm{E}\,\frac{\Gamma, \varphi \vee \psi \Rightarrow \Delta}{\Gamma, \psi \Rightarrow \Delta} \qquad \mathrm{R}\vee\mathrm{E}\,\frac{\Gamma \Rightarrow \varphi \vee \psi, \Delta}{\Gamma \Rightarrow \varphi, \psi, \Delta}$$

$$\mathrm{L}{\rightarrow}\mathrm{E}\,\frac{\Gamma, \varphi \rightarrow \psi \Rightarrow \Delta}{\Gamma \Rightarrow \varphi, \Delta} \qquad \mathrm{L}{\rightarrow}\mathrm{E}\,\frac{\Gamma, \varphi \rightarrow \psi \Rightarrow \Delta}{\Gamma, \psi \Rightarrow \Delta} \qquad \mathrm{R}{\rightarrow}\mathrm{E}\,\frac{\Gamma \Rightarrow \varphi \rightarrow \psi, \Delta}{\Gamma, \varphi \Rightarrow \psi, \Delta}$$

Notice that some of these elimination rules correspond to inverted rules of **LK**, while others do not. Intuitively, this is because not every rule in **LK** is invertible. First, the additive single-premise rules of **LK** are not invertible. As a result, among the single-premise rules of **LK**, only R→ (which is multiplicative) has an inverse that serves as an elimination rule: R→E. Second, the multiplicative two-premise rules are also not invertible. Consequently, L→ does not have its inverse as an elimination rule. To address this, we instead added the inverses of the additive rule for introducing the conditional on the left:

$$\mathrm{L}{\rightarrow}_a\,\frac{\Gamma \Rightarrow \varphi, \Delta \qquad \Gamma, \psi \Rightarrow \Delta}{\Gamma, \varphi \rightarrow \psi \Rightarrow \Delta}$$

thus obtaining the two L→E rules. Hence, the elimination rules of $\mathbf{LK}^{-c}_{+\mathrm{E}}$ fall into two categories: some of them are inverses of multiplicative single-premise rules, while the remaining ones are inverses of additive multi-premise rules.

As we will show in due course, with the Inferential Soundness and Completeness results from Sect. 3.2.3, the elimination rules do not allow us to prove any sequents that were not provable in $\mathbf{LK}^{-c}$. However, they do allow us to derive many new metainferences, and hence will play an essential role in the Metainferential Soundness and Completeness results from Sect. 4.2.3.

Exercises

Exercise 23 Prove Fact 3.2.2.

Exercise 24 Find derivations in $\mathbf{LK}^{-c}$ of the following sequents:

(a) $p \wedge (q \vee r) \Rightarrow (p \wedge q) \vee (p \wedge r)$
(b) $\Rightarrow p \vee \neg p$
(c) $p \wedge \neg p \Rightarrow$
(d) $p \vee (q \wedge \neg q) \Rightarrow (r \vee \neg r) \wedge p$

Exercise 25 Let $\mathbf{LK}_{+\mathrm{E}}^{-c^{\circ}}$ and $\mathbf{LK}^{-c^{\circ}}$ be the calculi that result from adding the axiom $\Rightarrow p \wedge \neg p$ to systems $\mathbf{LK}_{+\mathrm{E}}^{-c}$ and $\mathbf{LK}^{-c}$, respectively. Show that $\mathbf{LK}_{+\mathrm{E}}^{-c^{\circ}}$ proves the sequent $p \Rightarrow q$. Give an argument to explain why $\mathbf{LK}^{-c^{\circ}}$ does not prove this sequent.

3.2.2 *Trivaluations*

Throughout this chapter, we will work with the set of truth values $TV^3 = \{1, 1/2, 0\}$. Once again, values 1 and 0 are taken to represent Truth and Falsity, respectively. The non-classical value $1/2$ can be read in a number of ways; for instance, as a truth-value 'glut', viz. 'both true and false', or as a truth value 'gap', viz. 'neither true nor false'. We will call any function $v : FOR(\mathcal{L}) \to TV^3$ a *trivaluation*. We use V^3 to denote the class of all trivaluations. We start by defining formula satisfaction by a trivaluation:

Definition 3.2.4 A formula φ is *strictly satisfied* (from now on, s-satisfied) by a trivaluation $v \in V^3$ if and only if $v(\varphi) = 1$. A formula φ is *tolerantly satisfied* (t-satisfied) by a trivaluation $v \in V^3$ if and only if $v(\varphi) \neq 0$.

That is, a formula φ is strictly satisfied by trivaluation v when v assigns it the classical value True, and it is tolerantly satisfied by v when v does not assign it the classical value False.

Definition 3.2.5 For $x, y \in \{s, t\}$ and any valuation $v \in V^3$, we say that v *xy-satisfies* an inference $\Gamma \Rightarrow \Delta$ (written $v \vDash^{xy} \Gamma \Rightarrow \Delta$) just in case, if v x-satisfies every γ in Γ, then v y-satisfies some δ in Δ. If v does not xy-satisfy an inference $\Gamma \Rightarrow \Delta$, we say that it is an *$xy$-counterexample* to it (written $v \nvDash^{xy} \Gamma \Rightarrow \Delta$).

In turn, this definition enables us to define inferential validity in terms of inference satisfaction for every trivaluation in a given set $V \subseteq V^3$:

Definition 3.2.6 For $x, y \in \{s, t\}$ and any $V \subseteq V^3$, we say that an inference $\Gamma \Rightarrow \Delta$ is xy-*valid* in V (written $\Gamma \vDash^{xy}_V \Delta$) if and only if it is xy-satisfied by all $v \in V$. Equivalently, if there is no $v \in V$ that is an xy-counterexample to $\Gamma \Rightarrow \Delta$.

This definition yields four different notions of (inferential) validity, since there are four distinct ways of understanding inference satisfaction by a valuation v in a given set of trivaluations $V \subseteq V^3$: as ss-, tt-, st- or ts-satisfaction. We leave the cases of ss- and ts-satisfaction for the next chapter. Let us specify the notions of tt- and st-validity in a set $V \subseteq V^3$:

- An inference $\Gamma \Rightarrow \Delta$ is tt-valid in $V \subseteq V^3$ ($\Gamma \vDash^{tt}_V \Delta$) iff for all $v \in V$, if $v(\gamma) \in \{1, 1/2\}$ for every $\gamma \in \Gamma$, then $v(\delta) \in \{1, 1/2\}$ for some $\delta \in \Delta$.
- An inference $\Gamma \Rightarrow \Delta$ is st-valid in $V \subseteq V^3$ ($\Gamma \vDash^{st}_V \Delta$) iff for all $v \in V$, if $v(\gamma) = 1$ for every $\gamma \in \Gamma$, then $v(\delta) \in \{1, 1/2\}$ for some $\delta \in \Delta$.

It is worth noting that the notion of a trivaluation being an st-counterexample to an inference is identical to the corresponding notion for bivaluations. By this we mean that a valuation v is an st-counterexample to an inference $\Gamma \Rightarrow \Delta$ just in case it assigns the classical value 1 to all the premises and the classical value 0 to all the conclusions. Thus, st-validity can also be defined as:

- An inference $\Gamma \Rightarrow \Delta$ is st-valid in $V \subseteq V^3$ ($\Gamma \vDash^{st}_V \Delta$) iff there is no $v \in V$ such that $v(\gamma) = 1$ for every $\gamma \in \Gamma$ and $v(\delta) = 0$ for all $\delta \in \Delta$.

3.2.2.1 The Strong Kleene Valuations

The previous definitions were presented without any particular valuation scheme in mind. Before introducing the first three-valued scheme, notice that $V^2 \subseteq V^3$. Hence, in the limit case where $V = V^2$, the two notions of formula satisfaction (strict and tolerant) collapse, and so do the four notions of inference satisfaction (ss, tt, st and ts).

Figure 3.1 depicts the strong Kleene truth tables, first introduced by Kleene (1952). The *strong Kleene* valuation scheme SK for the language $\mathcal{L}$ is the set of functions encoded by those tables. We use V^{SK} to denote the set of all valuations that obey the strong Kleene valuation scheme.

Let us highlight a number of facts regarding the strong Kleene scheme. First, we say that a language $\mathcal{L}$ is *functionally complete* with respect to a scheme X (with truth values TV) whenever every truth-function on TV can be expressed in $\mathcal{L}$ by some

	$f^{\neg}_{\text{SK}}$
1	0
1/2	1/2
0	1

$f^{\wedge}_{\text{SK}}$	1	1/2	0
1	1	1/2	0
1/2	1/2	1/2	0
0	0	0	0

$f^{\vee}_{\text{SK}}$	1	1/2	0
1	1	1	1
1/2	1	1/2	1/2
0	1	1/2	0

$f^{\rightarrow}_{\text{SK}}$	1	1/2	0
1	1	1/2	0
1/2	1	1/2	1/2
0	1	1	1

Fig. 3.1 The strong Kleene truth tables

propositional formula, relative to the valuation scheme X. Namely, for any given truth table in TV, a functionally complete language is rich enough to enable us to construct a formula with exactly that truth table.

It is easy to observe that the language $\mathcal{L}$, interpreted by the strong Kleene scheme, is not rich enough to express every function in V^3. Case in point, there is no formula φ that takes the same truth value in all $v \in V^{\text{SK}}$. To see this, note that, since $V^{\text{B}} \subset V^{\text{SK}}$, there are valuations that make every formula either 1 or 0. Hence, no formula receives value $1/2$ in all valuations. On the other hand, there is a $v \in V^{\text{SK}}$ that assigns $1/2$ to every propositional variable; and by inspection of the truth tables, it is clear that said valuation also assigns $1/2$ to every compound formula. Hence, no formula receives the same classical value (1 or 0) in every valuation. Therefore, $\mathcal{L}$ is not functionally complete with respect to the set of strong Kleene valuations.

As a brief aside, it is important to note that some results presented in this chapter are specific to our current language and may not necessarily hold for extensions of it. For example, Theorem 3.2.8 fails to hold in a language enriched with a nullary connective $\curlywedge$ representing the value $1/2$, since this additional expressive power makes the set of st-valid inferences in V^{SK} no longer coextensional with the set of classically valid inferences. (For further illustration, see also Exercise 27).

For clarity, let us specify the notions of tt- and st-validity in the set of strong Kleene trivaluations:

- $\Gamma \vDash^{tt}_{V^{\text{SK}}} \Delta$ iff for all $v \in V^{\text{SK}}$, if $v(\gamma) \in \{1, 1/2\}$ for every $\gamma \in \Gamma$, then $v(\delta) \in \{1, 1/2\}$ for some $\delta \in \Delta$.
- $\Gamma \vDash^{st}_{V^{\text{SK}}} \Delta$ iff for all $v \in V^{\text{SK}}$, if $v(\gamma) = 1$ for every $\gamma \in \Gamma$, then $v(\delta) \in \{1, 1/2\}$ for some $\delta \in \Delta$.

Next, we present the most central result about the st notion of validity over the strong Kleene valuations, namely, that it has the same set of valid inferences as Classical Logic, $\mathbb{CL}$. For that, we will make use of the following Lemma:

Lemma 3.2.7 *Let $\Sigma \subseteq FOR(\mathcal{L})$. For every $v \in V^{\text{SK}}$ such that $v(\sigma) \in \{1, 0\}$ for all $\sigma \in \Sigma$, there is a $v' \in V^{\text{B}}$ that coincides with v on each $\sigma \in \Sigma$.*

Proof Let v be a valuation in V^{SK} such that $v(\sigma) \in \{1, 0\}$ for all $\sigma \in \Sigma$. Let v' be any Boolean valuation such that, for every propositional variable p that is a subformula of some formula in Σ, if $v(p) \in \{1, 0\}$ then $v'(p) = v(p)$. We show by induction on the complexity of the formula that for every $\sigma \in \Sigma$, $v(\sigma) = v'(\sigma)$.

- For the base case, $\sigma = p$, we have that $v(p) = v'(p)$ by definition of v'.
- For the inductive step, consider $\sigma = \neg\varphi$. If $v(\neg\varphi) = 1$ then, by the strong Kleene truth tables, $v(\varphi) = 0$. By inductive hypothesis, $v'(\varphi) = 0$. Then, by the Classical truth tables, $v'(\neg\varphi) = 1$. If $v(\neg\varphi) = 0$ then, by the strong Kleene truth tables, $v(\varphi) = 1$. By inductive hypothesis, $v'(\varphi) = 1$. Then, by the Classical truth tables, $v'(\neg\varphi) = 0$. Now, consider $\sigma = \varphi \wedge \psi$. Assume that $v(\varphi \wedge \psi) = 1$. Then, by the strong Kleene truth tables, $v(\varphi) = v(\psi) = 1$. By hypothesis, $v'(\varphi) = v'(\psi) = 1$. Then, by the Classical truth tables, $v'(\varphi \wedge \psi) = 1$. Assume that $v(\varphi \wedge \psi) = 0$. Then, by the strong Kleene truth tables, $v(\varphi) = 0$ or $v(\psi) = 0$. By hypothesis,

$v'(\varphi) = 0$ or $v'(\psi) = 0$. Then, by the Classical truth tables, $v'(\varphi \wedge \psi) = 0$. We leave the cases of disjunctive and conditional sentences to the reader.

□

Theorem 3.2.8 (Ripley, 2012) *An inference* $\Gamma \Rightarrow \Delta$ *is* st*-valid in* V^{SK} *iff* $\Gamma \Rightarrow \Delta$ *is valid in* V^{B}.

Proof We prove the left to right direction by contraposition. Assume that $\Gamma \nvDash_{V^{\mathrm{B}}} \Delta$. Then there is a valuation $v \in V^{\mathrm{B}}$ such that $v(\gamma) = 1$ for every $\gamma \in \Gamma$ and $v(\delta) = 0$ for every $\delta \in \Delta$. And since $V^{\mathrm{B}} \subseteq V^{\mathrm{SK}}$, v is also an st-counterexample. Hence, $\Gamma \nvDash^{st}_{V^{\mathrm{SK}}} \Delta$. For the right to left direction, we reason again by contraposition. Suppose that $\Gamma \nvDash^{st}_{V^{\mathrm{SK}}} \Delta$. Then there is a valuation v of V^{SK} in which $v(\gamma) = 1$ for every $\gamma \in \Gamma$ and $v(\delta) = 0$ for every $\delta \in \Delta$. According to Lemma 3.2.7, there is a $v' \in V^{\mathrm{B}}$ which evaluates the premises and conclusions just as v does. Hence, $\Gamma \nvDash_{V^{\mathrm{B}}} \Delta$. □

We say that a consequence relation xy for V is *paraconsistent* if $\varphi \wedge \neg\varphi \nvDash^{xy}_{V}$, and *paracomplete* if $\nvDash^{xy}_{V} \varphi \vee \neg\varphi$. It is worth noting that st-consequence on V^{SK} is neither (given Theorem 3.2.8 above), whereas tt-consequence on V^{SK} is not paracomplete (see Fact 3.2.11 below), but it is paraconsistent since it has no antitautologies:

Definition 3.2.9 A formula φ is a *tautology* of $\vDash^{xy}_{V}$ iff every valuation $v \in V$ y-satisfies it. And dually, a formula φ is an *antitautology* of $\vDash^{xy}_{V}$ iff there is no valuation $v \in V$ that x-satisfies it.

Fact 3.2.10 $\vDash^{tt}_{V^{\mathrm{SK}}}$ has no antitautologies.

Proof Consider a valuation $v \in V^{\mathrm{SK}}$, s.t. $v(p) = 1/2$ to every propositional variable $p \in FOR(\mathcal{L})$. By inspecting the way in which the strong Kleene connectives are defined, it is easy to check that v t-satisfies every formula. Therefore, no formula can be an antitautology of $\vDash^{tt}_{V^{\mathrm{SK}}}$. □

Fact 3.2.11 $\vDash^{tt}_{V^{\mathrm{SK}}}$ and $\vDash^{st}_{V^{\mathrm{SK}}}$ have the same tautologies as Classical Logic.

Proof Exercise 28. □

Exercises

Exercise 26 Let $\top$, $\bot$ and $\curlywedge$ be the propositional constants (viz. 0-ary connectives) denoting 1, 0 and 1/2, respectively. What is the least amount of constants that we need to add to $\mathcal{L}$ in order to be able to express all three? Which ones are included in this set? (There are two correct answers)

Exercise 27 Sequent calculus $\mathbf{LK}^{-c}_{+\curlywedge}$ is the result of extending **LK** with the axioms for $\curlywedge$:

$$\mathrm{R}\curlywedge \frac{}{\curlywedge \Rightarrow} \qquad \mathrm{L}\curlywedge \frac{}{\Rightarrow \curlywedge}$$

Show that Cut is not admissible in $\mathbf{LK}^{-c}_{+\curlywedge}$.

	$f^{\neg}_{\text{Sch}}$
1	0, 1/2
1/2	1/2
0	1, 1/2

$f^{\wedge}_{\text{Sch}}$	1	1/2	0
1	1, 1/2	1/2	0, 1/2
1/2	1/2	1/2	0, 1/2
0	0, 1/2	0, 1/2	0, 1/2

$f^{\vee}_{\text{Sch}}$	1	1/2	0
1	1, 1/2	1, 1/2	1, 1/2
1/2	1, 1/2	1/2	1/2
0	1, 1/2	1/2	0, 1/2

$f^{\rightarrow}_{\text{Sch}}$	1	1/2	0
1	1, 1/2	1/2	0, 1/2
1/2	1, 1/2	1/2	1/2
0	1, 1/2	1, 1/2	1, 1/2

Fig. 3.2 The Schütte truth tables

Exercise 28 Prove Fact 3.2.11. (Hint: use Lemma 3.2.7 in the proof!)

Exercise 29 Prove the following facts:

(a) st-consequence on V^{SK} is inferentially self-dual: $\Gamma \vDash^{st}_{V^{\text{SK}}} \Delta$ iff $\neg\Delta \vDash^{st}_{V^{\text{SK}}} \neg\Gamma$, where $\neg\Sigma = \{\neg\sigma \mid \sigma \in \Sigma\}$.
(b) st-consequence on V^{SK} satisfies the Deduction Theorem: $\vDash^{st}_{V^{\text{SK}}} \varphi \rightarrow \psi$ iff $\varphi \vDash^{st}_{V^{\text{SK}}} \psi$.

3.2.2.2 The Schütte Tables

So far, the truth functions we have been working with take n-tuples of truth values as inputs and deliver a single truth value as output. These are what we call *deterministic* truth functions. In what follows, we will also appeal to truth functions that take n-tuples of truth values as inputs and deliver a non-empty *set* of truth values as output. We call them *non-deterministic* truth functions.

Functions of this latter kind can also be depicted in truth tables. The difference, of course, is that now each output cell will contain a set of truth values instead of a single truth value. For instance, Fig. 3.2 depicts the tables of the the so-called *Schütte* non-deterministic truth functions (see, e.g. Girard, 1987). In the tables, we drop the braces for legibility: for example, $f^{\neg}_{\text{Sch}}(1) = \{0, 1/2\}$, but we write "0, 1/2" in the corresponding cell. An easy way to remember the Schütte tables is as follows: we just take the tables of strong Kleene, and add the value 1/2 to each and every cell! The Schütte tables will be useful later, when giving Metainferential Soundness and Completeness results for some of the calculi in this (and the next) chapter.

When we have a non-deterministic truth function $f^{\#}$ for each connective # in our language, by collecting these functions together in a set X we obtain what we call a *non-deterministic valuation scheme* for the language. The set of truth values associated with X is the union of the codomains of all the functions in X. So, for instance, we say that Sch is the non-deterministic valuation scheme for $\mathcal{L}$ whose functions are $f^{\neg}_{\text{Sch}}$, $f^{\wedge}_{\text{Sch}}$, $f^{\vee}_{\text{Sch}}$ and $f^{\rightarrow}_{\text{Sch}}$, and whose set of truth values is $\{1, 1/2, 0\}$. The notion of a valuation *based* on a non-deterministic valuation scheme is defined as follows:

Definition 3.2.12 Let X be a non-deterministic valuation scheme for a propositional language $\mathcal{L}$, with TV its associated set of truth values. A valuation for $\mathcal{L}$ *based* on X is a mapping v from $FOR(\mathcal{L})$ to TV such that, for every n-ary connective # of $\mathcal{L}$ and its associated function $f^{\#}$, it happens that $v(\#(\varphi_1, \ldots, \varphi_n)) \in f^{\#}(v(\varphi_1), \ldots, v(\varphi_n))$.

As in the case of valuation schemes *simpliciter*, given a non-deterministic valuation scheme X we write V^{X} to denote the set of all the valuations based on it. So, in particular, V^{Sch} is the set of all valuations based on Sch.

The notions we previously discussed of inference satisfaction generate four distinct consequence relations based on the Schütte scheme. As before, we leave the cases of ss- and ts- for later. The notions of tt- and st-validity over the set V^{Sch} are as follows:

- $\Gamma \vDash^{tt}_{V^{\text{Sch}}} \Delta$ iff for all $v \in V^{\text{Sch}}$, if $v(\gamma) \in \{1, 1/2\}$ for every $\gamma \in \Gamma$, then $v(\delta) \in \{1, 1/2\}$ for some $\delta \in \Delta$.
- $\Gamma \vDash^{st}_{V^{\text{Sch}}} \Delta$ iff for all $v \in V^{\text{Sch}}$, if $v(\gamma) = 1$ for every $\gamma \in \Gamma$, then $v(\delta) \in \{1, 1/2\}$ for some $\delta \in \Delta$.

Next, we outline some basic features of these consequence relations. Considering that $1/2$ appears on every cell, it is easy to observe that:

Fact 3.2.13 The consequence relation tt on V^{Sch} is paraconsistent and has no antitautologies.

Less obvious is the fact that st-valid inferences in V^{Sch} and st-valid inferences in V^{SK} coincide. In order to prove that, we make use of the next Lemma:

Lemma 3.2.14 *Let $\Sigma \subseteq FOR(\mathcal{L})$. For every $v \in V^{\text{Sch}}$ such that $v(\sigma) \in \{1, 0\}$ for all $\sigma \in \Sigma$, there is a $v' \in V^{\text{SK}}$ that coincides with v on each $\sigma \in \Sigma$.*

Proof Let v be a valuation in V^{Sch} such that $v(\sigma) \in \{1, 0\}$ for all $\sigma \in \Sigma$. Let v' be any strong Kleene valuation such that, for every propositional variable p that is a subformula of some $\sigma \in \Sigma$, if $v(p) \in \{1, 0\}$ then $v'(p) = v(p)$. We show by induction on the complexity of the formula that for every $\sigma \in \Sigma$, $v(\sigma) = v'(\sigma)$.

- For the base case, $\sigma = p$, we have that $v(p) = v'(p)$ by definition of v'.
- For the inductive step, consider $\sigma = \neg\varphi$. If $v(\neg\varphi) = 1$ then, by the Schütte truth tables, $v(\varphi) = 0$. By inductive hypothesis, $v'(\varphi) = 0$. Then, by the strong Kleene truth tables, $v'(\neg\varphi) = 1$. If $v(\neg\varphi) = 0$ then, by the Schütte truth tables, $v(\varphi) = 1$. By inductive hypothesis, $v'(\varphi) = 1$. Then, by the strong Kleene truth tables, $v'(\neg\varphi) = 0$. At last, consider $\sigma = \varphi \vee \psi$. Assume that $v(\varphi \vee \psi) = 1$. Then, by the Schütte truth tables, either $v(\varphi) = 1$ or $v(\psi) = 1$. By inductive hypothesis, $v'(\varphi)$ or $v'(\psi) = 1$. Then, by the strong Kleene truth tables, $v'(\varphi \vee \psi) = 1$. Assume that $v(\varphi \vee \psi) = 0$. Then, by the Schütte truth tables, $v(\varphi) = v(\psi) = 0$. By inductive hypothesis, $v'(\varphi) = v'(\psi) = 0$. Then, by the strong Kleene truth tables, $v'(\varphi \vee \psi) = 0$. We leave the cases of conjunctive and conditional sentences to the reader.

□

Fact 3.2.15 An inference $\Gamma \Rightarrow \Delta$ is st-valid in V^{Sch} iff it is st-valid in V^{SK}.

Proof For the left to right direction, assume $\Gamma \nvDash^{st}_{V^{\text{SK}}} \Delta$. Then there is a valuation $v \in V^{\text{SK}}$ such that $v(\gamma) = 1$ for every $\gamma \in \Gamma$ and $v(\delta) = 0$ for every $\delta \in \Delta$. And since $V^{\text{SK}} \subset V^{\text{Sch}}$, v also shows that $\Gamma \nvDash^{st}_{V^{\text{Sch}}} \Delta$. To establish the right to left direction, we proceed by contraposition once more: suppose $\Gamma \nvDash^{st}_{V^{\text{Sch}}} \Delta$. Then there is a valuation $v \in V^{\text{Sch}}$ such that $v(\gamma) = 1$ for every $\gamma \in \Gamma$ and $v(\delta) = 0$ for every $\delta \in \Delta$. According to Lemma 3.2.14, there is a $v' \in V^{\text{SK}}$ which evaluates the formulas in Γ and in Δ just as v does. Therefore, $\Gamma \nvDash^{st}_{V^{\text{SK}}} \Delta$. □

Something to notice about valuations based on non-deterministic schemes is that, intuitively, they can take different policies towards formulas of the same form. Thus, for instance, there are Schütte valuations v such that $v(p) = v(q) = v(r) = v(s) = 1$, but $v(p \wedge q) = 1$ and $v(r \wedge s) = 1/2$. Sometimes, however, we might want to focus only on the valuations that treat all formulas of the same form alike. We call them uniform valuations:

Definition 3.2.16 (Uniform Valuations) A valuation v for $\mathcal{L}$ is *uniform* just in case, for every n-ary connective # of $\mathcal{L}$, if $v(\varphi_1) = v(\psi_1), \ldots, v(\varphi_n) = v(\psi_n)$, then $v(\#(\varphi_1, \ldots, \varphi_n)) = v(\#(\psi_1, \ldots, \psi_n))$.

Given a non-deterministic valuation scheme X, we write V^{X}_u for the set of valuations in V^{X} that are uniform. Informally, each valuation in V^{X}_u chooses one of the possible ways of making the tables for X determinate (viz. with a singleton at every output cell) and follows the resulting tables throughout. In these valuations, knowing the values of all the subformulas of a complex formula is enough for knowing the value of the formula itself. This is of course not the case for valuations in V^{X} that are not in V^{X}_u. That is why we may think of the latter as the 'properly non-deterministic' valuations based on X.

Going back to the Schütte scheme, we have that $V^{\text{SK}} \subset V^{\text{Sch}}_u \subset V^{\text{Sch}}$. But we have seen that the st-valid inferences in V^{SK} and in V^{Sch} are the same. Hence,

Fact 3.2.17 An inference is st-valid in V^{Sch}_u iff it is st-valid in V^{Sch} (V^{SK}).

As we will see later, the metainferences that are st-valid in V^{Sch}_u and in V^{Sch} do not coincide.

Exercises

Exercise 30 Find an inference that is tt-valid in V^{Sch}_u but not in V^{Sch}. (Hint: use multiple conclusions.)

3.2.3 Inferential Soundness and Completeness

We start with the Soundness of the sequent calculi from Sect. 3.2.1. We will use the following lemma:

Lemma 3.2.18 *Every axiom and rule of* $\mathbf{LK}^{-c}_{+\text{E}}$ *preserves* st*-validity in* V^{SK}.

Proof The result is immediate for the axioms and rules of **LK**, since they preserve validity in V^{B}, and validity in V^{B} is coextensive with st-validity in V^{SK} by Theorem 3.2.8. So, we only need to check the result for the elimination rules of $\mathbf{LK}^{-c}_{+\mathrm{E}}$. We take care of a couple of cases:

- Rule L¬E. Suppose that $\Gamma, \neg\varphi \Rightarrow \Delta$ is st-valid in V^{SK}. Take any $v \in V^{\mathrm{SK}}$ and suppose that $v(\gamma) = 1$ for every $\gamma \in \Gamma$. If $v(\varphi) \in \{1, 1/2\}$, then v st-satisfies $\Gamma \Rightarrow \varphi, \Delta$. If $v(\varphi) = 0$, then the strong Kleene tables give $v(\neg\varphi) = 1$; it follows that $v(\delta) \in \{1, 1/2\}$ for some $\delta \in \Delta$, and again v st-satisfies $\Gamma \Rightarrow \varphi, \Delta$. Since v was arbitrary, we conclude that $\Gamma \Rightarrow \varphi, \Delta$ is st-valid in V^{SK}.
- Rules L→E. Suppose that $\Gamma, \varphi \to \psi \Rightarrow \Delta$ is st-valid in V^{SK}. Take any $v \in V^{\mathrm{SK}}$ and suppose that $v(\gamma) = 1$ for every $\gamma \in \Gamma$. If $v(\varphi) \in \{1, 1/2\}$, then v st-satisfies $\Gamma \Rightarrow \varphi, \Delta$. If $v(\varphi) = 0$, then the strong Kleene tables give $v(\varphi \to \psi) = 1$; it follows that $v(\delta) \in \{1, 1/2\}$ for some $\delta \in \Delta$, and again v st-satisfies $\Gamma \Rightarrow \varphi, \Delta$. Since v was arbitrary, we conclude that $\Gamma \Rightarrow \varphi, \Delta$ is st-valid in V^{SK}. The other L→E rule is dealt with analogously.

The remaining cases are left to the reader (see Exercise 31). □

Theorem 3.2.19 (Inferential Soundness $\mathbf{LK}^{-c}_{+\mathrm{E}}$) *If* $\Gamma \vdash_{\mathbf{LK}^{-c}_{+\mathrm{E}}} \Delta$, *then* $\Gamma \vDash^{st}_{V^{\mathrm{SK}}} \Delta$.

Proof If $\Gamma \vdash_{\mathbf{LK}^{-c}_{+\mathrm{E}}} \Delta$, then there are finite Γ' and Δ', such that $\Gamma' \subseteq \Gamma$ and $\Delta' \subseteq \Delta$ and $\Gamma' \Rightarrow \Delta'$ is provable in $\mathbf{LK}^{-c}_{+\mathrm{E}}$. To establish that $\Gamma' \Rightarrow \Delta'$ is st-valid in V^{SK}, we proceed by induction on the height of the derivation:

- The base case is trivial since the only axiomatic sequent is valid.
- For the inductive step, if $\Gamma' \Rightarrow \Delta'$ has a proof of height n, then the sequents immediately above it have proofs shorter than n, and by inductive hypothesis, they are st-valid in V^{SK}. Given that all rules of $\mathbf{LK}^{-c}_{+\mathrm{E}}$ preserve st-validity in V^{SK} (as shown in Lemma 3.2.18), $\Gamma' \Rightarrow \Delta'$ is also st-valid in V^{SK}.

Hence, $\Gamma' \vDash^{st}_{V^{\mathrm{SK}}} \Delta'$. By monotonicity, $\Gamma \vDash^{st}_{V^{\mathrm{SK}}} \Delta$. □

Since $\mathbf{LK}^{-c}$ has less rules than $\mathbf{LK}^{-c}_{+\mathrm{E}}$, we can automatically conclude that it is also sound with respect to st-consequence over V^{SK}.

Now we tackle Completeness, which follows easily from previous results:

Theorem 3.2.20 (Inferential Completeness $\mathbf{LK}^{-c}$) *If* $\Gamma \vDash^{st}_{V^{\mathrm{SK}}} \Delta$, *then* $\Gamma \vdash_{\mathbf{LK}^{-c}} \Delta$.

Proof Suppose $\Gamma \vDash^{st}_{V^{\mathrm{SK}}} \Delta$. Then $\Gamma \vDash_{V^{\mathrm{B}}} \Delta$ (by Theorem 3.2.8). Then $\Gamma \vdash_{\mathbf{LK}} \Delta$ (by Corollary 2.2.17). This means that, for some finite $\Gamma' \subseteq \Gamma$ and $\Delta' \subseteq \Delta$, the sequent $\Gamma' \Rightarrow \Delta'$ is provable in **LK**. But then $\Gamma' \Rightarrow \Delta'$ is also provable in $\mathbf{LK}^{-c}$ (by Corollary 2.4.8). This means that $\Gamma \vdash^{-c}_{\mathbf{LK}} \Delta$, and we are done. □

Since $\mathbf{LK}^{-c}_{+\mathrm{E}}$ has more rules than $\mathbf{LK}^{-c}$, we can automatically conclude that it is also complete with respect to st-consequence over V^{SK}.

Remember that Facts 3.2.15 and 3.2.17 established the coextensivity of st-consequence over the sets of valuations V^{SK}, V^{Sch} and V^{Sch}_{u}. Hence, we also have

Corollary 3.2.21 (Inferential Soundness and Completeness of $\mathbf{LK}^{-c}$ and $\mathbf{LK}^{-c}_{+\mathrm{E}}$) *Calculi* $\mathbf{LK}^{-c}$ *and* $\mathbf{LK}^{-c}_{+\mathrm{E}}$ *are both sound and complete with respect to the* st*-valid inferences in* V^{Sch} *(in* V_u^{Sch}*).*

Exercises

Exercise 31 Complete the proof of Lemma 3.2.18.

3.3 Metainferential Strict-Tolerant Logic

In this part we study the metainferential aspects of strict-tolerant logics. As before, we first we address the proof theory, then the model theory, and lastly the relationships between the two.

3.3.1 Derivability and Admissibility

This section briefly compares the admissibility and derivability of rules in calculi **LK**, $\mathbf{LK}^{-c}$ and $\mathbf{LK}^{-c}_{+\mathrm{E}}$. As a quick refresher (see Definitions 2.3.1 and 2.3.2) remember the following:

- A rule is admissible in a sequent calculus just in case, whenever the premise-sequents are provable, the conclusion-sequent is provable.
- A rule is derivable in a sequent calculus just in case, whenever the premise-sequents are added as new axioms, the conclusion-sequent becomes provable.

A very simple but important fact about admissibility is that, if two sequent calculi coincide in their provable sequents, then they also coincide in their admissible rules:

Fact 3.3.1 Suppose that two sequent calculi $\mathbf{S_1}$ and $\mathbf{S_2}$ prove the same sequents. Then, a rule R is admissible in $\mathbf{S_1}$ if and only if it is admissible in $\mathbf{S_2}$.

Proof Suppose that R is admissible in $\mathbf{S_1}$. Consider any instance of it. Either some premise-sequent is not provable in $\mathbf{S_1}$ or the conclusion-sequent is. If the former, some premise-sequent is not provable in $\mathbf{S_2}$. If the latter, the conclusion-sequent is provable in $\mathbf{S_2}$. Either way, this instance of R is admissible in $\mathbf{S_2}$. The reasoning for the converse inclusion is similar. □

We have seen that calculi **LK**, $\mathbf{LK}^{-c}$ and $\mathbf{LK}^{-c}_{+\mathrm{E}}$ all prove the same sequents. Hence, Fact 3.3.1 allows us to conclude

Corollary 3.3.2 *Calculi* **LK**, $\mathbf{LK}^{-c}$ *and* $\mathbf{LK}^{-c}_{+\mathrm{E}}$ *have the same admissible rules.*

Even though our three calculi coincide in the rules they can admit, they of course do not coincide in the rules they can derive. Remember that a sequent calculus is structurally complete when all the rules that are admissible in it are also derivable (Definition 2.3.5). Thus,

Fact 3.3.3 (Structural Incompleteness) $\mathbf{LK}^{-c}$ and $\mathbf{LK}^{-c}_{+\mathrm{E}}$ are not structurally complete. In particular, Cut is admissible in both calculi, but it is derivable in neither of them.

Proof Take the instance of Cut

$$\frac{\Rightarrow p \qquad p \Rightarrow}{\Rightarrow}$$

and consider the calculi that extend $\mathbf{LK}^{-c}$ and $\mathbf{LK}^{-c}_{+\mathrm{E}}$ with axioms $\Rightarrow p$ and $p \Rightarrow$. By inspection of the rules we know that the empty sequent $\Rightarrow$ is not provable in these calculi: it is not an axiom, and all the rules of the calculi are such that all their instances have a non-empty conclusion sequent. □

Now, we know by Fact 2.3.8 from Chap. 2 that **LK** *is* structurally complete. So, to have a better grasp of Fact 3.3.3, it will be useful to check where exactly the proof of Structural Completeness for **LK** breaks down in the cases of $\mathbf{LK}^{-c}$ and $\mathbf{LK}^{-c}_{+\mathrm{E}}$. Remember that the proof consisted of two steps. First, one shows a "Horizontalization" result (Lemma 2.3.7), which states, roughly, that if a rule is admissible in **LK**, then its inferential translation is provable. Second, one goes on to show that if the inferential translation of a rule is provable in **LK**, then the rule itself is derivable. Now, the first of these two results applies to our current calculi as well:

Lemma 3.3.4 (Horizontalization) *Let the following rule*

$$\frac{\Gamma_1 \Rightarrow \Delta_1 \qquad \dots \qquad \Gamma_n \Rightarrow \Delta_n}{\Sigma \Rightarrow \Pi}$$

be admissible in $\mathbf{LK}^{-c}$ *(resp.* $\mathbf{LK}^{-c}_{+\mathrm{E}}$*). Then, the following inference schema*

$$\bigwedge \Gamma_1 \to \bigvee \Delta_1, \dots, \bigwedge \Gamma_n \to \bigvee \Delta_n \Rightarrow \bigwedge \Sigma \to \bigvee \Pi$$

is provable in $\mathbf{LK}^{-c}$ *(resp.* $\mathbf{LK}^{-c}_{+\mathrm{E}}$*).*

(The proof is essentially as in Lemma 2.3.7.) So, where does the proof of Structural Completeness for $\mathbf{LK}^{-c}$ and $\mathbf{LK}^{-c}_{+\mathrm{E}}$ break down? It is in the second step. From the fact that the inferential translation of a rule is provable in these calculi, it does not follow that the rule itself is derivable: for the rule to be derived, the rule of Cut is needed, and this rule is not available now.

It is worth mentioning that, even though $\mathbf{LK}^{-c}$ and $\mathbf{LK}^{-c}_{+\mathrm{E}}$ are both structurally incomplete, the latter has strictly more derivable rules than the former; for instance, all the elimination rules of $\mathbf{LK}^{-c}_{+\mathrm{E}}$ are admissible but not derivable in $\mathbf{LK}^{-c}$. In the next section we will officially introduce three different logics, among which we find $\mathbb{ST}$, based on the strong Kleene valuations, and $\mathbb{ST}^{Sch}$, based on the Schütte valuations. These logics will be presented using model-theoretic machinery.

However, the Metainferential Soundness and Completeness results from Sect. 3.3.3 will show that, indeed, $\mathbb{ST}$ can also be characterized as the set of inferences provable and metainferences derivable in $\mathbf{LK}^{-c}_{+\mathrm{E}}$, while $\mathbb{ST}^{Sch}$ can be characterized as the set of inferences provable and metainferences derivable in $\mathbf{LK}^{-c}$.

Exercises

Exercise 32 Show Lemma 3.3.4.

Exercise 33 Let $\mathbf{LK}^{\star}$ be the calculus that results from $\mathbf{LK}$ by adding the axiom $\varphi \Rightarrow \varphi$ for every non-atomic φ. Prove that, for each connective of the language, its right (left) introduction rule is interderivable with its left (right) elimination rule (Přenosil, 2017).

Exercise 34 Show that rule (a) is derivable in $\mathbf{LK}^{-c}$, rule (b) is derivable in $\mathbf{LK}^{-c}_{+\mathrm{E}}$, and rule (c) is derivable in $\mathbf{LK}$:

$$(a)\ \frac{\varphi \Rightarrow \neg\varphi}{\Rightarrow \neg\varphi}$$

$$(b)\ \frac{\neg\varphi \Rightarrow \varphi}{\Rightarrow \varphi}$$

$$(c)\ \frac{\Rightarrow \varphi \to \psi \qquad \Rightarrow \varphi}{\Rightarrow \psi}\ \mathrm{MMP}$$

(Label 'MMP' stands for 'Meta Modus Ponens'.) Explain informally why (c) is not derivable in either $\mathbf{LK}^{-c}_{+\mathrm{E}}$ or $\mathbf{LK}^{-c}$, and why (b) is not derivable in $\mathbf{LK}^{-c}$.

3.3.2 *Global, Local and Absolute Global*

In this section we extend the work on local, global and absolute global validity made in Sect. 2.3.2 from a bi-valued to a three-valued setting. Basically, we will make various observations about the local, the global and the absolute global ranges in V^3 of calculi $\mathbf{LK}^{-c}$ and $\mathbf{LK}^{-c}_{+\mathrm{E}}$, in all cases relative to the consequence relation st. We will start with local validity.

3.3.2.1 Local Validity

In the bivalued context, we had only one reasonable notion of validity for inferences, namely, preservation of value 1. In the three-valued context, in contrast, we have at least four *prima facie* reasonable options: ss, tt, st and ts. So, when we define local validity for metainferences, we must have each of these options in mind. That motivates the following definition:

Definition 3.3.5 Let $x, y \in \{s, t\}$, and $V \subseteq V^3$. A metainference $S \Rrightarrow r$ is *locally xy-valid in V* just in case all V-valuations which xy-satisfy each $s \in S$ also xy-satisfy r.

As in the previous chapter, the definition extends to rules in the obvious way: a rule is locally xy-valid in V just in case each of its instances is.

Having defined local validity allows us to officially introduce the various non-transitive *logics* that are of our interest:

Definition 3.3.6

- $\mathbb{ST}^{\text{SK}}$ is the logic containing all and only the inferences and metainferences that are (locally) st-valid in V^{SK}.
- $\mathbb{ST}^{\text{Sch}}$ is the logic containing all and only the inferences and metainferences that are (locally) st-valid in V^{Sch}.
- $\mathbb{ST}_u^{\text{Sch}}$ is the logic containing all and only the inferences and metainferences that are (locally) st-valid in V_u^{Sch}.

Occasionally we will write "$\mathbb{ST}$" (without any superscript) as shorthand for $\mathbb{ST}^{\text{SK}}$. Of course, the three logics just introduced are different from $\mathbb{CL}$ because, even though they contain the same inferences, they do no contain the same metainferences. Unsurprisingly, they are also metainferentially different from each other. Indeed, they can be ordered by strict inclusion as follows:

Fact 3.3.7 $\mathbb{CL} \supset \mathbb{ST}^{\text{SK}} \supset \mathbb{ST}_u^{\text{Sch}} \supset \mathbb{ST}^{\text{Sch}}$

Proof The '$\supseteq$' claims are guaranteed by the fact that $V^{\text{B}} \subset V^{\text{SK}} \subset V_u^{\text{Sch}} \subset V^{\text{Sch}}$. Then, for the '$\not\subseteq$' claims, consider the metainferences

$$\frac{\Rightarrow p \qquad p \Rightarrow}{\Rightarrow} \qquad\qquad \frac{\Rightarrow p \wedge q}{\Rightarrow p} \qquad\qquad \frac{p \wedge q \Rightarrow}{r \wedge s, p, q, r, s \Rightarrow}$$

The one on the left belongs to $\mathbb{CL}$ but not to the other logics. The one in the middle belongs to $\mathbb{ST}^{\text{SK}}$ but not to $\mathbb{ST}_u^{\text{Sch}}$ or $\mathbb{ST}^{\text{Sch}}$. The one on the right belongs to $\mathbb{ST}_u^{\text{Sch}}$ but not to $\mathbb{ST}^{\text{Sch}}$. □

Incidentally, the above proof mentions that Cut is in none of our st-based logics. This observation can be generalized, as follows:

Fact 3.3.8 Let $V \subseteq V^3$ contain at least one valuation $v \notin V^2$. Then, Cut is not locally st-valid in V.

Proof Since $v \notin V^2$, there is a formula φ such that $v(\varphi) = 1/2$. Take the instance of Cut

$$\frac{\Rightarrow \varphi \qquad \varphi \Rightarrow}{\Rightarrow}$$

The premise-sequents are both st-satisfied at v. But the conclusion-sequent is not.

□

This makes clear in what sense all logics based on st-consequence can be said to be non-transitive—as long as they admit at least one valuation that is non-vacuously three-valued, viz. it assigns $1/2$ to at least one formula.

Next, we tackle the local ranges of $\mathbf{LK}^{-c}$ and $\mathbf{LK}^{-c}_{+\mathrm{E}}$. For this, we must first adjust our definition of local range to the three-valued setting:

Definition 3.3.9 Let $\mathbf{S}$ be a set of metainferences, $x, y \in \{s, t\}$, and $V \subseteq V^3$. The *local xy-range of* $\mathbf{S}$ *in* V, written $lr(\mathbf{S}, xy, V)$, is the maximal set of valuations $V_i \subseteq V$ such that each metainference in $\mathbf{S}$ is locally xy-valid in V_i.

Thus, now the local range of a calculus is defined not only relative to a set V of valuations, but also relative to a particular notion of consequence xy. But the idea of the definition is the same as in the bivalued case: the xy-range of a calculus $\mathbf{S}$ in a set of valuations V is the maximal subset of V in which the rules of $\mathbf{S}$ preserve xy-satisfaction.

Fact 3.3.10 (Local Range of $\mathbf{LK}^{-c}_{+\mathrm{E}}$) The local st-range in V^3 of $\mathbf{LK}^{-c}_{+\mathrm{E}}$ is V^{SK}.

Proof The procedure is similar to that of Fact 2.3.18. From right to left, we just check that each of the rules of $\mathbf{LK}^{-c}_{+\mathrm{E}}$ is locally st-valid in V^{SK}. We exhibit a couple of cases, and leave the remaining ones to the reader.

- Rule R$\wedge$. Suppose that a V^{SK}-valuation v st-satisfies both $\Gamma \Rightarrow \varphi, \Delta$ and $\Gamma \Rightarrow \psi, \Delta$. We want to show that v also st-satisfies $\Gamma \Rightarrow \varphi \wedge \psi, \Delta$. So, suppose further that $v(\gamma) = 1$ for every $\gamma \in \Gamma$. Then, either $v(\delta) \in \{1, 1/2\}$ for some $\delta \in \Delta$, or $v(\varphi), v(\psi) \in \{1, 1/2\}$—in which case $v(\varphi \wedge \psi) \in \{1, 1/2\}$ by definition of V^{SK}. Either way, we are done.
- Rules R$\wedge$E. Suppose that a V^{SK}-valuation v st-satisfies $\Gamma \Rightarrow \varphi \wedge \psi, \Delta$. We want to show that v st-satisfies both $\Gamma \Rightarrow \varphi, \Delta$ and $\Gamma \Rightarrow \psi, \Delta$. So, suppose further that $v(\gamma) = 1$ for every $\gamma \in \Gamma$. Then, either $v(\delta) \in \{1, 1/2\}$ for some $\delta \in \Delta$, or $v(\varphi \wedge \psi) \in \{1, 1/2\}$—in which case $v(\varphi), v(\psi) \in \{1, 1/2\}$ by definition of V^{SK}. Either way, we are done.

From left to right, suppose that $v \notin V^{\mathrm{SK}}$. We will show that v is also not in the local st-range of $\mathbf{LK}^{-c}_{+\mathrm{E}}$. There are four cases: v does not respect the strong Kleene tables for $\neg$, or for $\vee$, or for $\wedge$, or for $\rightarrow$. We tackle the first two cases, and leave the remaining ones to the reader.

- v is deviant for $\neg$. We distinguish four subcases:

  * There is a formula φ s.t. $v(\varphi) = 1$ but $v(\neg\varphi) \in \{1, 1/2\}$. Then, v is an st-counterexample to an instance of R$\neg$E of the following form:

$$\frac{\Rightarrow \neg\varphi}{\varphi \Rightarrow}$$

  * There is a formula φ s.t. $v(\varphi) = 0$ but $v(\neg\varphi) \in \{0, 1/2\}$. Then, v is an st-counterexample to an instance of L$\neg$E of the following form:

$$\frac{\neg\varphi \Rightarrow}{\Rightarrow \varphi}$$

* There is a formula φ s.t. $v(\varphi) = 1/2$ but $v(\neg\varphi) = 1$. Then, v is an *st*-counterexample to an instance of L$\neg$ of the following form:

$$\frac{\Rightarrow \varphi}{\neg\varphi \Rightarrow}$$

* There is a formula φ s.t. $v(\varphi) = 1/2$ but $v(\neg\varphi) = 0$. Then, v is an *st*-counterexample to an instance of R$\neg$ of the following form:

$$\frac{\varphi \Rightarrow}{\Rightarrow \neg\varphi}$$

- v is deviant for $\vee$. Again, we distinguish four subcases, all consisting in there being formulas φ and ψ satisfying certain conditions:

* $v(\varphi) = v(\psi) = 0$, but $v(\varphi \vee \psi) \in \{1, 1/2\}$. Then, v is an *st*-counterexample to an instance of R$\vee$E of the following form:

$$\frac{\Rightarrow \varphi \vee \psi}{\Rightarrow \varphi, \psi}$$

* $v(\varphi) = 1$ or $v(\psi) = 1$, but $v(\varphi \vee \psi) \in \{0, 1/2\}$. Then, v is an *st*-counterexample to an instance of L$\vee$E of one of the following forms:

$$\frac{\varphi \vee \psi \Rightarrow}{\varphi \Rightarrow} \qquad \frac{\varphi \vee \psi \Rightarrow}{\psi \Rightarrow}$$

* $v(\varphi) = 1/2$ or $v(\psi) = 1/2$, and $v(\varphi), v(\psi) \in \{0, 1/2\}$, but $v(\varphi \vee \psi) = 1$. Then, v is an *st*-counterexample to an instance of L$\vee$ of the following form:

$$\frac{\varphi \Rightarrow \qquad \psi \Rightarrow}{\varphi \vee \psi \Rightarrow}$$

* $v(\varphi) = 1/2$ or $v(\psi) = 1/2$, and $v(\varphi), v(\psi) \in \{0, 1/2\}$, but $v(\varphi \vee \psi) = 0$. Then, v is an *st*-counterexample to an instance of R$\vee$ of one of the following forms:

$$\frac{\Rightarrow \varphi}{\Rightarrow \varphi \vee \psi} \qquad \frac{\Rightarrow \psi}{\Rightarrow \varphi \vee \psi}$$

□

So, the local *st*-range in V^3 of calculus $\mathbf{LK}^{-c}_{+\mathrm{E}}$ is exactly the set of valuations used to define the logic $\mathbb{ST}^{\mathrm{SK}}$. As we have seen, $\mathbf{LK}^{-c}_{+\mathrm{E}}$ has strictly more derivable metainferences than $\mathbf{LK}^{-c}$. As expected, this correlates with the fact that the local *st*-range of the $\mathbf{LK}^{-c}$ has strictly more valuations than $\mathbf{LK}^{-c}_{+\mathrm{E}}$; the extra valuations in the local *st*-range of $\mathbf{LK}^{-c}$ counterexemplify metainferences that are derivable in $\mathbf{LK}^{-c}_{+\mathrm{E}}$ but not in $\mathbf{LK}^{-c}$.

Fact 3.3.11 (Local Range of $\mathbf{LK}^{-c}$) The local *st*-range in V^3 of $\mathbf{LK}^{-c}$ is V^{Sch}.

Proof The procedure is entirely similar to that of the previous result. From right to left, we again check that the rules of $\mathbf{LK}^{-c}$ are locally *st*-valid in V^{Sch}. From left to right, we assume that a valuation v is not in V^{Sch}, and show that v is an *st*-counterexample to the local *st*-validity of some rule of $\mathbf{LK}^{-c}$. We leave the proof to the reader (Exercise 37). □

So, the local st-range in V^3 of $\mathbf{LK}^{-c}$ is exactly the set of valuations that are used to define logic $\mathbb{ST}^{\mathrm{Sch}}$.

Exercises

Exercise 35 Give alternative examples of the '$\not\subseteq$' claims made in the proof of Fact 3.3.7. Use other connectives, such as the conditional and the negation.

Exercise 36 Prove the remaining cases of the proof of Fact 3.3.10.

Exercise 37 Sketch a proof of Fact 3.3.11.

3.3.2.2 Global Validity

Again, we begin by adapting our previous definitions to the three-valued framework:

Definition 3.3.12 Let $x, y \in \{s, t\}$, and $V \subseteq V^3$. A metainference $S \Rrightarrow r$ is *globally* xy*-valid in* V just in case, either some $s \in S$ is not xy-valid in V or r is.

The definition extends to metainference schemas in the obvious way. So, just as in the bivalued case, global validity can be understood in terms of preservation of validity.

Definition 3.3.13 Let $\mathbf{S}$ be a set of metainferences, $x, y \in \{s, t\}$, and $V \subseteq V^3$. The *global* xy*-range of* $\mathbf{S}$ *in* V, written $gr(\mathbf{S}, xy, V)$, is the maximal set of valuations $V_i \subseteq V$ such that each metainference in $\mathbf{S}$ is globally xy-valid in V_i.

So, in other words, the global st-range in V of a calculus $\mathbf{S}$ is the maximal subset V_i of V such that the rules of $\mathbf{S}$ preserve st-validity in V_i.

We present a couple of basic facts about the notions just presented, which will be useful for the proofs that follow.

Fact 3.3.14 Let $x, y \in \{s, t\}$ and $V_1, V_2 \subseteq V^3$. If V_1 and V_2 xy-validate the same inferences, then they globally xy-validate the same metainferences.

(The proof is left to the reader.) This can be seen as the semantic counterpart of Fact 3.3.1. The latter says, remember, that whenever two sequent calculi prove the same sequents, they have the same admissible rules. Leaving the relativization to xy implicit, Fact 3.3.14 says that, whenever two sets of valuations validate the same inferences, they also globally validate the same metainferences.

Fact 3.3.15 Let $\mathbf{S}$ be any sequent calculus, $x, y \in \{s, t\}$ and $V \subseteq V^3$. If a valuation v belongs to $gr(\mathbf{S}, xy, V)$, then v xy-satisfies all the sequents provable in $\mathbf{S}$.

Proof Suppose that v belongs to $gr(\mathbf{S}, xy, V)$. Let r be any sequent provable in $\mathbf{S}$, and let D be an $\mathbf{S}$-derivation of r. Since the axioms of D are rules with no premises, and all the rules in the derivation are globally xy-valid in $gr(\mathbf{S}, xy, V)$, it follows that each sequent in D must be xy-valid in $gr(\mathbf{S}, xy, V)$. Thus, v must xy-satisfy each sequent in D, and this applies in particular to r. □

So, leaving the relativization to xy and V implicit for a moment, we can say: for any sequent calculus whatsoever, if a valuation is in its global range, then it satisfies all of its provable sequents.

It is important to notice that the converse of Fact 3.3.15 does not hold in general. That is, it is not the case that, for any sequent calculus $\mathbf{S}$, and any set of valuations V, if a valuation in V xy-satisfies all the sequents provable in $\mathbf{S}$, then it belongs to the xy-global range of $\mathbf{S}$ in V. For a toy example, let $\mathbf{P}$ be the calculus whose only rules are

$$\text{Id}\ \frac{}{\varphi \Rightarrow \varphi} \qquad\qquad \text{R}\vee m\ \frac{\Gamma \Rightarrow \Delta, \varphi, \psi}{\Gamma \Rightarrow \Delta, \varphi \vee \psi}$$

Due to the lack of Weakening, $\mathbf{P}$ can only prove sequents of the form $\varphi \Rightarrow \varphi$. So, let v be a valuation in V^3 such that $v(p) = 1$ and $v(\varphi \vee \psi) = 0$ for any φ and ψ. On the one hand, v st-satisfies all the sequents that are provable in $\mathbf{P}$, since (of course!) it st-satisfies any sequent of the form $\varphi \Rightarrow \varphi$. On the other hand, v is not in $gr(\mathbf{P}, st, V^3)$. To see this, consider the following instance of R$\vee m$:

$$\frac{p \Rightarrow p, q}{p \Rightarrow p \vee q}$$

The premise-sequent is st-valid in V^3, and thus *a fortiori* it is st-valid in $gr(\mathbf{P}, st, V^3)$. It follows that the conclusion-sequent is st-valid in $gr(\mathbf{P}, st, V^3)$. But the mentioned sequent is not st-satisfied at v. Which implies that $v \notin gr(\mathbf{P}, st, V^3)$.

We are interested in the global st-ranges in V^3 of the calculi $\mathbf{LK}^{-c}$ and $\mathbf{LK}^{-c}_{+\text{E}}$. As it turns out, these ranges are identical: they are the set of all the valuations in V^3 that st-satisfy every sequent valid in V^{B}. Let us write CL^3_{st} to denote this set.

Fact 3.3.16 Let $\mathbf{S}$ be any sequent calculus that is sound and complete with respect to the inferences valid in V^{B}. Then, $gr(\mathbf{S}, st, V^3) = CL^3_{st}$.

Proof From right to left, we rely on the fact that the sequents valid in V^{B} are exactly the ones st-valid in CL^3_{st}. So, take any rule R of $\mathbf{S}$, and consider any instance of it. Suppose that the premise-sequents are all st-valid in CL^3_{st}. Then they are valid in V^{B}. By Completeness, they are provable in $\mathbf{S}$. Then, by R, the conclusion-sequent is provable in $\mathbf{S}$ as well. By Soundness, it is valid in V^{B}. Hence, it is also st-valid in CL^3_{st}. The upshot is that rules of $\mathbf{S}$ are all globally st-valid in CL^3_{st}. Which means that $CL^3_{st} \subseteq gr(\mathbf{S}, st, V)$. From left to right, let v be any valuation in $gr(\mathbf{S}, st, V)$. By Fact 3.3.15, v st-satisfies all the sequents that are provable in $\mathbf{S}$. But by Completeness, these include all the sequents that are valid in V^{B}. Thus, $v \in CL^3_{st}$. That is to say, $gr(\mathbf{S}, st, V^3) \subseteq CL^3_{st}$. □

Corollary 3.3.17 (Global Ranges) *The global st-range in V^3 of calculi* $\mathbf{LK}$, $\mathbf{LK}^{-c}$ *and* $\mathbf{LK}^{-c}_{+\text{E}}$ *is in all cases* CL^3_{st}.

So, we have a characterization result for the global st-range in V^3 of our calculi.[3] It should be mentioned that some seemingly plausible generalizations of Fact 3.3.16 fail. For instance, it is not the case that, whenever two calculi prove the same sequents, they have the same global xy-range in V for any $x, y \in \{s, t\}$ and $V \subseteq V^3$. To wit, compare calculus **P** from above with the calculus **Q** whose only rule is Id. Both calculi prove all and only the sequents of the form $\varphi \Rightarrow \varphi$. However, $gr(\mathbf{Q}, st, V^3) = V^3$. In contrast, we have seen that $gr(\mathbf{P}, st, V^3) \neq V^3$. So, the two calculi have different global st-ranges in V^3, in spite of proving the same sequents.

In Chap. 2, Fact 2.3.19 showed that local and global validity in V^{B} collapse for metainference schemas. Now, a similar collapse result fails for many distinguished sets of trivaluations. For instance, Cut is globally st-valid in V^{SK}, but locally st-invalid.[4] However, the coincidence between the two notions of validity can be restored if we switch to a sufficiently expressive language:

Fact 3.3.18 (Global and Local Collapse) Let $V \subseteq V^3$. Suppose that, for each $i \in \{1, 1/2, 0\}$, the language contains a formula φ_i such that, for every valuation $v \in V$, $v(\varphi_i) = i$. Let $x, y \in \{s, t\}$. Then, global and local xy-validity in V coincide for metainference schemas.

Proof The procedure is similar to that of Fact 2.3.19. Suppose that a metainference schema $S \Rrightarrow r$ is locally xy-invalid in V. Let $S' \Rrightarrow r'$ be an invalid instance, and $v \in V$ a witness to its invalidity. Replace each propositional variable p such that $v(p) = i$ by the formula φ_i. The result is a new instance of $S \Rrightarrow r$ that is globally xy-invalid in V. □

The result can be found for instance in Teijeiro (2021). We illustrate it by means of an example. Let $\mathcal{L}_{\bot\curlywedge}$ be the result of expanding $\mathcal{L}$ with the 0-ary constants $\bot$ and $\curlywedge$. Call $V^{\mathrm{SK}}_{\bot\curlywedge}$ the set of all and only the valuations $v : FOR(\mathcal{L}_{\bot\curlywedge}) \to \{1, 1/2, 0\}$ such

[3] The result is not as informative as we could wish, since it does not tell us much about what the valuations in CL^3_{st} look like. To approach this issue we could appeal to the work of Da Ré et al. (2024a). Using the authors' techniques, it can be shown that in V^3 there are 8.256 deterministic valuations schemes for $\mathcal{L}$ that st-validate exactly the inferences that are valid in V^{B}. Moreover, we know that they are all and only the deterministic schemes that satisfy the following two conditions: (a) they are Boolean normal (classical inputs yield the expected classical outputs), and (b) they are either (b.1) monotonic with respect to the information order (viz. the least partial order $\preceq$ on $\{1, 1/2, 0\}$ satisfying $1/2 \preceq 0$ and $1/2 \preceq 1$) or (b.2) 'collapsible', which means—intuitively—that value $1/2$ plays in them the same functional role as some of the classical values 1 or 0. Letting X be any of these schemes, we have $V^{\mathrm{X}} \subseteq CL^3_{st}$. However, not even this is enough for an alternative characterization result. The reason is that there are valuations that belong to CL^3_{st} in spite of being non-uniform (in the sense of Definition 3.2.16), and non-uniform valuations cannot belong to V^{X} for any of the deterministic schemes X just described. The question of how to better characterize the set CL^3_{st} remains an open problem.

[4] We focus here on well-known sets of valuations to simplify the discussion. But the same remark can be made by focusing on the global and local ranges of various calculi. For instance, consider the global and the local st-ranges of $\mathbf{LK}^{-c}$. Cut is globally st-valid in the former, but not st-locally valid in the latter.

that v behaves as a strong Kleene valuation for the constants of $\mathcal{L}$, and in addition $v(\perp) = 0$ and $v(\curlywedge) = 1/2$. Thus,

Corollary 3.3.19 *Let $x, y \in \{s, t\}$. Then, global and local xy-validity in $V^{\mathrm{SK}}_{\perp\curlywedge}$ coincide for metainference schemas in $\mathcal{L}_{\perp\curlywedge}$.*

Exercises

Exercise 38 Show Fact 3.3.14.

Exercise 39 Show the following weakened version of the converse of Fact 3.3.15: If $\mathbf{S}$ is complete with respect to the sequents that are xy-valid in $gr(\mathbf{S}, xy, V)$ then, if a valuation $v \in V$ xy-satisfies all the sequents provable in $\mathbf{S}$, v belongs to $gr(\mathbf{S}, xy, V)$.

3.3.2.3 Absolute Global Validity

Now, we turn to absolute global validity. Once again, we start with the appropriate definitions:

Definition 3.3.20 Let $x, y \in \{s, t\}$, and $\mathcal{V} \subseteq \mathcal{P}(V^3)$. A metainference is *absolutely globally xy-valid in $\mathcal{V}$* just in case it is globally xy-valid in V for every $V \in \mathcal{V}$.

Once more, the definition extends to schemas in the obvious way.

Definition 3.3.21 Let $\mathbf{S}$ be a set of metainferences, $x, y \in \{s, t\}$, and $V \subseteq V^3$. The *absolute global xy-range of* $\mathbf{S}$ *in* V, written $agr(\mathbf{S}, xy, V)$, is the maximal set $\mathcal{V} \subseteq \mathcal{P}(V)$ such that each metainference in $\mathbf{S}$ is absolutely globally xy-valid in $\mathcal{V}$.

We are interested in the absolute global st-ranges in V^3 of $\mathbf{LK}^{-c}$ and $\mathbf{LK}^{-c}_{+\mathrm{E}}$. As of now we do not have a proper characterization result for these ranges. We do know some things about them, however, which we proceed to mention.

Our first observation appeals to Fact 2.3.23. It states, remember, that for any calculus $\mathbf{S}$, and any valuation v in its local range, the singleton $\{v\}$ is in its absolute global range. Given this, Fact 3.3.10 (local st-range of $\mathbf{LK}^{-c}_{+\mathrm{E}}$) implies that for any $v \in V^{\mathrm{SK}}$, $\{v\}$ belongs to $agr(\mathbf{LK}^{-c}_{+\mathrm{E}}, st, V^3)$. Similarly, Fact 3.3.11 (local st-range of $\mathbf{LK}^{-c}$) implies that for any $v \in V^{\mathrm{Sch}}$, $\{v\}$ belongs to $agr(\mathbf{LK}^{-c}, st, V^3)$. And we can strengthen this kind of result:

Fact 3.3.22 Let $V \subseteq V^3$. A metainference is locally xy-valid in V if and only if it is absolutely globally xy-valid in $\mathcal{P}(V)$.

Proof From left to right, suppose that a metainference $S \Rrightarrow r$ is not absolutely globally xy-valid in $\mathcal{P}(V)$. Then, there is a $V_i \in \mathcal{P}(V)$ such that $S \Rrightarrow r$ is not globally xy-valid in V_i. This means that each $s \in S$ is xy-valid but r is not, which implies that at least one $v \in V_i$ is a local xy-counterexample to $S \Rrightarrow r$. Hence $S \Rrightarrow r$ is not locally xy-valid in V. We leave the right to left direction to the reader. □

As a consequence we have that, first, for each $V \subseteq V^{\text{SK}}$, V belongs to the absolute global st-range of $\mathbf{LK}^{-c}_{+\text{E}}$, and second, for each $V \subseteq V^{\text{Sch}}$, V belongs to the absolute global st-range of $\mathbf{LK}^{-c}$. More succinctly,

Fact 3.3.23 $\mathcal{P}(V^{\text{Sch}}) \subseteq agr(\mathbf{LK}^{-c}, st, V^3)$, and $\mathcal{P}(V^{\text{SK}}) \subseteq agr(\mathbf{LK}^{-c}_{+\text{E}}, st, V^3)$

Our second observation is that the above inclusions are proper.[5] Intuitively, this is because all the valuations that belong to the global st-range in V^3 of our calculi (which, remember, is CL^3_{st}) but do not belong to the respective local st-ranges will enter the stage again, and in fact belong to many elements of the absolute global st-ranges. Formally, we will rely on the following fact:

Fact 3.3.24 Let $\mathbf{S}$ be a sequent calculus, $x, y \in \{s, t\}$ and $V \subseteq V^3$. Suppose that $\mathbf{S}$ is complete with respect to the inferences xy-valid in V. Then, for any $V \subseteq W \subseteq gr(\mathbf{S}, xy, V^3)$, $W \in agr(\mathbf{S}, xy, V^3)$.

Proof Take any rule R of $\mathbf{S}$, and consider any instance of it. Suppose that the premise-sequents are all xy-valid in W. Then *a fortiori* they are xy-valid in $V \subseteq W$. Hence by Completeness they are provable in $\mathbf{S}$. Thus, by R the conclusion-sequent is provable in $\mathbf{S}$ as well. Therefore, by Fact 3.3.15 it is xy-valid in $gr(\mathbf{S}, xy, V^3)$. Then, *a fortiori* it is xy-valid in $W \subseteq gr(\mathbf{S}, xy, V^3)$. The conclusion is that the rules of $\mathbf{S}$ are all globally xy-valid in W. That is, $W \in agr(\mathbf{S}, xy, V^3)$. □

Corollary 3.3.25 *For any* $V^{\text{B}} \subseteq W \subseteq CL^3_{st}$, *$W$ belongs to the absolute global st-ranges of* **LK**, $\mathbf{LK}^{-c}$ *and* $\mathbf{LK}^{-c}_{+\text{E}}$.

Now, let v_1 be any valuation such that (a) it assigns $1/2$ to at least one propositional variable, and (b) it behaves exactly as an SK valuation, except that for every formula φ, if $v_1(\varphi) = 1/2$ then $v_1(\neg\varphi) = 1$.[6] It is routine to show that $v_1 \in CL^3_{st}$ (see Exercise 41). So, consider now the set $W = V^{\text{B}} \cup \{v_1\}$. By Corollary 3.3.25, W belongs to the absolute global st-ranges of $\mathbf{LK}^{-c}$ and $\mathbf{LK}^{-c}_{+\text{E}}$. However, W is not a subset of either V^{SK} or V^{Sch}. Thus, $agr(\mathbf{LK}^{-c}, st, V^3) \not\subseteq \mathcal{P}(V^{\text{Sch}})$, and $agr(\mathbf{LK}^{-c}_{+\text{E}}, st, V^3) \not\subseteq \mathcal{P}(V^{\text{SK}})$.

The third observation is that even Fact 3.3.23 and Corollary 3.3.25 together are not enough to characterize the absolute global st-ranges in V^3 of our calculi. That is, there are sets in these ranges that are neither a subset of the corresponding local range, nor a set extending V^{B} but included in CL^3_{st}. As an example, consider the set of valuations $\{v_1, v_2\}$ where v_1 is as before, and v_2 is the SK valuation defined by stipulating that, for every propositional variable p,

---

[5] It is worth mentioning that this observation corrects a mistake in Da Ré et al. (2021). There, the authors claim that $agr(\mathbf{LK}^{-c}_{+\text{E}}, st, V^3) = \mathcal{P}(V^{\text{SK}})$.

[6] Following the terminology of Da Ré et al. (2024a), we may say that v_1 is a '0-collapsing' valuation: a valuation where value $1/2$ plays the same functional role as value 0.

$$v_2(p) = \begin{cases} 1 & \text{if } v_1(p) = 1 \\ 0 & \text{if } v_1(p) \in \{1/2, 0\} \end{cases}$$

It is easy to show (by induction on the complexity of the formula) that (i) for every φ, if $v_1(\varphi) = 1$ then $v_2(\varphi) = 1$, and if $v_1(\varphi) \in \{1/2, 0\}$ then $v_2(\varphi) = 0$. This is enough to establish that $\{v_1, v_2\}$ is in the absolute global st-ranges in V^3 of both $\mathbf{LK}^{-c}$ and $\mathbf{LK}^{-c}_{+\mathrm{E}}$. Roughly, the reasoning goes as follows. Suppose that an instance of a rule R of one of these calculi is globally st-invalid in $\{v_1, v_2\}$. Then, either v_1 or v_2 must be a local st-counterexample to it. But it cannot be v_2, because it is in the local st-range in V^3 of both calculi. So it must be v_1. Hence, v_1 st-satisfies all the premise-sequents but does not st-satisfy the conclusion-sequent. Applying (i) we obtain that v_2 does not st-satisfy the conclusion-sequent either. But then, v_2 also fails to st-satisfy some of the premise sequents, which will not be st-valid in $\{v_1, v_2\}$. As a consequence, the instance of the rule R is globally st-valid in $\{v_1, v_2\}$, contrary to our assumption. This example gives the reader a flavor of how unmanageable the absolute global st-ranges of our calculi are!

Lastly, we note that the absolute global st-range in V^3 of $\mathbf{LK}^{-c}_{+\mathrm{E}}$ is strictly included in the corresponding range of $\mathbf{LK}^{-c}$. That is,

Fact 3.3.26 $agr(\mathbf{LK}^{-c}_{+\mathrm{E}}, st, V^3) \subset agr(\mathbf{LK}^{-c}, st, V^3)$

Proof The '$\subseteq$' claim follows from the fact that $\mathbf{LK}^{-c}$ is a subsystem of $\mathbf{LK}^{-c}_{+\mathrm{E}}$. The '$\not\supseteq$' claim can be shown by taking any Schütte valuation v such that $v(p) = 0$ but $v(\neg p) = 1/2$. This valuation is a local st-counterexample of the metainference

$$\frac{\neg p \Rightarrow}{\Rightarrow p}$$

But this metainference is an instance of the rule L$\neg$E of $\mathbf{LK}^{-c}_{+\mathrm{E}}$. Thus, the set $\{v\}$ is not in $agr(\mathbf{LK}^{-c}_{+\mathrm{E}}, st, V^3)$. Moreover, since v is a Schütte valuation, by Fact 3.3.11 it belongs to the local st-range in V^3 of $\mathbf{LK}^{-c}$, which by Fact 3.3.23 means that $\{v\}$ is in $agr(\mathbf{LK}^{-c}, st, V^3)$. □

As explained at the end of Sect. 2.3.2, given a particular sequent calculus **S**, the metainferences that are locally valid in **S**'s local range do not always coincide with the metainferences that are absolutely globally valid in **S**'s absolute global range. In Sect. 2.3.2, the phenomenon was exemplified by taking the corresponding ranges in V^2. But the same is true when we consider the various possible xy-ranges in V^3. Here, we exemplify the case of the st-range, because we are interested in non-transitive systems.

So, let $\mathbf{S_2}$ be a sequent calculus containing all the structural rules except for Cut, and the following rules for disjunction:

$$(1a)\ \frac{\Rightarrow \varphi}{\Rightarrow \varphi \vee \psi} \qquad (1b)\ \frac{\Rightarrow \psi}{\Rightarrow \varphi \vee \psi}$$

$$(2a)\ \frac{\varphi \vee \psi \Rightarrow}{\varphi \Rightarrow} \qquad (2b)\ \frac{\varphi \vee \psi \Rightarrow}{\psi \Rightarrow}$$

$$(3)\ \frac{\varphi \Rightarrow \qquad \psi \Rightarrow}{\varphi \vee \psi \Rightarrow} \qquad (4)\ \frac{\varphi \vee \psi \Rightarrow \qquad \Rightarrow \varphi \vee \psi}{\Rightarrow \varphi, \psi}$$

It is easy to check that the valuations belonging to the local st-range in V^3 of $\mathbf{S_2}$ are all and only those satisfying the following conditions:

- If $v(\varphi \vee \psi) = 0$, then $v(\varphi) = v(\psi) = 0$ [rules (1a) and (1b)]
- If $v(\varphi) = 1$ or $v(\psi) = 1$, then $v(\varphi \vee \psi) = 1$ [rules (2a) and (2b)]
- If $v(\varphi \vee \psi) = 1$, then $v(\varphi) = 1$ or $v(\psi) = 1$ [rule (3)]
- If $v(\varphi) = v(\psi) = 0$, then $v(\varphi \vee \psi) \in \{0, 1\}$ [rule (4)]

Moreover, the valuations in V^3 satisfying these conditions are exactly those which respect the strong Kleene table for disjunction. Now, take the rule

$$(*)\ \frac{\Rightarrow \varphi \vee \psi}{\Rightarrow \varphi, \psi}$$

This rule is locally st-valid in the strong Kleene valuations. Hence, it is locally st-valid in the local st-range of $\mathbf{S_2}$. However, consider the set of valuations $\{v_1, v_2\}$ where v_1 and v_2 behave as follows:

$$v_1(\varphi) = 1 \text{ for every } \varphi \qquad v_2(\varphi) = \begin{cases} 1/2 & \text{if } \varphi \text{ is of the form } \delta \vee \gamma \\ 0 & \text{otherwise} \end{cases}$$

The set $\{v_1, v_2\}$ is in the absolute global st-range of $\mathbf{S_2}$, as each of the rules (1a)–(4) is globally st-valid in it. However, rule $(*)$ is not globally st-valid in $\{v_1, v_2\}$. (For example, its instance

$$\frac{\Rightarrow p \vee q}{\Rightarrow p, q}$$

is such that the premise-sequent is st-valid in $\{v_1, v_2\}$, but the conclusion sequent is not.) As a consequence, $(*)$ is not absolutely globally st-valid in the absolute global st-range of $\mathbf{S_2}$.

Exercises

Exercise 40 Prove the right to left direction of Fact 3.3.22.

Exercise 41 Show claim (i) concerning valuations v_1 and v_2 from above. Then, use it to show that $v_1 \in CL^3_{st}$.

Exercise 42 Explain why, for every $W \subseteq V^{\mathrm{B}}$, W belongs to the absolute global st-ranges of both $\mathbf{LK}^{-c}$ and $\mathbf{LK}^{-c}_{+\mathrm{E}}$.

Exercise 43 Suppose $W \in agr(\mathbf{S}, xy, V)$. Does it follow that $W' \in agr(\mathbf{S}, xy, V)$, for every $W' \subseteq W$?

3.3.3 Metainferential Soundness and Completeness

We begin by stating a couple of general facts about the relationship between our syntactic and our semantic notions of metainferential validity. For one thing, as in the bivalued case, global validity coincides with admissibility under the assumption of inferential Soundness and Completeness:

Fact 3.3.27 (Admissibility and Global Validity) Let $\mathbf{S}$ be a sequent calculus, $x, y \in \{s, t\}$ and $V \subseteq V^3$. If $\mathbf{S}$ is sound and complete with respect to the inferences xy-valid in V, then a rule is admissible in $\mathbf{S}$ if and only if it is globally xy-valid in V.

(The proof is simple, and we leave it as an exercise to the reader.)

For another thing, derivability coincides with absolute global validity, in the following sense:

Theorem 3.3.28 (Derivability and Absolute Global Validity) *Let* $\mathbf{S}$ *be a sequent calculus with all the structural rules except for Cut. Suppose that* $\mathcal{V} \subseteq \mathcal{P}(V^3)$ *is the absolute global* st*-range in* V^3 *of* $\mathbf{S}$*. Then, a rule is derivable in* $\mathbf{S}$ *if and only if it is absolutely globally* st*-valid in* $\mathcal{V}$*.*

The result was first shown by Da Ré et al. (2021). (The proof is less than trivial, and we omit it for the sake of conciseness and readability.)

Now, we will give two results that are particular to calculi $\mathbf{LK}^{-c}$ and $\mathbf{LK}^{-c}_{+\mathrm{E}}$, and concern the relation between derivability and local validity. The first result states that a sequent rule is derivable in $\mathbf{LK}^{-c}$ just in case it is st-valid in V^{Sch}. The result was first shown by Girard (1987).

Theorem 3.3.29 (Metainferential Adequacy $\mathbf{LK}^{-c}$) *A rule is derivable in* $\mathbf{LK}^{-c}$ *if and only if it is locally* st*-valid in* V^{Sch}*.*

Proof Soundness is an easy consequence of Fact 3.3.11, which states that the local st-range in V^3 of $\mathbf{LK}^{-c}$ is V^{Sch}. So we leave it to the reader, and focus on Completeness instead. We present a slight variation of Girard's proof. The strategy is similar to that of Theorem 2.2.12, which shows the finite Completeness of $\mathbf{LK}$ with respect to the inferences valid in V^{B}. Take any instance $S \Rrightarrow r$ of the relevant rule. We build a reduction tree for the sequent r. Then, we show that if the tree satisfies certain closure conditions, r is derivable from S in $\mathbf{LK}^{-c}$, and if the tree does *not* satisfy those conditions, there is a Schütte valuation v that st-satisfies each sequent in S but does not st-satisfy r. Crucially, the closure conditions for the tree will not only allow leaves that can be obtained by Weakening applied on axioms, but also leaves that can be obtained by Weakening applied on sequents of S.

Stage 0. Let $r = \Gamma_0 \Rightarrow \Delta_0$ be the root of the tree, and mark all the non-atomic formulas in it as non-reduced. A branch of the tree is *closed* if its leaf can be obtained by (perhaps 0) applications of Weakening and/or Contraction on some sequent $s \in S$ or some sequent of the form $\varphi \Rightarrow \varphi$; if the branch is not closed, it is *open*.

Stage n + 1. If either all the branches of the tree are closed, or all the non-atomic formulas have been marked as reduced in previous stages, then halt the process. Otherwise, take any of the open branches. Call its leaf $\Gamma_n \Rightarrow \Delta_n$. Take any formula φ that has not been marked as reduced yet, and reduce it as follows:

Case (a) φ is a negation, or a conjunction in Δ_n, or a disjunction in Γ_n, or a conditional in Δ_n. First, extend the branch by adding an additional copy of φ to the corresponding multiset. Then, extend it further by applying the inverse of the rule corresponding to φ's form and position. Thus, for instance, if φ is a conjunction $\psi \wedge \rho$ in Δ_n, then the result looks like this:

$$\dfrac{\dfrac{\Gamma_n \Rightarrow \Delta_n, \psi \qquad \Gamma_n \Rightarrow \Delta_n, \rho}{\Gamma_n \Rightarrow \Delta_n, \psi \wedge \rho}}{\Gamma_n \Rightarrow \Delta_n}$$
$$\vdots$$

Case (b) φ is a conjunction in Γ_n or a disjunction in Δ_n. First, extend the branch by adding, one at a time, two extra copies of φ to the corresponding multiset. Then, extend it further by applying the inverses of the two rules corresponding to φ's form and position. Thus, for instance, if φ is a conjunction $\psi \wedge \rho$ in Γ_n, the result looks like this:

$$\dfrac{\dfrac{\dfrac{\dfrac{\psi, \rho, \Gamma_n \Rightarrow \Delta_n}{\psi, \psi \wedge \rho, \Gamma_n \Rightarrow \Delta_n}}{\psi \wedge \rho, \psi \wedge \rho, \Gamma_n \Rightarrow \Delta_n}}{\psi \wedge \rho, \Gamma_n \Rightarrow \Delta_n}}{\Gamma_n \Rightarrow \Delta_n}$$
$$\vdots$$

Case (c) φ is a conditional in Γ_n. First, extend the branch by adding, one at a time, two extra copies of φ and one extra copy of each of the other formulas in $\Gamma_n \Rightarrow \Delta_n$. Then, extend it further by applying the inverse of L$\rightarrow$; use Γ_n and Δ_n as the left and right contexts, respectively, of the premises. Letting φ be $\psi \rightarrow \rho$, the result looks as follows:

$$\dfrac{\rho, \Gamma_n \Rightarrow \Delta_n \qquad \Gamma_n \Rightarrow \Delta_n, \psi}{\Gamma_n, \Gamma_n, \psi \rightarrow \rho \Rightarrow \Delta_n, \Delta_n}$$
$$\vdots$$
$$\Gamma_n \Rightarrow \Delta_n$$
$$\vdots$$

In all cases, we obtain leaves of the form $\Sigma, \Gamma_n \Rightarrow \Delta_n, \Pi$, where Σ and Π only contain immediate subformulas of φ. Mark the non-atomic formulas in Σ and in Π as non-reduced. Mark φ as already reduced; all the other formulas in Γ_n and in Δ_n inherit their marks from sequent $\Gamma_n \Rightarrow \Delta_n$.

That ends the description of the tree. Note that a difference with Theorem 2.2.12 is that here the tree is cumulative, in the sense that, if a formula occurs at a given node, then it also occurs at all the nodes above. We say that the tree is *closed* just in case all its branches are closed; otherwise it is *open*. It is easy to check that, if the tree is closed, then we can construct a derivation of r from S in $\mathbf{LK}^{-c}$: we just take each branch and extend it with the (perhaps 0) applications of Weakening and/or Contraction needed to reach either a sequent $\varphi \Rightarrow \varphi$ or a sequent s in S; if we reached a sequent $\varphi \Rightarrow \varphi$ with φ non-atomic, we further extend the branch with a derivation of it. As a consequence, if r is *not* derivable from S, there will be an open branch $\Gamma_0 \Rightarrow \Delta_0, \ldots, \Gamma_n \Rightarrow \Delta_n$. We define the valuation v as follows:

$$v(\varphi) = \begin{cases} 1 & \text{if } \varphi \in \Gamma_n \\ 0 & \text{if } \varphi \in \Delta_n \\ 1/2 & \text{otherwise} \end{cases}$$

Notice that conditions $\varphi \in \Gamma_n$ and $\varphi \in \Delta_n$ are mutually exclusive, since if φ occurred both in Γ_n and in Δ_n, then the branch would be closed contrary to our assumption. It remains to show three things: (a) v does not st-satisfy $\Gamma_0 \Rightarrow \Delta_0$, (b) v st-satisfies every sequent in S, and (c) v is a Schütte valuation, viz. $v \in V^{\mathrm{Sch}}$. Claim (a) follows from the definition of v and the fact that Γ_0 and Δ_0 are submultisets of Γ_n and Δ_n, respectively. As for (b), suppose that v does not st-satisfy a sequent $\Sigma \Rightarrow \Pi$ in S. This means that $v(\sigma) = 1$ for each $\sigma \in \Sigma$ and $v(\pi) = 0$ for each $\pi \in \Pi$. It follows that each $\sigma \in \Sigma$ occurs in Γ_n and each $\pi \in \Pi$ occurs in Δ_n. But this makes $\Gamma_n \Rightarrow \Delta_n$ derivable from $\Sigma \Rightarrow \Pi$ by applications of Weakening and Contraction, which contradicts again our assumption that the branch is open. Lastly, to show (c) we must consider each of the possible shapes of a non-atomic formula. We consider two of the cases, and leave the others to the reader.

- Negation: v is a Schütte valuation for $\neg$ just in case (i) if $v(\neg\varphi) = 1$ then $v(\varphi) = 0$, and (ii) if $v(\neg\varphi) = 0$ then $v(\varphi) = 1$. To establish (i), suppose $v(\neg\varphi) = 1$. Then, $\neg\varphi \in \Gamma_n$. Hence, by design of the tree, $\varphi \in \Delta_n$. Thus, $v(\varphi) = 0$, as desired. The reasoning for (ii) is symmetric.
- Conjunction: v is a Schütte valuation for $\wedge$ just in case (i) if $v(\varphi \wedge \psi) = 1$ then $v(\varphi) = v(\psi) = 1$, and (ii) if $v(\varphi \wedge \psi) = 0$ then $v(\varphi) = 0$ or $v(\psi) = 0$. To establish (i), suppose $v(\varphi \wedge \psi) = 1$. Then, $\varphi \wedge \psi \in \Gamma_n$. Hence, by design of the tree, $\varphi \in \Gamma_n$ and $\psi \in \Gamma_n$. Thus, $v(\varphi) = v(\psi) = 1$, as desired. To establish (ii), suppose $v(\varphi \wedge \psi) = 0$. Then, $\varphi \wedge \psi \in \Delta_n$. Hence, by design of the tree, $\varphi \in \Delta_n$ or $\psi \in \Delta_n$. Either case, $v(\varphi) = 0$ or $v(\psi) = 0$, as desired.

□

A couple of comments on the above proof. First, the reader might wonder why we defined valuation v for all formulas 'in one go', when the usual procedure (see, e.g., Theorem 2.2.12 in Chap. 2) is to define it for propositional variables only, and let the truth tables do the rest of the work. The reason stems from the shape of the Schütte tables: if we defined v by letting it be any valuation in V^{Sch} that assigns 1 to the variables in Γ_n, 0 to the variables in Δ_n and $1/2$ to the variables neither in Γ_n nor Δ_n, then we would have no guarantee that v extends to complex formulas in the way we want, namely, by assigning 1 to everything in Γ_n and 0 to everything in Δ_n. So, we had to first impose on v the behaviour that all in all we want from it, and then show that it indeed belongs to V^{Sch}.

Second, the reader might wonder why in this proof the tree must be cumulative, viz. such that complex formulas do not disappear from it as they get reduced. In few words, cumulativity ensures that we do not get rid of formulas that we may need later to arrive at the Weakening closure of some sequent. For example, consider the metainference

$$(*)\ \frac{\Rightarrow p \vee q, q}{\Rightarrow p \vee q}$$

The reduction procedure from the proof delivers the tree

$$\mathrm{RC}\ \dfrac{\mathrm{RC}\ \dfrac{\mathrm{R}\vee\ \dfrac{\mathrm{R}\vee\ \dfrac{\Rightarrow p \vee q, p, q}{\Rightarrow p \vee q, p \vee q, q}}{\Rightarrow p \vee q, p \vee q, p \vee q}}{\Rightarrow p \vee q, p \vee q}}{\Rightarrow p \vee q}$$

The topmost sequent $\Rightarrow p \vee q, p, q$ is in the Weakening closure of the premise $\Rightarrow p \vee q, q$. Hence, the tree is closed, and we know that $(*)$ is derivable. However, if we did not add additional copies of complex formulas before reducing them, and instead followed the procedure from Theorem 2.2.12, the result would be

$$\mathrm{RC}\ \dfrac{\mathrm{R}\vee\ \dfrac{\mathrm{R}\vee\ \dfrac{\Rightarrow p, q}{\Rightarrow p \vee q, q}}{\Rightarrow p \vee q, p \vee q}}{\Rightarrow p \vee q}$$

The topmost sequent $\Rightarrow p, q$ is not in the Weakening closure of $\Rightarrow p \vee q, q$ because of the formula $p \vee q$ which is in the latter but not in the former. Hence the tree is open, contrary to the fact that $(*)$ is indeed derivable in $\mathbf{LK}^{-c}$. In other words, then, cumulativity ensures that Weakening remains applicable in all cases where it *should* be applicable.

Our Theorem establishes that calculus $\mathbf{LK}^{-c}$ is metainferentially sound and complete with respect to the rules that are valid in $\mathbb{ST}^{\mathrm{Sch}}$. Bear in mind that $\mathbb{ST}^{\mathrm{Sch}}$ is based on the set V^{Sch} of all Schütte valuations, and hence it has strictly less metainferences than logic $\mathbb{ST}_u^{\mathrm{Sch}}$, which is based on the set V_u^{Sch} of the Schütte valuations that are uniform. This means that $\mathbf{LK}^{-c}$ is not metainferentially complete with respect to $\mathbb{ST}_u^{\mathrm{Sch}}$. For instance, take again the metainference

$$\frac{p \wedge q \Rightarrow}{r \wedge s, p, q, r, s \Rightarrow}$$

from the proof of Fact 3.3.7. This metainference belongs to $\mathbb{ST}_u^{\mathrm{Sch}}$, but it is not derivable in $\mathbf{LK}^{-c}$, as the calculus has no elimination rules to get rid of the formula $p \wedge q$ that is present in the premise but not in the conclusion.

Lastly, at the end of Sect. 3.3.2.3 we mentioned that local validity in the local range of a calculus does not always coincide with absolute global validity in the absolute global range of the same calculus. Well, an immediate consequence of Theorems 3.3.28 and 3.3.29 is that these two things coincide in the case of $\mathbf{LK}^{-c}$:

Corollary 3.3.30 *A rule is locally st-valid in $lr(\mathbf{LK}^{-c}, st, V^3)$ if and only if it is absolutely globally st-valid in $agr(\mathbf{LK}^{-c}, st, V^3)$.*

Alternatively, we can also say that, in the case of calculus $\mathbf{LK}^{-c}$, *derivability coincides with local validity*—with the proviso, of course, that we talk about rules, and that local validity is relativized to the appropriate set of valuations and notion of consequence.

Now, the second result. It states that a sequent rule is derivable in $\mathbf{LK}^{-c}_{+\mathrm{E}}$ just in case it is st-valid in V^{SK}. This was first shown by Dicher and Paoli (2019).

Theorem 3.3.31 (Metainferential Adequacy $\mathbf{LK}^{-c}_{+\mathrm{E}}$) *A rule $S \Rrightarrow r$ is derivable in $\mathbf{LK}^{-c}_{+\mathrm{E}}$ if and only if it is locally st-valid in V^{SK}.*

Proof Soundness is an easy consequence of Fact 3.3.10, which states that the local st-range in V^3 of $\mathbf{LK}^{-c}_{+\mathrm{E}}$ is V^{SK}. So we focus on Completeness. We extend the strategies used by Hösli and Jäger (1994) and Da Ré et al. (2024b) to apply the technique of tree reduction even with calculi that contain elimination rules. Take any instance $S \Rrightarrow r$ of the relevant rule. The idea is to construct a reduction tree for r where, following some enumeration of the language, we progressively consider all the ways in which the formulas in the tree could have been obtained by applying the elimination rules of the calculus. So, in what follows, let $\xi_1, \ldots, \xi_i, \ldots$ be some fixed enumeration of the formulas in the language. We start the construction with stage $n = 0$. Then, each successor stage $n + 1$ is either even (that is, of the form $2m$) or odd (that is, of the form $2m + 1$). We have different instructions for each case; intuitively, even cases take care of elimination rules, while odd cases address introduction rules.

Stage $\mathbf{n = 0}$. Let $r = \Gamma_0 \Rightarrow \Delta_0$ be the root of the tree. Also, let $\theta_0, \ldots, \theta_k$ be any listing of the non-atomic formulas occurring in r. As before, a branch of the tree is *closed* if its leaf can be obtained by Weakening and/or Contraction from a sequent $s \in S$ or a sequent $\varphi \Rightarrow \varphi$; otherwise it is *open*.

Stage $\mathbf{n + 1 = 2m}$. If all the branches are closed, halt the process. Otherwise, take any of the open branches. Call its leaf $\Gamma_n \Rightarrow \Delta_n$. Extend the branch with the sequent $\Gamma_{n+1} \Rightarrow \Delta_{n+1} = \Gamma_n, \Gamma'_n \Rightarrow \Delta_n, \Delta'_n$, where Γ'_n and Δ'_n are as follows:

- For each φ and ψ occurring in Γ_n at least once, Γ'_n contains $\varphi \wedge \psi$ and $\psi \wedge \varphi$.
- For each $\varphi \in \Gamma_n$ and ξ_i with $i \leq m$, Γ'_n contains $\varphi \vee \xi_i$ and $\xi_i \vee \varphi$
- For each $\varphi \in \Delta_n$, Γ'_n contains $\neg\varphi$.
- For each $\varphi \in \Gamma_n$, $\psi \in \Delta_n$ and ξ_i with $i \leq m$, Γ'_n contains $\xi_i \rightarrow \varphi$ and $\psi \rightarrow \xi_i$.

- For each φ and ψ occurring in Δ_n at least once, Δ'_n contains $\varphi \vee \psi$ and $\psi \vee \varphi$.
- For each $\varphi \in \Delta_n$ and ξ_i with $i \leq m$, Δ'_n contains $\varphi \wedge \xi_i$ and $\xi_i \wedge \varphi$.
- For each $\varphi \in \Gamma_n$, Δ'_n contains $\neg\varphi$.
- For each $\varphi \in \Gamma_n$ and $\psi \in \Delta_n$, Δ'_n contains $\varphi \to \psi$.
- Γ'_n and Δ'_n contain no other formulas.

The listing of the non-atomic formulas of $\Gamma_{n+1} \Rightarrow \Delta_{n+1}$ is $\theta_0, \ldots, \theta_k, \rho_0, \ldots, \rho_j$, where $\theta_0, \ldots, \theta_k$ is the inherited listing of the non-atomic formulas of $\Gamma_n \Rightarrow \Delta_n$ and $\rho_0, \ldots, \rho_j$ is any listing of the formulas in $\Gamma'_n \cup \Delta'_n$.

Stage $\mathbf{n + 1 = 2m + 1}$. If all the branches are closed, then halt the process. Otherwise, take any of the open branches. Call its leaf $\Gamma_n \Rightarrow \Delta_n$, and suppose that the listing of its non-atomic formulas is $\theta_0, \ldots, \theta_k$. Then take formula θ_0 and reduce it following Cases (a)–(c) from the tree construction in Theorem 3.3.29. For each of the newly obtained topmost sequents, the listing of the non-atomic formulas starts with $\theta_1, \ldots, \theta_k$, continues with the auxiliary formula(s) of the case, and ends with θ_0. Thus, for example, if θ_0 is $\varphi \to \psi$ and occurs in Γ_n, we apply instruction (c) and obtain the topmost sequents $\psi, \Gamma_n \Rightarrow \Delta_n$ and $\Gamma_n \Rightarrow \Delta_n, \varphi$; the listing of the former goes $\theta_1, \ldots, \theta_k, \psi, \varphi \to \psi$, and the listing of the the latter goes $\theta_1, \ldots, \theta_k, \varphi, \varphi \to \psi$.

That ends the description of the tree. A couple of things to notice: first, in stages $n + 1 = 2m$, the items defining Γ'_n and Δ'_n track one by one the elimination rules of $\mathbf{LK}^{-c}_{+\mathrm{E}}$. Second, an obvious difference with Theorem 3.3.29 is that, here, open branches are always infinite. As before, a tree is *closed* just in case all its branches are. Again, if a tree is closed, we can construct a derivation of r from S; the only novelty is that whenever a node of the tree was generated by a stage $n+1 = 2m$, we must replace it with a number of applications of elimination rules and Contraction. It follows that, if r is *not* derivable from S, there will be an open branch $\Gamma_0 \Rightarrow \Delta_0, \ldots, \Gamma_i \Rightarrow \Delta_i, \ldots$. Let Γ_u and Δ_u be the multiset unions of all the Γ_is and Δ_is, respectively. By the cumulativity of the tree and the fact that the branch is open we know that if $\varphi \in \Gamma_u$ then $\varphi \notin \Delta_u$. Hence we know that the valuation v defined as[7]

$$v(\varphi) = \begin{cases} 1 & \text{if } \varphi \in \Gamma_u \\ 0 & \text{if } \varphi \in \Delta_u \\ 1/2 & \text{otherwise} \end{cases}$$

exists. The claim that (a) v does not st-satisfy $\Gamma_0 \Rightarrow \Delta_0$ again follows from the definition of v and the fact that Γ_0 and Δ_0 are submultisets of Γ_u and Δ_u. As for the claim that (b) v st-satisfies every sequent in S, suppose, for contradiction, that it does not. Then, there is a sequent $\Sigma \Rightarrow \Pi$ in S that is st-counterexemplified by v. This means that $v(\sigma) = 1$ for every $\sigma \in \Sigma$ and $v(\pi) = 0$ for every $\pi \in \Pi$. By the

[7] Since SK valuations are deterministic, here we *could* have defined v as an SK valuation assigning 1 to every variable in Γ_u, 0 to every variable in Δ_u and 1/2 to every other variable. We proceed as we do to restrict the differences between the proofs of Theorems 3.3.29 and 3.3.31 to the essential ones only.

definition of v, the cumulativity of the tree, and the fact that Σ and Π are finite, it follows that there is some $\Gamma_k \Rightarrow \Delta_k$ such that $\sigma \in \Gamma_k$ for every $\sigma \in \Sigma$, and $\pi \in \Delta_k$ for every $\pi \in \Pi$. But this makes $\Gamma_k \Rightarrow \Delta_k$ derivable from $\Sigma \Rightarrow \Pi$ by applications of Weakening and Contraction, contrary to the assumption that the branch is open. Finally, we must show that (c) v is an SK valuation. We consider two of the possible shapes of a non-atomic formula, and leave the others to the reader.

- Negation: v is an SK valuation for $\neg$ just in case (i) if $v(\neg\varphi) = 1$ then $v(\varphi) = 0$, (ii) if $v(\neg\varphi) = 0$ then $v(\varphi) = 1$, and (iii) if $v(\neg\varphi) = 1/2$ then $v(\varphi) = 1/2$. Items (i) and (ii) are secured as in Theorem 3.3.29, that is, by the facts that if $\neg\varphi \in \Gamma_u$ then $\varphi \in \Delta_u$, and if $\neg\varphi \in \Delta_u$ then $\varphi \in \Gamma_u$. Item (iii) is secured by the fact (which does not hold for the tree construction in Theorem 3.3.29) that if $\neg\varphi \notin \Gamma_u \cup \Delta_u$, then $\varphi \notin \Gamma_u \cup \Delta_u$.
- Conjunction: v is an SK valuation for $\wedge$ just in case (i) if $v(\varphi \wedge \psi) = 1$ then $v(\varphi) = v(\psi) = 1$, (ii) if $v(\varphi \wedge \psi) = 0$ then $v(\varphi) = 0$ or $v(\psi) = 0$, and (iii) if $v(\varphi \wedge \psi) = 1/2$, then either $v(\varphi) = v(\psi) = 1/2$, or $v(\varphi) = 1/2$ and $v(\psi) = 1$, or $v(\varphi) = 1$ and $v(\psi) = 1/2$. Items (i) and (ii) are secured as in Theorem 3.3.29, that is, by the facts that if $\varphi \wedge \psi \in \Gamma_u$ then $\varphi, \psi \in \Gamma_u$, and if $\varphi \wedge \psi \in \Delta_u$ then either $\varphi \in \Delta_u$ or $\psi \in \Delta_u$, respectively. Item (iii) is secured by the fact (which does not hold for the tree construction in Theorem 3.3.29) that if $\varphi \wedge \psi \notin \Gamma_u \cup \Delta_u$, then none of φ and ψ can occur in Δ_u, and at least one of them must not occur in Γ_u.

□

So, our Theorem establishes that calculus $\mathbf{LK}^{-c}_{+\mathrm{E}}$ is metainferentially sound and complete with respect to the rules that are valid in $\mathbb{ST}^{\mathrm{SK}}$.[8] Just as before, an immediate consequence of Theorems 3.3.28 and 3.3.31 is the following:

Corollary 3.3.32 *A rule is locally st-valid in* $lr(\mathbf{LK}^{-c}_{+\mathrm{E}}, st, V^3)$ *if and only if it is absolutely globally st-valid in* $agr(\mathbf{LK}^{-c}_{+\mathrm{E}}, st, V^3)$.

Again, we can say that, in the case of $\mathbf{LK}^{-c}$, derivability coincides with local validity—provided we talk about rules, and local validity is relativized to the appropriate set of valuations and notion of consequence.

Exercises

Exercise 44 Proof Fact 3.3.27.

Exercise 45 Complete the proof of Theorem 3.3.29 by considering, in the argument to show that v is a Schütte valuation, the cases where φ is a disjunction and a conditional.

Exercise 46 Complete the proof of Theorem 3.3.31 by considering, in the argument to show that v is a strong Kleene valuation, the cases where φ is a disjunction and a conditional.

[8] The reader may well wonder what a metainferentially adequate calculus for logic $\mathbb{ST}^{\mathrm{Sch}}_u$ looks like. That remains an open question.

Exercise 47 Prove the Soundness part of either Theorems 3.3.29 or 3.3.31.

3.4 A Wolf in Sheep's Clothing

As we have seen, in **LK**, from the fact that the horizontalization of a given rule is provable, it follows that the rule itself is derivable (see Fact 2.3.8). We also mentioned that this is no longer the case in calculi $\mathbf{LK}^{-c}$ and $\mathbf{LK}^{-c}_{+\mathrm{E}}$. In this section we explore some surprising facts related to this phenomenon.

A number of authors—among them Pynko (2010), Barrio et al. (2015) and Dicher and Paoli (2021)—have shown that there is a close relationship between the non-transitive logics that we explore in this chapter, on the one hand, and certain paraconsistent logics on the other. This relationship can be presented from a model-theoretic or from a proof-theoretic perspective. We shall take the first path, although nothing of relevance hinges on this choice. The paraconsistent logics that we will be concerned with are the following:

Definition 3.4.1

- $\mathbb{LP}^{\mathrm{SK}}$ (or just $\mathbb{LP}$, for short) is the logic containing all and only the inferences and metainferences that are (locally) tt-valid in V^{SK}
- $\mathbb{LP}^{\mathrm{Sch}}$ is the logic containing all and only the inferences and metainferences that are (locally) tt-valid in V^{Sch}
- $\mathbb{LP}^{\mathrm{Sch}}_u$ is the logic containing all and only the inferences and metainferences that are (locally) tt-valid in V^{Sch}_u

The phenomenon we are going to study arises in full generality when the language is slightly stronger than our default language $\mathcal{L}$; in particular, the language in question must contain at least one formula that receives the same classical value at every valuation. Thus, let $\mathcal{L}_\perp$ be the language that results by expanding $\mathcal{L}$ with a 0-ary constant $\perp$, and assume that $\perp$ receives value 0 at every valuation in V^3. Throughout this section, all the objects we manipulate—(meta)inferences, logical systems, and so on—are to be read as being couched in $\mathcal{L}_\perp$.

Our first result relates logics $\mathbb{ST}^{\mathrm{SK}}$ and $\mathbb{LP}^{\mathrm{SK}}$; roughly, it says that the sequent-made metainferences locally valid in the former are, *modulo* translation, exactly the inferences valid in the latter. To state our result precisely, we introduce some terminology.

Definition 3.4.2 Given a sequent $\Gamma \Rightarrow \Delta$, we define its *formula-translation* $t(\Gamma \Rightarrow \Delta)$ as follows:[9]

[9] This definition is similar but non-identical to that of the horizontalization of a sequent (Definition 2.3.6). There, the constant $\perp$ was not needed, since in the bivalued context the empty sequent can be assigned any classical contradiction.

$$\tau(\Gamma \Rightarrow \Delta) = \begin{cases} \bigwedge \Gamma \to \bigvee \Delta & \text{if } \Gamma, \Delta \neq \varnothing \\ \bigvee \Delta & \text{if } \Gamma = \varnothing, \Delta \neq \varnothing \\ \neg \bigwedge \Gamma & \text{if } \Gamma \neq \varnothing, \Delta = \varnothing \\ \bot & \text{if } \Gamma, \Delta = \varnothing \end{cases}$$

If S is a set of sequents, $\tau(S)$ denotes the set $\{\tau(s) \mid s \in S\}$.

Definition 3.4.3 Given a metainference $S \Rrightarrow r$ that is made out of sequents,[10] we define its *inference-translation* $\tau(S \Rrightarrow r)$ as the inference that goes from the formula-translations of the premises to the formula-translation of the conclusion:

$$\tau(S \Rrightarrow r) = \tau(S) \Rightarrow \tau(r)$$

Now we can present the result:

Theorem 3.4.4 *Let $S \Rrightarrow r$ be a metainference made out of sequents. Then, $S \Rrightarrow r$ is locally st-valid in V^{SK} if and only if the inference $\tau(S \Rrightarrow r)$ is tt-valid in V^{SK}.*

Thus, we can write: $S \Rrightarrow r \in \mathbb{ST}^{\mathrm{SK}}$ if and only if $\tau(S \Rrightarrow r) \in \mathbb{LP}^{\mathrm{SK}}$.

Proof From left to right, we assume $S \Rrightarrow r \in \mathbb{ST}^{\mathrm{SK}}$, and argue that $\tau(S) \Rightarrow \tau(r) \in \mathbb{LP}^{\mathrm{SK}}$. So, consider any valuation $v \in V^{\mathrm{SK}}$ such that, for every $\tau(s) \in \tau(S)$, $v(\tau(s)) \in \{1, 1/2\}$. Take any $s \in S$, and suppose $s = \Gamma \Rightarrow \Delta$. Since $v(\bot) = 0$ and $v(\tau(\Gamma \Rightarrow \Delta)) \in \{1, 1/2\}$, we know that $\tau(\Gamma \Rightarrow \Delta)$ cannot be $\bot$. So, it must be either $\neg \bigwedge \Gamma$, or $\bigvee \Delta$, or $\bigwedge \Gamma \to \bigvee \Delta$. By the SK tables it follows that:

- If $\tau(\Gamma \Rightarrow \Delta)$ is $\neg \bigwedge \Gamma$, then $v(\gamma) \in \{1/2, 0\}$ for some $\gamma \in \Gamma$.
- If $\tau(\Gamma \Rightarrow \Delta)$ is $\bigvee \Delta$, then $v(\delta) \in \{1, 1/2\}$ for some $\delta \in \Delta$.
- If $\tau(\Gamma \Rightarrow \Delta)$ is $\bigwedge \Gamma \to \bigvee \Delta$, then either $v(\gamma) \in \{1/2, 0\}$ for some $\gamma \in \Gamma$ or $v(\delta) \in \{1, 1/2\}$ for some $\delta \in \Delta$.

In all three cases, $s = \Gamma \Rightarrow \Delta$ is st-satisfied at v. Now, suppose $r = \Sigma \Rightarrow \Pi$. Since $S \Rrightarrow r \in \mathbb{ST}^{\mathrm{SK}}$ and every sequent in S is st-satisfied at v, $\Sigma \Rightarrow \Pi$ is st-satisfied at v. This means that either $v(\sigma) \in \{1/2, 0\}$ for some $\sigma \in \Sigma$ or $v(\pi) \in \{1, 1/2\}$ for some $\pi \in \Pi$. By the SK tables, in the former case $v(\bigwedge \Sigma) \in \{1/2, 0\}$, and in the latter case $v(\bigvee \Pi) \in \{1, 1/2\}$. Either way, $v(\tau(\Sigma \Rightarrow \Pi)) \in \{1, 1/2\}$, for any of the available shapes of $\tau(\Sigma \Rightarrow \Pi)$. Thus, $\tau(S) \Rightarrow \tau(r) \in \mathbb{LP}^{\mathrm{SK}}$.

From right to left the reasoning is similar: we assume that $\tau(S) \Rightarrow \tau(r) \in \mathbb{LP}^{\mathrm{SK}}$, and also that each $s \in S$ is st-satisfied at a valuation v. Then, using the SK tables we show that $v(\tau(s)) \in \{1, 1/2\}$ for each $s \in S$, and hence $v(\tau(r)) \in \{1, 1/2\}$. We apply the properties of the SK tables once more and conclude that r is st-satisfied at v, which shows that $S \Rrightarrow r \in \mathbb{ST}^{\mathrm{SK}}$. (We leave the details for Exercise 48). □

[10] Notice that we have not defined the notion of formula-translation for inferences that are not sequents—that is, inferences with infinitely many premises or conclusions.

Theorem 3.4.4 reveals that there is a deep connection between the material conditional of strong Kleene tables and the notion of consequence st. The connection is further illustrated by the following simple fact:

Fact 3.4.5 Let v be an SK valuation. $\varphi \Rightarrow \psi$ is st-satisfied at v if and only if $v(\varphi \to \psi) \in \{1, 1/2\}$.

(We leave the proof to the reader.) Notice that $\{1, 1/2\}$ is the standard for conclusions of logics $\mathbb{LP}^{\mathrm{SK}}$ and $\mathbb{ST}^{\mathrm{SK}}$. Hence, the fact tells us that an inference is st-satisfied at a strong Kleene valuation just in case the corresponding conditional is satisfied *qua* conclusion according to these two logics.

When $\mathbb{ST}^{\mathrm{SK}}$ made its first appearances on the logical stage, its promoters talked *as if* it was just a different presentation of Classical Logic. However, results of the above kind motivated the criticism that $\mathbb{ST}^{\mathrm{SK}}$ does not improve substantially over the weaker logic $\mathbb{LP}^{\mathrm{SK}}$: these logics are close relatives and, arguably, theories based upon them will share many of their virtues and drawbacks. Following a metaphor by Cobreros et al. (2020), the idea is that, far from being a mere presentation of Classical Logic, $\mathbb{ST}^{\mathrm{SK}}$ is nothing more than $\mathbb{LP}^{\mathrm{SK}}$ in sheep's clothing.

Our second series of results is rather negative, and concerns the systems based on the Schütte valuations. In Theorem 3.4.4 we have seen that, when we focus on the strong Kleene scheme, the sequent-made metainferences that are locally st-valid coincide *modulo* translation with the inferences that are tt-valid. The natural conjecture, then, would be that a similar relationship between the notions of consequence st and tt will carry over to other valuation schemes, such as for instance the Schütte scheme. However, at least when we use the standard translation procedures, this is not the case.

As mentioned earlier in this chapter, when we work with the Schütte scheme, we have to choose whether to focus on the set V^{Sch} of all the Schütte valuations, or on the set V_u^{Sch} of all the Schütte valuations that in addition are uniform. Of course, we want to check whether the relationship between st and tt obtains for any of these options. Hence, we have to compare, on the one hand, logics $\mathbb{ST}^{\mathrm{Sch}}$ and $\mathbb{LP}^{\mathrm{Sch}}$ based on V^{Sch}, and on the other, logics $\mathbb{ST}_u^{\mathrm{Sch}}$ and $\mathbb{LP}_u^{\mathrm{Sch}}$ based on V_u^{Sch}.

Now, using the translation procedure specified above, neither of these two pairs of systems exemplifies the relationship between st and tt that we conjectured. This can be shown with a simple example. Consider for instance the metainference

$$\frac{\Rightarrow p \to q}{p \Rightarrow q}$$

Take a valuation v such that $v(p) = 1$, $v(q) = 0$, and for any formula $\varphi \to \psi$, if $v(\varphi) = 1$ and $v(\psi) = 0$, then $v(\varphi \to \psi) = 1/2$. These constraints are compatible with v belonging to V_u^{Sch}. And of course, $V_u^{\mathrm{Sch}} \subseteq V^{\mathrm{Sch}}$. Thus, v shows that the metainference does not belong to either $\mathbb{ST}^{\mathrm{Sch}}$ or $\mathbb{ST}_u^{\mathrm{Sch}}$. However, its inferential translation is

$$p \to q \Rightarrow p \to q$$

and this inference is valid in both $\mathbb{LP}^{\mathrm{Sch}}$ and $\mathbb{LP}_u^{\mathrm{Sch}}$—indeed, it is tt-valid in any set $V \subseteq V^3$. So, an analogous of Theorem 3.4.4 fails both for $\mathbb{ST}^{\mathrm{Sch}}$ and $\mathbb{LP}^{\mathrm{Sch}}$ and for $\mathbb{ST}_u^{\mathrm{Sch}}$ and $\mathbb{LP}_u^{\mathrm{Sch}}$.

As we mentioned, the example appeals to the standard translation procedure. Da Ré et al. (2021) put forward the more general question of whether there is *any* function from sequents to formulas that allows to make the connection between st and tt in the Schütte framework. Now we will give a partial negative answer to this question. We will show that, for any function f from sequents to formulas, if f satisfies the constraint that, for any formula φ, $f(\Rightarrow \varphi) = \varphi$, then the connection between st and tt fails. The result is relevant because the constraint mentioned is a quite plausible one: it arguably expresses the common idea that logical truths can be understood as valid arguments with no premises.

First, a preliminary fact that generalizes our previous example—and contrasts with Fact 3.4.5 about the strong Kleene framework. We take a particular sequent, say $p \Rightarrow q$, and show that there is *no* formula φ that satisfies certain conditions with respect to that sequent:

Fact 3.4.6 There is no formula φ of $\mathcal{L}_\bot$ such that, for every valuation $v \in V_u^{\mathrm{Sch}}$ (or V^{Sch}), $p \Rightarrow q$ is st-satisfied at v if and only if $v(\varphi) \in \{1, 1/2\}$.

Proof We prove the case of V_u^{Sch}. We partition our language in two: $\mathcal{L}_\bot/\{q, \bot\}$ and $\{q, \bot\}$. First, take the valuation v_1 such that $v_1(p) = 1$, $v_1(q) = 0 = v_1(\bot)$ and $v_1(\varphi) = 1/2$ for every other formula φ. It is easy to see that $v_1 \in V_u^{\mathrm{Sch}}$. Sequent $p \Rightarrow q$ is not st-satisfied at v_1, even though $v_1(\varphi) \in \{1, 1/2\}$ for every formula φ in $\mathcal{L}_\bot/\{q, \bot\}$. Now, take any Boolean valuation v_2 such that $v_2(p) = v_2(q) = v(\bot) = 0$. Sequent $p \Rightarrow q$ is st-satisfied at v_2, even though $v(q), v(\bot) \notin \{1, 1/2\}$. The case of V^{Sch} follows immediately from the fact that $V_u^{\mathrm{Sch}} \subset V^{\mathrm{Sch}}$. □

To make things more intuitive, given a formula ξ let us say that it *st-flattens* a sequent $\varphi \Rightarrow \psi$ at a valuation v just in case the biconditional " $\Rightarrow \xi$ is st-satisfied at v if and only if $\varphi \Rightarrow \psi$ is st-satisfied at v" holds. Then, Fact 3.4.6 can be equivalently restated by saying that there is *no* formula ξ that st-flattens sequent $p \Rightarrow q$ at every valuation $v \in V_u^{\mathrm{Sch}}$. This allows us to prove the result that we announced. In the following statement, $f(S \Rrightarrow r) = f(S) \Rightarrow f(r)$, where in turn $f(S) = \{f(s) \mid s \in S\})$.

Corollary 3.4.7 *Let f be a function from sequents to formulas of $\mathcal{L}_\bot$ such that, for any formula φ, $f(\Rightarrow \varphi) = \varphi$. Then, it is not the case that, for every metainference $S \Rrightarrow r$ made out of sequents, $S \Rrightarrow r$ belongs to $\mathbb{ST}^{\mathrm{Sch}}$ ($\mathbb{ST}_u^{\mathrm{Sch}}$) if and only if $f(S \Rrightarrow r)$ belongs to $\mathbb{LP}^{\mathrm{Sch}}$ ($\mathbb{LP}_u^{\mathrm{Sch}}$).*

Proof We focus on $\mathbb{ST}_u^{\mathrm{Sch}}$ and $\mathbb{LP}_u^{\mathrm{Sch}}$. Take any function f of the kind specified. Consider the following pair of metainferences:

$$\text{(i)}\ \frac{p \Rightarrow q}{\Rightarrow f(p \Rightarrow q)} \qquad\qquad \text{(ii)}\ \frac{\Rightarrow f(p \Rightarrow q)}{p \Rightarrow q}$$

By Fact 3.4.6, we know that either there is a valuation $v \in V_u^{\text{Sch}}$ such that $p \Rightarrow q$ is st-satisfied at v but $v(f(p \Rightarrow q)) = 0$, or there is a valuation $v \in V_u^{\text{Sch}}$ such that $p \Rightarrow q$ is not st-satisfied at v but $v(f(p \Rightarrow q)) \in \{1, 1/2\}$. In the former case, (i) $\notin \mathbb{ST}_u^{\text{Sch}}$. In the latter case, (ii) $\notin \mathbb{ST}_u^{\text{Sch}}$. But by the definition of f, the translation of these metainferences is in both cases $f(p \Rightarrow q) \Rightarrow f(p \Rightarrow q)$, and this inference belongs to $\mathbb{LP}_u^{\text{Sch}}$. The argument for $\mathbb{ST}^{\text{Sch}}$ and $\mathbb{LP}^{\text{Sch}}$ is similar. □

So, at the very least, the translations that can possibly witness the connection between st and tt in the Schütte framework will have to give up the idea that a logical truth φ can be understood in term of the argument $\Rightarrow \varphi$.

Our third and last result shows that there is weaker sense in which the connection between st and tt can be brought into light without using translations at all. This is achieved by restricting our attention to metainferences of a certain form, namely, metainferences such that (i) the conclusion-inference has no premises, and (ii) the premise-inferences have no premises and exactly one conclusion. Thus, we have metainferences of the form $\{\varnothing \Rightarrow \gamma_1, \ldots, \varnothing \Rightarrow \gamma_n, \ldots\} \Rrightarrow \varnothing \Rightarrow \Delta$.[11] Next, we show that metainferences of this form allow us to reformulate the connection between st and tt in such a way that it holds both for the strong Kleene and the Schütte schemes.

Fact 3.4.8 Let $\Gamma = \{\gamma_1, \ldots, \gamma_n, \ldots\}$. A metainference of the form

$$\frac{\Rightarrow \gamma_1 \qquad \ldots \qquad \Rightarrow \gamma_n \qquad \ldots}{\Rightarrow \Delta}$$

belongs to $\mathbb{ST}^{\text{SK}}$ (resp. $\mathbb{ST}^{\text{Sch}}$, $\mathbb{ST}_u^{\text{Sch}}$) if and only if the inference $\Gamma \Rightarrow \Delta$ belongs to $\mathbb{LP}^{\text{SK}}$ (resp. $\mathbb{LP}^{\text{Sch}}$, $\mathbb{LP}_u^{\text{Sch}}$).

Proof We just prove the case of $\mathbb{ST}^{\text{Sch}}$ and $\mathbb{LP}^{\text{Sch}}$—the other cases are similar. From left to right, suppose that the metainference that concerns us belongs to $\mathbb{ST}^{\text{Sch}}$. Assume further that a valuation v in V^{Sch} is such that $v(\gamma) \in \{1, 1/2\}$ for each $\gamma \in \Gamma$. Then, each premise of the metainference is st-satisfied at v, and hence the conclusion $\Rightarrow \Delta$ is st-satisfied at v as well. But that means that $v(\delta) \in \{1, 1/2\}$ for some $\delta \in \Delta$. Which implies that inference $\Gamma \Rightarrow \Delta$ is tt-satisfied at v. Hence, $\Gamma \Rightarrow \Delta$ belongs to $\mathbb{LP}^{\text{Sch}}$. The converse is obtained by analogous reasoning. □

Exercises

Exercise 48 Fill in the details of the right to left side of the proof of Theorem 3.4.4.

Exercise 49 Proof Fact 3.4.5.

[11] The metainferential consequence relation that one obtains when one focuses only on metainferences of this form can be understood as a semantic counterpart of Avron's (1988) syntactic notion of *external consequence*. Roughly, a formula φ is an external consequence of a collection of formulas Γ in a sequent calculus **S** just in case the sequent $\Rightarrow \varphi$ is derivable in the calculus that extends **S** with all the sequents $\Rightarrow \gamma$ for $\gamma \in \Gamma$.

Exercise 50 We gave an example of a metainference that is *invalid* in $\mathbb{ST}^{\text{Sch}}$ and $\mathbb{ST}_u^{\text{Sch}}$, even though its standard translation is valid in $\mathbb{LP}^{\text{Sch}}$ and $\mathbb{LP}_u^{\text{Sch}}$. Provide an example of a metainference that is *valid* $\mathbb{ST}^{\text{Sch}}$ and $\mathbb{ST}_u^{\text{Sch}}$, even though its translation is invalid in $\mathbb{LP}^{\text{Sch}}$ and $\mathbb{LP}_u^{\text{Sch}}$.

3.5 The ST Phenomenon Generalized

As we have seen, $\mathbb{ST}$ is a logic that has the same inferential validities as $\mathbb{CL}$, though it differs from it at the metainferential level by not containing Cut. Now, one could wonder if a similar phenomenon arises with respect to other logical systems: given some logic $\mathbb{L}$, is it possible to define a new system, $\mathbb{L}^{nt}$, which has the same inferential validities as $\mathbb{L}$, but is non-transitive at the metainferential level? The answer is positive.

In the literature, one can find two distinct strategies for generalizing the strict-tolerant phenomenon to other (non-classical) logics. The first approach can be found in a series of papers by Fitting (2020, 2021c,a,b), who employs bilattice theory to define what he terms the 'strict-tolerant counterpart' of several well-known logics (among them, $\mathbb{CL}$ and $\mathbb{LP}$, which we already presented, but also the paracomplete logic $\mathbb{K3}$ by Kleene (1952), to be seen in Chap. 4, as well as the relevant logic $\mathbb{FDE}$ by Belnap (1977) and Dunn (1976)). A second method was developed by Szmuc (2024), where a key result due to Wójcicki (1988) along with the notion of *infectious extension of an algebra* allow the author to define what he calls the 'non-transitive counterpart' of a logic. In this section, we will summarize the latter strategy.

Following Szmuc (2024) and taking into account the amendments made later by Da Ré and Fiore (2025), we will show that every non-trivial Tarskian Logic $\mathbb{L}$ has *at least one* non-transitive counterpart, $\mathbb{L}^{nt}$, having the same valid inferences as $\mathbb{L}$ but different valid metainferences, with the absence of Cut salient among them. Interestingly enough, the non-transitive counterpart of $\mathbb{CL}$ obtained by following this method won't be $\mathbb{ST}$, but rather the logic $\mathbb{wST}$ (Szmuc & Ferguson, 2021; Ferguson, 2023). In that sense, $\mathbb{ST}$ is just one of the many possible non-transitive counterparts of $\mathbb{CL}$.[12]

The remainder of this section is structured as follows: in the first part we will introduce some technical preliminaries, and in the second one, we will reconstruct Szmuc's method for obtaining the non-transitive counterpart of any (non-trivial) Tarskian logic.

[12] Indeed, the results by Da Ré et al. (2024a) (see Footnote 3 above) indicate that $\mathbb{ST}$ and $\mathbb{wST}$ are just two among the 8.256 three-valued, locally non-transitive counterparts of $\mathbb{CL}$ having a strict-tolerant notion of consequence and a deterministic valuation scheme!

3.5.1 Technical Preliminaries

The results we are going to present concern logics based on sets of formulas instead of multisets. So, throughout this section, $\Gamma, \Delta, \ldots$ will denote *sets* of formulas. All the relevant theoretical notions, such as 'inference', 'metainference', 'logic', and the like are defined exactly as before, only that replacing multisets with sets.

We start by explaining what Tarskian logical consequence relations are:

Definition 3.5.1 A *Tarskian logical consequence relation* is a binary relation $\Rightarrow \subseteq \mathcal{P}(FOR(\mathcal{L})) \times FOR(\mathcal{L})$ that satisfies the following conditions for every $\Gamma, \Delta \subseteq FOR(\mathcal{L})$ and all $\varphi, \psi \in FOR(\mathcal{L})$:

(Reflexivity) $\varphi \Rightarrow \varphi$
(Monotonicity) If $\Gamma \Rightarrow \varphi$, then $\Gamma, \psi \Rightarrow \varphi$
(Transitivity) If $\Gamma \Rightarrow \psi$ for every $\psi \in \Delta$ and $\Delta \Rightarrow \varphi$, then $\Gamma \Rightarrow \varphi$
(Invariance) If $\Gamma \Rightarrow \varphi$, then $\sigma(\Gamma) \Rightarrow \sigma(\varphi)$ for every substitution σ

where $\sigma(\Gamma)$ denotes the set $\{\sigma(\psi) : \psi \in \Gamma\}$.

Given any logic $\mathbb{L}$, let $\Rightarrow_{\mathbb{L}} \subseteq \mathcal{P}(FOR(\mathcal{L})) \times \mathcal{P}(FOR(\mathcal{L}))$ be the relation defined as follows:

$$\Gamma \Rightarrow_{\mathbb{L}} \Delta \text{ iff } \Gamma \Rightarrow \Delta \in \mathbb{L}$$

then, we say that $\mathbb{L}$ is a *Tarskian logic* just in case $\Rightarrow_{\mathbb{L}}$ is Tarskian, that is, it is single-conclusioned and satisfies the conditions in Definition 3.5.1.[13]

In this section, we will work with so-called logical matrices. These structures are defined using algebras, so let's introduce the latter first.[14] We define an *algebra* as a structure $\mathcal{A} = \langle A, \langle f_{\mathcal{A}} \rangle \rangle$, where A is a non-empty set, called the *universe* or *carrier set*, and $\langle f_{\mathcal{A}} \rangle$ is a sequence of finitary operations on elements of A.[15] The *type* of $\mathcal{A}$ is the sequence $\langle \text{ar}(f_{\mathcal{A}}) \rangle$ of the arities of the operations in $\mathcal{A}$. In this section, our language $\mathcal{L}$ is itself understood as an algebra: it is the algebra $\mathcal{L} = \langle FOR(\mathcal{L}), \langle \neg, \wedge, \vee, \rightarrow \rangle \rangle$ of type $\langle 1, 2, 2, 2 \rangle$. Other algebras are used to 'interpret' our language, and their universes are read as sets of 'truth values'. For instance, we have the Boolean algebra $\mathcal{B} = \langle \{1, 0\}, \langle f_{\mathcal{B}}^{\neg}, f_{\mathcal{B}}^{\wedge}, f_{\mathcal{B}}^{\vee}, f_{\mathcal{B}}^{\rightarrow} \rangle \rangle$, whose operations are those of Fig. 2.1, and the strong Kleene algebra $\mathcal{SK} = \langle \{1, 1/2, 0\}, \langle f_{\mathcal{SK}}^{\neg}, f_{\mathcal{SK}}^{\wedge}, f_{\mathcal{SK}}^{\vee}, f_{\mathcal{SK}}^{\rightarrow} \rangle \rangle$, where operations are those of Fig. 3.1.

[13] Strictly speaking, given our definition of inference (as a pair of sets of formulas), $\Rightarrow_{\mathbb{L}}$ will never have as its codomain the set $FOR(\mathcal{L})$ of all formulas (which is what the definition of Tarskian logical consequence relation requires); rather, it may have as it codomain the set of all singletons of formulas, i.e. things of the form $\{\varphi\}$. But we are assuming that, for all practical purposes, a formula φ and its corresponding singleton $\{\varphi\}$ are the same thing.

[14] For an introduction to algebraic logic, see, e.g. Font (2016).

[15] Nullary operations are taken to be identical to their image.

Definition 3.5.2 A *logical matrix* $\mathfrak{M}$ for a language $\mathcal{L}$ is a structure $\langle \mathcal{A}, D \rangle$ where $\mathcal{A}$ is an algebra of the same type as $\mathcal{L}$ and D is a subset of A, called the set of *designated values*.

An $\mathfrak{M}$-*valuation* for $\mathcal{L}$ is a homomorphism from $\mathcal{L}$ to $\mathcal{A}$; namely, a function $v_{\mathfrak{M}} : FOR(\mathcal{L}) \to A$ such that, for every n-ary constant # of $\mathcal{L}$, and formulas $\varphi_1, \ldots, \varphi_n$, we have that $v_{\mathfrak{M}}(\#(\varphi_1, \ldots, \varphi_n)) = \#^{\mathcal{A}}(v_{\mathfrak{M}}(\varphi_1), \ldots, v_{\mathfrak{M}}(\varphi_n))$. When no confusion arises, we drop the subindex and just write v.

It is crucial to observe that logical matrices enable us to define logical consequence relations. We say that the logical matrix $\mathfrak{M}$ *induces* the consequence relation $\Rightarrow_{\mathfrak{M}} \subseteq \mathcal{P}(FOR(\mathcal{L})) \times FOR(\mathcal{L})$ defined as:

$$\Gamma \Rightarrow_{\mathfrak{M}} \varphi \quad \text{iff} \quad \text{for all } v_{\mathfrak{M}} : \text{ if } v_{\mathfrak{M}}(\Gamma) \subseteq D, \text{ then } v_{\mathfrak{M}}(\varphi) \in D$$

where $\Gamma \cup \{\varphi\} \subseteq FOR(\mathcal{L})$ and $v_{\mathfrak{M}}(\Gamma) = \{v_{\mathfrak{M}}(\gamma) : \gamma \in \Gamma\}$.[16] We call any valuation $v_{\mathfrak{M}}$ such that $v_{\mathfrak{M}}(\gamma) \in D$ for every $\gamma \in \Gamma$ but $v_{\mathfrak{M}}(\varphi) \notin D$ an $\mathfrak{M}$-*counterexample* to the inference $\Gamma \Rightarrow \varphi$. We can also say that a *set* of logical matrices $\boldsymbol{\mathfrak{M}}$ *induces* the consequence relation $\Rightarrow_{\boldsymbol{\mathfrak{M}}}$, where $\Rightarrow_{\boldsymbol{\mathfrak{M}}} = \bigcap\{\Rightarrow_{\mathfrak{M}} \mid \mathfrak{M} \in \boldsymbol{\mathfrak{M}}\}$.

As for metainferences, we stipulate that a metainference $S \Rrightarrow r$ is *locally valid* in the logical matrix $\mathfrak{M}$ if and only if for every $v_{\mathfrak{M}}$, if $v_{\mathfrak{M}}$ is not an $\mathfrak{M}$-counterexample to any inference $s \in S$ then it is not an $\mathfrak{M}$-counterexample to r (and it is locally valid in a set of matrices $\boldsymbol{\mathfrak{M}}$ just in case it is locally valid in every $\mathfrak{M} \in \boldsymbol{\mathfrak{M}}$). Finally, given any logic $\mathbb{L}$, let $\Rrightarrow_{\mathbb{L}}$ be the relation between sets of inferences and inferences defined as follows:

$$S \Rrightarrow_{\mathbb{L}} r \text{ iff } S \Rrightarrow r \in \mathbb{L}$$

Then, given any matrix $\mathfrak{M}$, we say that the logic *induced* by it is the (single-conclusion) logic $\mathbb{L}$ defined as follows:

$$\begin{array}{ll} \Gamma \Rightarrow_{\mathbb{L}} \varphi \text{ iff} & \Gamma \Rightarrow_{\mathfrak{M}} \varphi \\ S \Rrightarrow_{\mathbb{L}} r \text{ iff} & S \Rrightarrow r \text{ is locally valid in } \mathfrak{M} \end{array}$$

Thus, $\langle \mathcal{B}, \{1\} \rangle$ induces (the single-conclusion fragment of) $\mathbb{CL}$, whereas $\langle \mathcal{SK}, \{1, 1/2\} \rangle$ induces (the single-conclusion fragment of) $\mathbb{LP}$. The notion of the logic induced by a set $\boldsymbol{\mathfrak{M}}$ of logical matrices is defined in the expected way.

A key result that will allow us to define the non-transitive counterpart of Tarskian logics is the next Theorem due to Wójcicki:

[16] In this section we assume a single-conclusion framework, because Tarskian consequence relations, and hence also Wójcicki's theorem about them, are formulated with single-conclusions. However, the results of this section can be easily generalized to multiple conclusions. (For a multiple-conclusion version of Wójcicki's theorem, see, e.g. Chemla & Egré, 2019.)

Theorem 3.5.3 (Wójcicki, 1988) *For every Tarskian logic* $\mathbb{L}$ *there is a class of logical matrices* $\mathfrak{M}$ *such that* $\Gamma \Rightarrow_{\mathbb{L}} \varphi$ *just in case* $\Gamma \Rightarrow_{\mathfrak{M}} \varphi$.

Proof We use $\Gamma_{\mathbb{L}}$ to denote the closure of Γ under $\mathbb{L}$, that is, the set $\{\varphi : \Gamma \Rightarrow_{\mathbb{L}} \varphi\}$. For every set of formulas Γ, let $\mathfrak{M}_\Gamma$ be the matrix $= \langle \mathcal{L}, \Gamma_{\mathbb{L}} \rangle$. Observe that an $\mathfrak{M}_\Gamma$-valuation v is an homomorphism from $\mathcal{L}$ to $\mathcal{L}$, that is, a substitution. We show that the set of matrices $\mathfrak{M}_{\mathbb{L}} = \{\mathfrak{M}_\Gamma : \Gamma \subseteq FOR(\mathcal{L})\}$, known as the *Lindenbaum bundle of* $\mathbb{L}$, witnesses the theorem:

- Left to right. Assume that $\Gamma \Rightarrow_{\mathbb{L}} \varphi$. Consider any matrix $\mathfrak{M}_\Delta$ and any valuation v over it. Suppose that $v(\Gamma) \subseteq \Delta_{\mathbb{L}}$; this means that $\Delta \Rightarrow_{\mathbb{L}} v(\gamma)$ for every $v(\gamma) \in v(\Gamma)$. Since v is a substitution, by (Invariance) we also have $v(\Gamma) \Rightarrow_{\mathbb{L}} v(\varphi)$. Then, by (Transitivity), we have $\Delta \Rightarrow_{\mathbb{L}} v(\varphi)$, which means that $v(\varphi) \in \Delta_{\mathbb{L}}$. Since v and $\mathfrak{M}_\Delta$ where arbitrary, we successively conclude $\Gamma \Rightarrow_{\mathfrak{M}_\Delta} \varphi$ and $\Gamma \Rightarrow_{\mathfrak{M}_{\mathbb{L}}} \varphi$
- Right to left. Suppose that $\Gamma \not\Rightarrow_{\mathbb{L}} \varphi$. Take the matrix $\mathfrak{M}_\Gamma$ and the identity valuation v. We have $v(\Gamma) = \Gamma$, and by (Reflexivity), $\Gamma \subseteq \Gamma_{\mathbb{L}}$; thus, $v(\Gamma) \subseteq \Gamma_{\mathbb{L}}$. But $v(\varphi) = \varphi$, and $\varphi \notin \Gamma_{\mathbb{L}}$; hence, $v(\varphi) \notin \Gamma_{\mathbb{L}}$, which means that $\Gamma \not\Rightarrow_{\mathfrak{M}_\Gamma} \varphi$. Hence, $\Gamma \not\Rightarrow_{\mathfrak{M}_{\mathbb{L}}} \varphi$.

□

Intuitively, this Theorem guarantees that for every Tarskian logic $\mathbb{L}$ there is a class of matrices $\mathfrak{M}$ that validates exactly the inferences in $\mathbb{L}$. But, let us say that a matrix $\mathfrak{M}$ is *trivial* just in case its set of designated values is identical to the universe of the corresponding algebra. Then, the Theorem is compatible with the following two situations: first, the case where the class $\mathfrak{M}$ is empty; second, the case where $\mathfrak{M}$ contains some trivial matrix. To prove the main result of this section we will appeal to a refined version of Wójcicki's theorem. Let us say that a Tarskian logic $\mathbb{L}$ is *non-trivial* if there is some $\Gamma \subseteq FOR(\mathcal{L})$ and some $\varphi \in FOR(\mathcal{L})$ such that $\Gamma \not\Rightarrow_{\mathbb{L}} \varphi$; otherwise, $\mathbb{L}$ it is trivial. Then, we have:

Theorem 3.5.4 (Wójcicki Refined) *For every non-trivial Tarskian logic* $\mathbb{L}$ *there is a non-empty class* $\mathfrak{M}$ *of non-trivial logical matrices such that* $\Gamma \Rightarrow_{\mathbb{L}} \varphi$ *just in case* $\Gamma \Rightarrow_{\mathfrak{M}} \varphi$.

Proof Just as in Theorem 3.5.3, with the only difference that we take the class of matrices $\{\mathfrak{M}_\Gamma : \Gamma \subset FOR(\mathcal{L})\}$. In the right to left direction, after assuming $\Gamma \not\Rightarrow_{\mathbb{L}} \varphi$ we note that, by (Reflexivity) and (Monotonicity), it follows that $\Gamma \neq FOR(\mathcal{L})$, and hence the matrix $\mathfrak{M}_\Gamma$ is non-trivial. □

Next, we are going to introduce the notion of logical p-matrices, due to Frankowski (2011, 2004). They can be understood as structures that induce logics where (Transitivity) might fail.

Definition 3.5.5 A *logical p-matrix* $\mathfrak{M}$ for a language $\mathcal{L}$ is a structure $\langle \mathcal{A}, D^p, D^c \rangle$ where $\mathcal{A}$ is an algebra of the same type as $\mathcal{L}$, and D^p and D^c are subsets of A such that $D^p \subseteq D^c$.

It is easy to check that p-matrices give us a generalization of the regular logical matrices described in Definition 3.5.2: the latter can be represented as an instance of the former where $D^p = D^c$. We say that a logical p-matrix *induces* the p-consequence relation $\Rightarrow_{\mathfrak{M}} \subseteq \mathcal{P}(FOR(\mathcal{L})) \times FOR(\mathcal{L})$ defined as:

$$\Gamma \Rightarrow_{\mathfrak{M}} \varphi \qquad \text{iff} \qquad \text{for all } v_{\mathfrak{M}}\text{: if } v_{\mathfrak{M}}(\Gamma) \subseteq D^p, \text{ then } v_{\mathfrak{M}}(\varphi) \in D^c$$

The notion of an $\mathfrak{M}$-counterexample, and the consequence relation $\Rightarrow_{\mathfrak{M}}$ induced by a set $\mathfrak{M}$ of p-matrices are defined similarly as before. The same goes for the notions of local validity over a p-matrix (or set of p-matrices), and the logic induced by a p-matrix (or set of p-matrices). Notice that in logics induced by p-matrices, both (Reflexivity) and (Monotonicity) hold, as in Tarskian logical consequence relations, but this isn't necessarily the case for (Transitivity).

The last notion that we need for establishing the desired result is that of *infectious extension* of an algebra. Let us start by defining what an infectious element is:

Definition 3.5.6 An algebra $\mathcal{A}$ has an *infectious element* e if and only if for every n-ary operation # of $\mathcal{A}$ and every $\{a_1, \ldots, a_n\} \subseteq A$: if $e \in \{a_1, \ldots, a_n\}$, then $\#^{\mathcal{A}}(a_1, \ldots, a_n) = e$.

This truth value is called an 'infectious element' since it is a zero (aka. absorbing) element for all the operations in the algebra. By the same token, logics that feature this kind of semantic value are usually called 'infectious logics'.

Definition 3.5.7 Given an algebra $\mathcal{A}$ and an element $e \notin A$, its *infectious extension with* e, denoted as $\mathcal{A}[e]$, is the algebra with universe $A \cup \{e\}$ whose operations are defined by:

$$\#^{\mathcal{A}[e]}(a_1, \ldots, a_n) = \begin{cases} e & \text{if } e \in \{a_1, \ldots, a_n\} \\ \#^{\mathcal{A}}(a_1, \ldots, a_n) & \text{otherwise} \end{cases}$$

Moreover, given any logical matrix $\mathfrak{M} = \langle \mathcal{A}, D \rangle$, and any element $e \notin A$, the *infectious extension of* $\mathfrak{M}$ *with* e is the p-matrix $\mathfrak{M}[e] = \langle \mathcal{A}[e], D^p, D^c \rangle$, where $D^p = D$ and $D^c = D \cup \{e\}$. It is easy to establish the following fact about $\mathfrak{M}[e]$-valuations:

Fact 3.5.8 Let $Var(\varphi)$ be the set of propositional variables occurring in φ. For every $\mathfrak{M}[e]$-valuation v and every formula φ, if $v(p) = e$ for some variable $p \in Var(\varphi)$, then $v(\varphi) = e$.

Proof Exercise 51. □

When all operations are captured by truth tables, extending a logic with an infectious value is equivalent to appending an additional row and column to each truth table, where all entries in this new row and column are filled with the new value e.

	$f^{\neg}_{\mathcal{WK}}$
1	0
1/2	1/2
0	1

$f^{\wedge}_{\mathcal{WK}}$	1	1/2	0
1	1	1/2	0
1/2	1/2	1/2	1/2
0	0	1/2	0

$f^{\vee}_{\mathcal{WK}}$	1	1/2	0
1	1	1/2	1
1/2	1/2	1/2	1/2
0	1	1/2	0

$f^{\rightarrow}_{\mathcal{WK}}$	1	1/2	0
1	1	1/2	0
1/2	1/2	1/2	1/2
0	1	1/2	1

Fig. 3.3 The weak Kleene truth tables

To finish, we present the weak Kleene algebra $\mathcal{WK}$, which notably coincides with the infectious extension of the Boolean algebra, providing a concrete example of the previous definitions:

Definition 3.5.9 The weak Kleene algebra $\mathcal{WK}$ is the structure $\mathcal{WK} = \langle\{1, 1/2, 0\}, \langle f^{\neg}_{\mathcal{WK}}, f^{\wedge}_{\mathcal{WK}}, f^{\vee}_{\mathcal{WK}}, f^{\rightarrow}_{\mathcal{WK}}\rangle\rangle$, whose operations are depicted in Fig. 3.3.

With this algebra we can define the p-matrix $\langle\mathcal{WK}, \{1\}, \{1, 1/2\}\rangle$, which induces the logic $\mathbb{wST}$ already mentioned before.

Exercises

Exercise 51 Prove Fact 3.5.8 by induction on the complexity of the formula.

Exercise 52 Prove that, for all inferences $\Gamma \Rightarrow \varphi$, $\Gamma \Rightarrow \varphi$ is valid in $\mathbb{wST}$ iff $\Gamma \Rightarrow \varphi$ is valid in $\mathbb{CL}$.

3.5.2 *The Generalization*

With the above tools, we can now demonstrate how to obtain the non-transitive counterpart of any non-trivial Tarskian logic $\mathbb{L}$. The strategy consists in, first, finding a set of p-matrices $\mathfrak{M}$ such that $\Gamma \Rightarrow_{\mathbb{L}} \varphi$ if and only if $\Gamma \Rightarrow_{\mathfrak{M}} \varphi$, and second, extending each matrix $\mathfrak{M}$ in $\mathfrak{M}$ with an infectious element e that is designated for conclusions but not for premises. The new class of matrices will validate the same inferences as $\mathfrak{M}$ but locally invalidate Cut; hence, it will induce a new logic, here called $\mathbb{L}^{nt}$, which has the same inferences as $\mathbb{L}$ but different metainferences.

Theorem 3.5.10 *Let $\mathfrak{M} = \langle\mathcal{A}, D\rangle$ be any logical matrix, and $e \notin A$. Then, its infectious extension with e, $\mathfrak{M}[e]$, is such that $\Gamma \Rightarrow_{\mathfrak{M}} \varphi$ if and only if $\Gamma \Rightarrow_{\mathfrak{M}[e]} \varphi$.*

Proof We proceed by contraposition in both directions:

- Right to left. Assume that $\Gamma \not\Rightarrow_{\mathfrak{M}} \varphi$. Thus, there's an $\mathfrak{M}$-counterexample given by an $\mathfrak{M}$-valuation v such that $v(\gamma) \in D$ for every $\gamma \in \Gamma$ and also $v(\varphi) \notin D$. But every $\mathfrak{M}$-valuation is also a $\mathfrak{M}[e]$-valuation, and so v is also a $\mathfrak{M}[e]$-counterexample. Therefore, $\Gamma \not\Rightarrow_{\mathfrak{M}[e]} \varphi$.
- Left to right. Suppose that $\Gamma \not\Rightarrow_{\mathfrak{M}[e]} \varphi$. Then there's an $\mathfrak{M}[e]$-counterexample given by an $\mathfrak{M}[e]$-valuation v such that $v(\gamma) \in D$ for all $\gamma \in \Gamma$ and also $v(\varphi) \notin D \cup \{e\}$. Furthermore, by the infectious character of e (Fact 3.5.8), we can infer

that for all propositional variables p appearing in $\Gamma \cup \{\varphi\}$, $v(p) \neq e$. Thus, v coincides with an $\mathfrak{M}$-valuation v' on the set of sentences $\Gamma \cup \{\varphi\}$. Hence, $\Gamma \not\Rightarrow_{\mathfrak{M}} \varphi$.

□

Now, let $\mathfrak{M}$ be any class of logical matrices, and let e be such that, for every $\mathfrak{M} = \langle \mathcal{A}, D\rangle$ in $\mathfrak{M}$, $e \notin A$. The above result easily extends to the class of p-matrices $\mathfrak{M}[e] = \{\mathfrak{M}[e] : \mathfrak{M} \in \mathfrak{M}\}$. Also, by Theorem 3.5.4 we had that, for every non-trivial Tarskian logic $\mathbb{L}$ there is a non-empty class of non-trivial logical matrices $\mathfrak{M}$ such that $\Gamma \Rightarrow_{\mathbb{L}} \varphi$ if and only if $\Gamma \Rightarrow_{\mathfrak{M}} \varphi$.

Corollary 3.5.11 *For every non-trivial Tarskian logic* $\mathbb{L}$ *there is a non-empty class of p-matrices* $\mathfrak{M}[e]$ *such that* $\Gamma \Rightarrow_{\mathbb{L}} \varphi$ *if and only if* $\Gamma \Rightarrow_{\mathfrak{M}[e]} \varphi$.

Moreover, it is easy to check that each matrix $\mathfrak{M}[e]$ in $\mathfrak{M}[e]$ locally invalidates Cut. To wit, just consider the instance

$$\frac{\Rightarrow q \qquad q \Rightarrow r}{\Rightarrow r}$$

Since $\mathfrak{M}$ is non-trivial, we know that there is at least one truth value $\dagger \in A$ such that $\dagger \notin D$. So, take the $\mathfrak{M}[e]$-valuation v in which $v(q) = e$ and $v(r) = \dagger$. This valuation shows that the instance of Cut in question is locally invalid in $\mathfrak{M}[e]$.

All in all, letting $\mathbb{L}^{nt}$ be the logic induced by the class of p-matrices $\mathfrak{M}[e]$, we have the main result of this section:

Theorem 3.5.12 (Szmuc, 2024; Da Ré and Fiore, 2025) *For every non-trivial Tarskian logic* $\mathbb{L}$, *there is a logic* $\mathbb{L}^{nt}$ *that has the same inferences but does not contain Cut.*

As noted by Da Ré and Fiore (2025), the restriction to *non-trivial* logics is essential for the result to hold. In Szmuc's paper, Theorem 3.5.12 is mistakenly formulated for any Tarskian logic whatsoever. But consider any trivial logic $\mathbb{T}$.[17] The inferences of such a logic are characterized either by the empty class of logical matrices, $\varnothing$, or by any class of logical matrices whose elements are all trivial matrices. Now, let $\mathfrak{M}_{\mathbb{T}}$ be any class of matrices with either of these traits, and let e be such that, for every $\mathfrak{M} = \langle \mathcal{A}, D\rangle$ in $\mathfrak{M}_{\mathbb{T}}$, $e \notin A$. It is easy to check that, even though the class of p-matrices $\mathfrak{M}_{\mathbb{T}}[e]$ validates the same inferences as $\mathfrak{M}_{\mathbb{T}}$, it does not locally invalidate (the single-conclusion restriction of) Cut: it does not counterexemplify any inference, and so it also does not counterexemplify any instance of the rule. Anyway, we can regard trivial logics as a limit (if not degenerate) case, which does not affect the philosophical import of the result itself.

[17] We say "any" because, following our previous definition, there are *many* trivial logics: they all coincide in their valid inferences, but differ in their valid metainferences.

Bibliography

Avron, A. (1988). The semantics and proof theory of linear logic. *Theoretical Computer Science, 57*(2–3), 161–184.

Barrio, E. A,, Rosenblatt, L., & Tajer, D. (2015). The logics of strict-tolerant logic. *Journal of Philosophical Logic, 44*(5), 551–571.

Belnap, N. (1977). A useful four-valued logic. In *Modern uses of multiple-valued logic* (pp. 5–37). Springer.

Chemla, E., & Egré, P. (2019). Suszko's problem: Mixed consequence and compositionality. *The Review of Symbolic Logic, 12*(4), 736–767.

Cobreros, P., Egré, P., Ripley, E., & van Rooij, R. (2020). Inferences and Metainferences in ST. *Journal of Philosophical Logic, 49*(6), 1057–1077.

Da Ré, B., & Fiore, C. (2025). Maximally substructural classical logic. Manuscript.

Da Ré, B., Szmuc, D., & Teijeiro, P. (2021). Derivability and metainferential validity. *Journal of Philosophical Logic, 51*(6), 1521–1547.

Da Ré, B., Szmuc, D., Chemla, E., & Egré, P. (2024a). On three-valued presentations of classical logic. *The Review of Symbolic Logic, 17*(3), 682–704.

Da Ré, B., Szmuc, D., & Corbalán, M. I. (2024b). Non-reflexive nonsense: proof theory of paracomplete weak Kleene logic. *Studia Logica, 112*(6), 1243–1259.

Dicher, B., & Paoli, F. (2019). ST, LP, and tolerant metainferences. In C. Başkent, & T. M. Ferguson (Eds.), *Graham priest on dialetheism and paraconsistency* (pp. 383–407). Springer Verlag.

Dicher, B., & Paoli, F. (2021). The original sin of proof-theoretic semantics. *Synthese, 198*(1), 615–640.

Dunn, J. M. (1976). Intuitive semantics for first-degree entailments and 'coupled trees'. *Philosophical Studies, 29*(3), 149–168.

Ferguson, T. M. (2023). Monstrous content and the bounds of discourse. *Journal of Philosophical Logic, 52*(1), 111–143.

Fitting, M. (2020). A family of strict/tolerant logics. *Journal of Philosophical Logic, 50*(2), 363–394.

Fitting, M. (2021a). Strict/tolerant logics built using generalized weak Kleene logics. *The Australasian Journal of Logic, 18*(2), 73–92.

Fitting, M. (2021b). The strict/tolerant family continued: quantifiers and modalities. *The Australasian Journal of Logic, 18*(6), 616–644.

Fitting, M. (2021c). The strict/tolerant idea and bilattices. In O. Arieli, & A. Zamansky (Eds.), *Arnon avron on semantics and proof theory of non-classical logics* (Chap. 8, pp. 167–191). Springer.

Font, J. (2016). *Abstract algebraic logic: An introductory textbook*. College Publications.

Frankowski, S. (2004). Formalization of a plausible inference. *Bulletin of the Section of Logic, 33*(1), 41–52.

Frankowski, S. (2011). Syntactic properties of p-consequence. *Logic and Logical Philosophy, 20*(4), 285–295.

Girard, J.-Y. (1987). *Proof theory and logical complexity*. Bibliopolis.

Hösli, B., & Jäger, G. (1994). About some symmetries of negation. *The Journal of Symbolic Logic, 59*(2), 473–485.

Kleene, S. C. (1952). *Introduction to metamathematics*. North-Holland.

Přenosil, A. (2017). Cut elimination, identity elimination, and interpolation in super-belnap logics. *Studia Logica, 105*(6), 1255–1289.

Pynko, A. (2010). Gentzen's cut-free calculus versus the logic of paradox. *Bulletin of the Section of Logic, 39*(1/2), 35–42.

Ripley, E. (2012). Conservatively extending classical logic with transparent truth. *Review of Symbolic Logic, 5*(2), 354–378.

Ripley, E. (2013a). Paradoxes and failures of cut. *Australasian Journal of Philosophy, 91*(1), 139–164.
Ripley, E. (2013b). Revising up: Strengthening classical logic in the face of paradox. *Philosophers' Imprint, 13*(5), 1–13.
Szmuc, D. E. (2024). Non-transitive counterparts of every tarskian logic. *Analysis, 84*(2), 320–326.
Szmuc, D., & Ferguson, T. M. (2021). Meaningless divisions. *Notre Dame Journal of Formal Logic, 62*(3), 399–424.
Teijeiro, P. (2021). Strength and stability. *Análisis Filosófico, 41*(2), 337–349.
Wójcicki, R. (1988). *Theory of logical calculi: basic theory of consequence operations* (Vol. 199). Springer Science & Business Media.

Chapter 4
Non-Reflexive Logics

Joaquín S. Toranzo Calderón and Martina Zirattu

Abstract This chapter is about non-reflexive logics in the "tolerant-strict" family. We present them both from a proof-theoretic and a model theoretic point of view. For the first, we define new sequent calculi in which identity is not admissible nor derivable, that are obtained by removing the identity axiom from the calculus **LK** and possibly adding or removing other rules. For the second, we build on some notions already introduced in the previous chapter and present different semantics, which are then related to the calculi through soundness and completeness results both at the inferential and metainferential level. The level of metainferences is further discussed showing that the non-reflexive calculi are structurally incomplete and then proving some results regarding their global, local and absolute global ranges. Next, the non-reflexive logic $\mathbb{TS}$ is related to other logics introduced in this and the previous chapters, namely $\mathbb{CL}$, $\mathbb{ST}$, $\mathbb{LP}$ and $\mathbb{K}3$. In particular, several notions of translation and duality which can be found in the literature are presented and compared. Some space is also devoted to the introduction of metainferential mixed logics. Finally, we build on the notion of anti-inference central to the discussion on dualities and show that, given a Tarskian logic, one can construct a non-reflexive logic with the same anti-inferences. We conclude by briefly presenting some three-sided sequent calculi used in the literature to characterize some of the logics discussed here and in the previous chapters.

4.1 Inferential Tolerant-Strict Logics

In this chapter, we present some logics in which Identity (Reflexivity) fails, that is, the inference $\varphi \Rightarrow \varphi$ is not in the extension of these logics. This can be done in several ways, similar to the strategies used for non-transitive logics, which result in

J. S. Toranzo Calderón
UBA/UTN/IIF-SADAF-CONICET, Buenos Aires, Argentina

M. Zirattu (✉)
Department of Philosophy and Education Sciences, University of Turin, Torino, Italy

P. Teijeiro, E. A. Barrio (eds.), *Metainferences in Substructural Logics*, Synthese Library 528, https://doi.org/10.1007/978-3-032-05449-4_4

some interesting features for non-reflexive logics, like the possibility of extending its calculus with a truth predicate without running into troubles. But some of their most interesting features are probably those that arise from their close relationship with other logics, especially but not restricted to symmetries with non-transitive ones. For showing those relations we use the same semantic framework from the previous chapter for providing non-reflexive systems but here based on tolerant-strict standards of satisfaction.

4.1.1 Proof Theory

There are several proof systems in the literature for non-reflexive logics. However, here we will only focus on the presentation of the sequent calculi $\mathbf{LK}^{-i}$ and $\mathbf{LK}^{-i}_{+\mathrm{E}}$ introduced by French (2016), and on $\mathbf{LK}^{-i}_{+\mathrm{E}-\mathrm{I}}$. These three calculi, as we will show, determine three different logics. As in the previous chapter, we choose these sequent calculi since they are natural candidates for comparisons with **LK**, given that they add or drop rules from the initial sequent calculus. There is a symmetry between the calculi presented in this chapter and $\mathbf{LK}^{-c}$ and $\mathbf{LK}^{-c}_{+\mathrm{E}}$, which will be explored in Sect. 4.3. Other systems, like those presented by Da Ré and Pailos (2022), make the comparison more difficult, although the results are quite similar.

Our first sequent calculus for a non-reflexive logic can be obtained from that for Classical Logic in Chap. 2, by removing the purportedly unsound structural rule (in our case, Id, i.e. reflexivity) as done by French (2016).

Definition 4.1.1 The sequent calculus $\mathbf{LK}^{-i}$ results from **LK** (see Definition 3.2.1) by removing Id.

Since the rule Id is missing (and we have no other axiom), it is not possible to start any proof. Hence, dropping Identity as a structural rule gives a logic that is empty at the inferential level.

French (2016) also introduces the sequent calculus $\mathbf{LK}^{-i}_{+\mathrm{E}}$, which is just the result of adding all the elimination rules to $\mathbf{LK}^{-i}$. Those are the operational rules read bottom-up, with some exceptions, as discussed in Sect. 3.2.1.

Definition 4.1.2 The sequent calculus $\mathbf{LK}^{-i}_{+\mathrm{E}}$ results from $\mathbf{LK}^{-i}$ by adding the elimination rules defined in Definition 3.2.3.

It is easy to see that adding the elimination rules does not affect the provable inferences of $\mathbf{LK}^{-i}$: $\mathbf{LK}^{-i}_{+\mathrm{E}}$ is still inferentially empty since, once more, no proof can start without Id. Nevertheless, this addition has some interesting effects on the derivable metainferences. In fact, at the level of metainferences $\mathbf{LK}^{-i}_{+\mathrm{E}}$ is a significantly stronger system than $\mathbf{LK}^{-i}$. We show this in Sect. 4.2.1.

Notice that in the previous chapter, adding the elimination rules was prompted by the need to recover some of the derivability power that was lost with the departure of Cut. However, it remained unanswered what would happen if, instead of only having the introduction rules, the calculus only had the elimination ones. Instead, we do consider this option and will then briefly explain why it is explored here and not in Chap. 3.

Definition 4.1.3 The sequent calculus $\mathbf{LK}^{-i}_{+\mathrm{E}-\mathrm{I}}$ results from $\mathbf{LK}^{-i}$ by removing the operational rules and adding the elimination rules.

In the case of the Cut-free system, we face the following situation. The Identity axiom was formulated for atoms, because that is enough to derive the complex instances, given the corresponding introduction rules for their main connective. Moreover, the ability to prove full Identity from its atomic instances is commonly considered to be a desirable property for a calculus. This property, known as 'Identity elimination', can be taken to be a test regarding how well the introduction rules work as definitions of the connectives (see Hacking, 1979).

To show the role of introduction rules in the derivation of Identity for complex formulas in $\mathbf{LK}^{-c}$ consider for instance the case of negated formulas:

$$\dfrac{\dfrac{\overline{\varphi \Rightarrow \varphi}\ \text{Id up to complexity of } \varphi}{\Rightarrow \varphi, \neg\varphi}\ \mathrm{R}\neg}{\neg\varphi \Rightarrow \neg\varphi}\ \mathrm{L}\neg$$

Thus, in the absence of introduction rules, Identity for complex formulas is no longer derivable. This means that the substitution of introduction rules with elimination rules in $\mathbf{LK}^{-c}$ would have left us with a non-transitive *and* non-reflexive logic. At that point, we were only exploring non-transitive systems and thus wanted to wait to a further point to entertain this idea.

However, in the case of $\mathbf{LK}^{-i}$, the slot for a system with only primitive elimination rules remains open, since Identity need not be proved anymore. Moreover, the non-reflexive character of this system does not come with a failure of transitivity. Thus, in the context of a chapter devoted to non-reflexive logics, it seems appropriate to study systems without Identity and with only elimination rules, rather than the analogous in a Cut-free system.

In Sect. 4.2.1 we show the effect of adding or dropping introduction or elimination rules for the metainferential aspect of the resulting logics.

Exercises

Exercise 53 Elimination rules are admissible in **LK**. Prove that elimination rules are also admissible in $\mathbf{LK}^{-i}$.

4.1.2 Model Theory

In the context of three-valued semantics, some notions were defined in Sect. 3.2.2, mainly the *tt*- and the *st*-validity notions. We continue with the *ss*- and the *ts*-validity notions, the latter being more important for our purposes in this section, because it can be the basis to define some interesting non-reflexive logics. It will be useful to recall the notions of ts-satisfaction and ts-validity and, for later use, the same for ss-satisfaction and ss-validity (implied by Definitions 3.2.4, 3.2.5 and 3.2.6):

- $\Gamma \vDash_V^{ss} \Delta$ iff for all v in V, if $v(\gamma) \in \{1\}$ for every $\gamma \in \Gamma$, then $v(\delta) \in \{1\}$ for some $\delta \in \Delta$.
- $\Gamma \vDash_V^{ts} \Delta$ iff for all v in V, if $v(\gamma) \in \{1/2, 1\}$ for every $\gamma \in \Gamma$, then $v(\delta) \in \{1\}$ for some $\delta \in \Delta$.

Before presenting some specific semantics, we want to show that the notion of ts-validity can render non-reflexive logics under a very minimal condition.

Fact 4.1.4 Let V be a set of valuations for which there is a valuation $v \in V$ such that, for a formula $\varphi \in \mathcal{L}$, $v(\varphi) = 1/2$. The inference $\varphi \Rightarrow \varphi$ is not ts-valid in V.

Proof The valuation v clearly t-satisfies φ but does not s-satisfy it. Thus, v does not ts-satisfy $\varphi \Rightarrow \varphi$, which makes it ts-invalid. □

This valuation can be such that it only affects one and only one formula, or it could affect an arbitrary subset of the language. In what follows, we present some semantics for which there is a valuation v that assigns value $1/2$ to every formula $\varphi \in \mathcal{L}$, which is a stronger condition given that it makes Identity fail for every φ; we will call this valuation $v_{1/2}$.

4.1.2.1 Strong Kleene Valuations

The first logic is based on the Strong Kleene valuations, denoted by V^{SK} and defined in Sect. 3.2.2.1. With V^{SK} as the set of valuations, we can show that the resulting logic is empty, and therefore it is also non-reflexive.[1]

Fact 4.1.5 There are no inferences $\Gamma \Rightarrow \Delta$ which are ts-valid in V^{SK}.

Proof It follows directly from the fact that the valuation $v_{1/2} \in V^{\mathrm{SK}}$, that is, for every formula $\varphi \in \mathcal{L}$, $v_{1/2}(\varphi) = 1/2$ (see Exercise 54). That means that, for an arbitrary inference $\Gamma \Rightarrow \Delta$, the valuation $v_{1/2}$ t-satisfies every $\gamma \in \Gamma$ but s-satisfies no $\delta \in \Delta$. Thus, $v_{1/2}$ ts-satisfies no $\Gamma \Rightarrow \Delta$, and therefore every $\Gamma \Rightarrow \Delta$ is not ts-valid. □

As a result of this emptiness, the logic does not validate inferences like $\varphi \wedge \neg\varphi \Rightarrow$ (with emphasis in the conclusions being the empty set) nor $\Rightarrow \varphi \vee \neg\varphi$ (with emphasis in the premises being the empty set). The former fact gives us a paraconsistent system, and the latter a paracomplete system (see Sect. 3.2.2.1). Even more, it has no tautologies, that is, there are no formulas φ such that every $v \in V^{\mathrm{SK}}$ s-satisfies them. This is equivalent to saying that there are no ts-valid

[1] This last feature should be of no surprise, since V^{SK} already complies with the condition for V in Fact 4.1.4.

	$f^{\neg}_{\text{SSch}}$
1	0
1/2	0, 1/2, 1
0	1

$f^{\wedge}_{\text{SSch}}$	1	1/2	0
1	1	0, 1/2, 1	0
1/2	0, 1/2, 1	0, 1/2, 1	0
0	0	0	0

$f^{\vee}_{\text{SSch}}$	1	1/2	0
1	1	1	1
1/2	1	0, 1/2, 1	0, 1/2, 1
0	1	0, 1/2, 1	0

$f^{\rightarrow}_{\text{SSch}}$	1	1/2	0
1	1	0, 1/2, 1	0
1/2	1	0, 1/2, 1	0, 1/2, 1
0	1	1	1

Fig. 4.1 The Strong Schütte truth-tables

inferences of the form $\Rightarrow \varphi$, for which paracompleteness is a particular case. The logic has no antitautologies (see Sect. 3.2.2.1 again) either, i.e. formulas φ such that no $v \in V^{\text{SK}}$ t-satisfies them.[2] This is equivalent to saying that there are no ts-valid inferences of the form $\varphi \Rightarrow$, of which paraconsistency is a particular case.

Exercises

Exercise 54 Show that a valuation $v \in V^{\text{SK}}$ that assigns value 1/2 to every atomic formula, also assigns value 1/2 to every formula $\varphi \in \mathcal{L}$.

Exercise 55 Explain why there are neither tautologies nor antitautologies in V^{SK}.

4.1.2.2 Schütte Valuations

In Sect. 4.1.3 we will show a result on inferential Soundness and Completeness, but when we get to the metainferential aspect of our logics, Strong Kleene valuations will only be useful to the calculus $\mathbf{LK}^{-i}_{+\text{E}}$. For the metainferential aspect of the calculi $\mathbf{LK}^{-i}$ and $\mathbf{LK}^{-i}_{+\text{E}-\text{I}}$ we need to recall the Schütte tables (Fig. 3.2), and introduce the Strong Schütte tables (Fig. 4.1). We denote the sets of valuations which they represent with V^{Sch} and V^{SSch}, respectively.

Strong Schütte tables are non-deterministic extensions of the Strong Kleene tables (Fig. 3.1). In particular, the previous placement of value 1/2 now involves a choice from the set {0, 1/2, 1}.

Since both V^{SSch} and V^{Sch} are extensions of V^{SK}, they also have the valuation $v_{1/2}$ that assigns value 1/2 to every formula. Therefore, we may establish that the resulting logics are also empty.

Fact 4.1.6 There are no ts-valid inferences either in V^{SSch} or in V^{Sch}.

Exercises

Exercise 56 Show that there is a valuation $v \in V^{\text{SSch}}$ such that it assigns value 1/2 to every formula $\varphi \in \mathcal{L}$.

Exercise 57 Show that the consequence relation ts in V^{SSch} has neither tautologies nor antitautologies.

---

[2] Cfr. the notion of *antitheorem* given by Wintein (2016).

4.1.3 Inferential Soundness and Completeness

As mentioned before in Sect. 4.1.1, the sequent calculi presented have no axioms, and without them, no proof can be found for any given sequent. Something similar happens when we consider the semantics presented in Sect. 4.1.2, since the valuation that assigns value 1/2 to every formula invalidates (in terms of ts-satisfaction) every inference. So we can claim Fact 4.1.7.

Fact 4.1.7 $\Gamma \vdash_{\mathbf{LK}^{-i}} \Delta$ iff $\Gamma \vdash_{\mathbf{LK}^{-i}_{+\mathrm{E}}} \Delta$ iff $\Gamma \vdash_{\mathbf{LK}^{-i}_{+\mathrm{E}-\mathrm{I}}} \Delta$ iff $\Gamma \vDash_{ts} \Delta$ in $V \subseteq V^3$, such that $v_{1/2} \in V$.

One could think that, since they are all empty, they are useless consequence relations, but they have received some interesting applications, for instance when dealing with semantic paradoxes (French, 2016). In Sect. 4.2, we show that when considering the metainferential aspect of a logic, more interesting things about these empty logics can be said.

4.2 Metainferential Tolerant-Strict Logics

As it was stated in previous chapters, metainferences are interesting on their own, but in the case of non-reflexive logics they are more so than regular inferences. Indeed, in a non-reflexive setting and, perhaps more importantly, in a context with no valid inferences, the focus naturally shifts to the transitions between inferences that are hypothetically assumed to be valid and those that follow from them. In other words, to metainferences. In this section, we present some distinctions and relations that can be established when working with metainferences.

This section will follow the same order of reviewing the proof-theoretical and then the semantic approach to metainferential validity.

4.2.1 Admissibility and Derivability

The major feature of the calculi we presented above is that they are inferentially empty. However, they determine quite different logics, though these differences can only be seen at the level of the metainferences. Here we show these differences under both admissibility and derivability notions.

Similar to Fact 3.3.3, that states that calculi $\mathbf{LK}^{-c}$ and $\mathbf{LK}^{-c}_{+\mathrm{E}}$ are not structurally complete, Fact 4.2.1 states that the non-reflexive calculi are not structurally complete either.

Fact 4.2.1 Calculi $\mathbf{LK}^{-i}$ and $\mathbf{LK}^{-i}_{+\mathrm{E}}$ are not structurally complete.

Proof It is sufficient to present a metainference schema admissible in the calculi, but not derivable in them. Consider the following metainference:

$$\frac{\varphi \Rightarrow \psi}{\sigma \Rightarrow \pi}$$

Since $\varphi \Rightarrow \psi$ has no provable instances in either $\mathbf{LK}^{-i}$ nor $\mathbf{LK}^{-i}_{+\mathrm{E}}$, all instances of the metainference that have them as a premise are admissible. However, adding some instance of $\varphi \Rightarrow \psi$ (for example $p \Rightarrow q$) as an axiom to the calculi is not enough to prove some instance of $\sigma \Rightarrow \pi$ (for example $r \Rightarrow s$). Thus, there is an underivable instance and the schema is underivable. □

The proof rests on the fact that every metainference with non-empty premises is admissible in both systems, because no inference is provable, and therefore the admissibility condition would be trivially satisfied (recall Definition 2.3.1). This is not the case for derivability, since there are metainferences with non-empty premises that are not derivable. For the case with no premises, admissibility and derivability collapse, given that no such metainference is either admissible or derivable.

However, $\mathbf{LK}^{-i}$ is strictly weaker than $\mathbf{LK}^{-i}_{+\mathrm{E}}$. Inferentially they prove nothing, and they have the same admissible metainferences. But $\mathbf{LK}^{-i}_{+\mathrm{E}}$ has more derivable metainferences than $\mathbf{LK}^{-i}$. This comes from the additional (elimination) rules that we add to $\mathbf{LK}^{-i}$. For example, the following metainferences are derivable in $\mathbf{LK}^{-i}_{+\mathrm{E}}$, but not in $\mathbf{LK}^{-i}$.

$$\frac{\Rightarrow \neg\varphi}{\varphi \Rightarrow} \qquad\qquad \frac{\varphi \Rightarrow \psi \qquad \varphi \Rightarrow \psi \to \rho}{\Rightarrow \varphi \to \rho}$$

Notice that the first example corresponds to the elimination rule for negation on the right. The reason why this and the other elimination rules are not derivable in $\mathbf{LK}^{-i}$, depends on the lack of Id. The derivation of the elimination rule for negation on the right would go as follows:

$$\cfrac{{}^{[*]} \Rightarrow \neg\varphi \qquad \cfrac{\cfrac{}{\varphi \Rightarrow \varphi}\,\mathrm{Id}}{{}^{[**]}\varphi, \neg\varphi \Rightarrow}\,\mathrm{L}\neg}{\varphi \Rightarrow}\,\mathrm{Cut}$$

Similarly, for the derivation of all elimination rules, both Cut, which we have, and Id, which we do not have, are needed. The former is necessary since it is the only rule which eliminates formulas. The latter is necessary for the application of Cut, which requires two premises. One is the premise of the elimination rule, that is assumed (see [*] in the example above). The other premise has to be derived (see [**] in the example above). And, since we do not have axioms, i.e. Id, we cannot obtain it. Therefore, if elimination rules are added, as in $\mathbf{LK}^{-i}_{+\mathrm{E}}$, the previous derivations can be performed thanks to the rule of negation elimination on the right, and the elimination of the conditional on the right (see Exercise 60), respectively.

Exercises

Exercise 58 Prove that $\mathbf{LK}^{-i}_{+\mathrm{E}-\mathrm{I}}$ is not structurally complete.

Exercise 59 Prove that $\mathbf{LK}^{-i}_{+\mathrm{E}-\mathrm{I}}$ is strictly weaker than $\mathbf{LK}^{-i}_{+\mathrm{E}}$, and that it is incomparable to $\mathbf{LK}^{-i}$.

Exercise 60 Show that assuming $\varphi \Rightarrow \psi$ and $\varphi \Rightarrow \psi \rightarrow \rho$, there is a proof of $\Rightarrow \varphi \rightarrow \rho$ in $\mathbf{LK}^{-i}_{+\mathrm{E}}$.

4.2.2 Global, Local and Absolute Global Validity

The work done in Sect. 2.3.2 for local, global and absolute global validity was extended in Sect. 3.3.2 from two-valued to three-valued logics. There the extension was for non-transitive logics, but here we further extend it to explore non-reflexive logics.

4.2.2.1 Global Validity

Let us start with the notion of global validity. Global validity expresses whether a metainference preserves the validity from the premises to its conclusion. Definition 4.2.2 is the corresponding instance of Definition 3.3.12 for ts, i.e. the global ts-validity notion.

Definition 4.2.2 Let $V \subseteq V^3$. A metainference $S \Rrightarrow r$ is *globally ts-valid in V* just in case some $s \in S$ is not ts-valid in V or r is.

The set of valuations V mentioned in Definition 4.2.2 can be properly selected when we look at the global range of the sequent calculus of our interest, defined in Definition 4.2.3.

Definition 4.2.3 Let $\mathbf{S}$ be a set of metainferences, and $V \subseteq V^3$. The *global ts-range of a set of metainferences $\mathbf{S}$ in V*, abbreviated $gr(\mathbf{S}, ts, V)$, is the maximal set of valuations $V_i \subseteq V$ such that each metainference in $\mathbf{S}$ is globally ts-valid in V_i.

The logics obtained in Sect. 4.1.2 have no valid inferences. So, every metainference with non-empty premises is globally valid, independently of the premises and the conclusion. However, these logics are not globally trivial, since metainferences with no premises are globally invalid, no matter the conclusion they have. Notice that the same metainferences are admissible in $\mathbf{LK}^{-i}$, $\mathbf{LK}^{-i}_{+\mathrm{E}}$ and $\mathbf{LK}^{-i}_{+\mathrm{E}-\mathrm{I}}$. As a result, we can state which is the global ts-range in V^3 of $\mathbf{LK}^{-i}$, $\mathbf{LK}^{-i}_{+\mathrm{E}}$ and $\mathbf{LK}^{-i}_{+\mathrm{E}-\mathrm{I}}$ (Fact 4.2.4).

Fact 4.2.4 The global ts-range in V^3 of $\mathbf{LK}^{-i}$, $\mathbf{LK}^{-i}_{+\mathrm{E}}$ and $\mathbf{LK}^{-i}_{+\mathrm{E}-\mathrm{I}}$ is V^3.

Proof Recall that $v_{1/2} \in V^3$, which, by definition, assigns value $1/2$ to every formula $\varphi \in \mathcal{L}$, rendering every inference ts-invalid, since every set of premises is t-satisfied but the conclusion is not s-satisfied by $v_{1/2}$. Every rule in $\mathbf{LK}^{-i}$ (and in $\mathbf{LK}^{-i}_{+\mathrm{E}}$ and in $\mathbf{LK}^{-i}_{+\mathrm{E}-\mathrm{I}}$) has a non-empty set of premises which are all ts-invalid in V^3, and therefore it is globally ts-valid in V^3, and V^3 is maximal in relation with itself. Therefore V^3 is the maximal set of valuations such that each metainference in $\mathbf{LK}^{-i}$ (and in $\mathbf{LK}^{-i}_{+\mathrm{E}}$ and in $\mathbf{LK}^{-i}_{+\mathrm{E}-\mathrm{I}}$) is globally ts-valid. □

4.2.2.2 Local Validity

As expected, local and global ts-validity do not collapse. From Definitions 3.3.5 and 3.3.9, we can obtain their instances for local ts-validity (Definition 4.2.5) and local ts-range (Definition 4.2.6).

Definition 4.2.5 Let $V \subseteq V^3$. A metainference $S \Rrightarrow r$ is *locally ts-valid in V* just in case all V-valuations which ts-satisfy each $s \in S$ also ts-satisfy r.

Definition 4.2.6 Let $V \subseteq V^3$. The *local ts-range of a set of metainferences* $\mathbf{S}$ *in* V, abbreviated $lr(\mathbf{S}, ts, V)$, is the maximal set of valuations $V_i \subseteq V$ such that each metainference in $\mathbf{S}$ is locally ts-valid in V_i.

Take then, for instance, the metainference schema with premise $\varphi \Rightarrow \varphi$ and conclusion $\psi \Rightarrow \psi$. It is globally ts-valid since every instance of the premise is ts-invalid, and therefore the metainference is trivially globally ts-valid. But some of its instances, say the one from $p \Rightarrow p$ to $q \Rightarrow q$, are not locally ts-valid, since if we consider the valuation v such that $v(p) = 1$ and $v(q) = 1/2$, we have that, being the premise ts-satisfied, the conclusion is not ts-satisfied, so it is locally ts-invalid.

With Definitions 4.2.5 and 4.2.6 at hand, we can characterize the local ts-ranges of $\mathbf{LK}^{-i}$ and $\mathbf{LK}^{-i}_{+\mathrm{E}}$, starting with the latter (Fact 4.2.7).[3]

Fact 4.2.7 The local ts-range in V^3 of $\mathbf{LK}^{-i}_{+\mathrm{E}}$ is V^{SK}.

Proof The proof follows the one for Fact 3.3.10. From right to left, we just check that each of the rules of $\mathbf{LK}^{-i}_{+\mathrm{E}}$ are locally ts-valid in V^{SK}. We show it for a couple of rules, but we leave the rest of them as exercises.

- Rule R$\wedge$. Suppose that a valuation $v \in V^{\mathrm{SK}}$ ts-satisfies both $\Gamma \Rightarrow \varphi, \Delta$ and $\Gamma \Rightarrow \psi, \Delta$. We have to show then that v also ts-satisfies $\Gamma \Rightarrow \varphi \wedge \psi, \Delta$. Suppose now that for every $\gamma \in \Gamma$, $v(\gamma) \in \{1/2, 1\}$, that is, every γ is t-satisfied. Then, either there is a $\delta \in \Delta$ such that $v(\delta) = 1$, or $v(\varphi) = v(\psi) = 1$ (equivalently, they are s-satisfied). The latter disjunct means that $v(\varphi \wedge \psi) = 1$ by the Strong Kleene tables. Therefore, in any case v also ts-satisfies the conclusion of the rule.

[3] For an original statement of this fact, see (Da Ré et al., 2021).

- Rules R$\wedge$E. Suppose that a valuation $v \in V^{\text{SK}}$ ts-satisfies $\Gamma \Rightarrow \varphi \wedge \psi, \Delta$. We have to show that v ts-satisfies both $\Gamma \Rightarrow \varphi, \Delta$ and $\Gamma \Rightarrow \psi, \Delta$ (the former for one elimination rule, the latter for the other one). Suppose then that for every $\gamma \in \Gamma$, $v(\gamma) \in \{1/2, 1\}$, i.e. every γ is ts-satisfied. Then, either there is a $\delta \in \Delta$ such that $v(\delta) = 1$, or $v(\varphi \wedge \psi) = 1$ (again, they are s-satisfied). The conjunction receives value 1 just in case $v(\varphi) = v(\psi) = 1$ given the Strong Kleene tables. Therefore, v also ts-satisfies the conclusion of each rule.

From left to right, suppose that $v \notin V^{\text{SK}}$. We show that v is therefore not in the local ts-range of $\mathbf{LK}^{-i}_{+\text{E}}$. There is one case for each connective for which v does not respect the Strong Kleene tables. We work the cases for $\neg$ and $\vee$, leaving the case for $\wedge$ as an exercise.

- v is deviant for $\neg$. We distinguish four subcases:
  * There is a formula φ such that $v(\varphi) = 1$ but $v(\neg\varphi) \in \{1/2, 1\}$. Then, v is a ts-counterexample to an instance of L$\neg$ of the following form:

$$\frac{\Rightarrow \varphi}{\neg\varphi \Rightarrow}$$

  * There is a formula φ such that $v(\varphi) = 0$ but $v(\neg\varphi) \in \{0, 1/2\}$. Then, v is a ts-counterexample to an instance of R$\neg$ of the following form:

$$\frac{\varphi \Rightarrow}{\Rightarrow \neg\varphi}$$

  * There is a formula φ such that $v(\varphi) = 1/2$ but $v(\neg\varphi) = 1$. Then, v is a ts-counterexample to an instance of R$\neg$E of the following form:

$$\frac{\Rightarrow \neg\varphi}{\varphi \Rightarrow}$$

  * There is a formula φ such that $v(\varphi) = 1/2$ but $v(\neg\varphi) = 0$. Then, v is a ts-counterexample to an instance of L$\neg$E of the following form:

$$\frac{\neg\varphi \Rightarrow}{\Rightarrow \varphi}$$

- v is deviant for $\vee$. We distinguish four subcases:
  * $v(\varphi) = 0$ and $v(\psi) = 0$ but $v(\varphi \vee \psi) \in \{1/2, 1\}$. Then, v is a ts-counterexample to an instance of L$\vee$ of the following form:

$$\frac{\varphi \Rightarrow \qquad \psi \Rightarrow}{\varphi \vee \psi \Rightarrow}$$

  * $v(\varphi) = 1$ or $v(\psi) = 1$, but $v(\varphi \vee \psi) \in \{0, 1/2\}$. Then, v is a ts-counterexample to an instance of R$\vee$ of one of the following forms:

$$\frac{\Rightarrow \varphi}{\Rightarrow \varphi \vee \psi} \qquad\qquad \frac{\Rightarrow \psi}{\Rightarrow \varphi \vee \psi}$$

* $v(\varphi) \in \{1/2, 1\}$ or $v(\psi) \in \{1/2, 1\}$, but $v(\varphi \vee \psi) = 0$. Then, v is a ts-counterexample to an instance of L$\vee$E of one of the following forms:

$$\frac{\varphi \vee \psi \Rightarrow}{\varphi \Rightarrow} \qquad \frac{\varphi \vee \psi \Rightarrow}{\psi \Rightarrow}$$

* $v(\varphi) \in \{0, 1/2\}$ and $v(\psi) \in \{0, 1/2\}$, but $v(\varphi \vee \psi) = 1$. Then, v is a ts-counterexample to an instance of R$\vee$E of the following form:

$$\frac{\Rightarrow \varphi \vee \psi}{\Rightarrow \varphi, \psi}$$

□

Regarding the local ts-range of $\mathbf{LK}^{-i}$, we have Fact 4.2.8.[4]

Fact 4.2.8 The local ts-range in V^3 of $\mathbf{LK}^{-i}$ is V^{SSch}.

Proof Again, the strategy is as in Fact 4.2.7. □

Now it is easy to see that the local ts-range of $\mathbf{LK}^{-i}$ has more valuations than the local ts-range of $\mathbf{LK}^{-i}_{+\text{E}}$. These additional valuations serve as counterexamples to some metainferences that are locally ts-valid in V^{SK} but are not in V^{SSch}, such as an instance of R$\neg$E of the following form:

$$\frac{\Rightarrow \neg\varphi}{\varphi \Rightarrow}$$

Finally, we want to introduce the following novel result about $\mathbf{LK}^{-i}_{+\text{E}-\text{I}}$:

Fact 4.2.9 The local ts-range in V^3 of $\mathbf{LK}^{-i}_{+\text{E}-\text{I}}$ is V^{Sch}.

Proof It follows the same strategy that Fact 4.2.8. □

It is worth highlighting some remarks. First, notice that $\mathbf{LK}^{-i}_{+\text{E}}$ has the rules from both $\mathbf{LK}^{-i}$ and $\mathbf{LK}^{-i}_{+\text{E}-\text{I}}$, and that V^{SK} is the intersection between V^{SSch} and V^{Sch}, which are all three, respectively, the local ts-ranges of those calculi. By having more rules, we have more restrictions on the set of valuations we get. By having fewer rules we allow for more valuations, though the lack of the introduction rules and the lack of elimination rules extend the valuations set differently.

The effect these rules have on the set of valuations can be better seen if we compare what we have discovered in Sect. 3.3.2 regarding the local st-ranges. First (leaving every other structural rule fixed), Cut and introduction rules, in one hand, and Id and elimination rules, on the other hand, determine both the Strong Schütte valuations, though the former do it as the local ts-range and the latter as the local st-range. Similarly, Cut and elimination rules, on one side, and Id and introduction rules, on the other, determine Schütte valuations as the local ts-range and the local st-range, respectively. Finally, in the presence of both introduction and elimination rules, Strong Kleene valuations are determined as the local ts-range when we have

[4] Proved in (Hösli & Jäger, 1994).

Cut and the local st-range when we have Id. If we have only the introduction rules, as in $\mathbf{LK}^{-i}$ and $\mathbf{LK}^{-c}$, the superset of V^{SK} that we obtain is different depending on whether we want either the local ts-range or the local st-range. There is a kind of symmetry here: if we have only the elimination rules, as in $\mathbf{LK}^{-i}_{+\mathrm{E}-\mathrm{I}}$ and $\mathbf{LK}^{-c}_{+\mathrm{E}-\mathrm{I}}$, the extension from V^{SK} that we obtain is the opposite—that is, V^{Sch} instead of V^{SSch} and vice versa.

Exercises

Exercise 61 Prove that

(i) every rule in $\mathbf{LK}^{-i}_{+\mathrm{E}}$ is locally ts-valid in V^{SK};
(ii) every rule in $\mathbf{LK}^{-i}$ is locally ts-valid in V^{SSch};
(iii) every rule in $\mathbf{LK}^{-i}_{+\mathrm{E}-\mathrm{I}}$ is locally ts-valid in V^{Sch}.

Exercise 62 Show that

(i) $v \notin V^{\mathrm{SK}}$, then v is deviant for $\wedge$ in $\mathbf{LK}^{-i}_{+\mathrm{E}}$;
(ii) $v \notin V^{\mathrm{SSch}}$, then v is deviant for $\wedge$ in $\mathbf{LK}^{-i}$;
(iii) $v \notin V^{\mathrm{Sch}}$, then v is deviant for $\wedge$ in $\mathbf{LK}^{-i}_{+\mathrm{E}-\mathrm{I}}$.

Exercise 63 Show that

(i) every elimination rule in $\mathbf{LK}^{-i}_{+\mathrm{E}}$ is locally ts-invalid in V^{SSch};
(ii) every introduction rule in $\mathbf{LK}^{-i}_{+\mathrm{E}}$ is locally ts-invalid in V^{Sch}.

4.2.2.3 Absolute Global Validity

For the absolute global ts-validity we recall Definitions 3.3.20 and 3.3.21 as a base for Definitions 4.2.10 and 4.2.11.

Definition 4.2.10 Let $\mathcal{V} \subseteq \mathcal{P}(V^3)$. A metainference is *absolutely globally ts-valid in $\mathcal{V}$* just in case it is globally ts-valid in V for every $V \in \mathcal{V}$.

Definition 4.2.11 Let $V \subseteq V^3$. The *absolute global ts-range of* **S** *in* V, abbreviated $agr(\mathbf{S}, ts, V)$, is the maximal set $\mathcal{V} \subseteq \mathcal{P}(V)$ such that each metainference in **S** is absolutely globally ts-valid in $\mathcal{V}$.

We would want to identify the absolute global ts-ranges in V^3 of our calculi $\mathbf{LK}^{-i}$, $\mathbf{LK}^{-i}_{+\mathrm{E}}$ and $\mathbf{LK}^{-i}_{+\mathrm{E}-\mathrm{I}}$, but, similarly to what was showed in Sect. 3.3.2.3, those results are not yet available. However, we can present some facts about them.

Firstly, Fact 2.3.23 lets us know some elements from the absolute global ts-ranges. Together with it, Fact 4.2.8 implies that for any v in V^{SSch}, $\{v\}$ is in the $agr(\mathbf{LK}^{-i}, ts, V^3)$, Fact 4.2.7 implies that for any v in V^{SK}, $\{v\}$ is in the $agr(\mathbf{LK}^{-i}_{+\mathrm{E}}, ts, V^3)$, and Fact 4.2.9 implies that for any v in V^{Sch}, $\{v\}$ is in the $agr(\mathbf{LK}^{-i}_{+\mathrm{E}-\mathrm{I}}, ts, V^3)$. A more general result is Fact 4.2.12, obtained with the aid of Fact 3.3.22, but applied to the ts-consequence relations.

Fact 4.2.12

(i) $\mathcal{P}(V^{\mathrm{SSch}}) \subseteq agr(\mathbf{LK}^{-i}, ts, V^3)$
(ii) $\mathcal{P}(V^{\mathrm{SK}}) \subseteq agr(\mathbf{LK}^{-i}_{+\mathrm{E}}, ts, V^3)$
(iii) $\mathcal{P}(V^{\mathrm{Sch}}) \subseteq agr(\mathbf{LK}^{-i}_{+\mathrm{E}-\mathrm{I}}, ts, V^3)$

However, the inclusions mentioned in Fact 4.2.12 are proper: there are sets of valuations in the absolute global ts-ranges outside the corresponding powerset.[5]

Fact 4.2.13

(i) $agr(\mathbf{LK}^{-i}, ts, V^3) \nsubseteq \mathcal{P}(V^{\mathrm{SSch}})$
(ii) $agr(\mathbf{LK}^{-i}_{+\mathrm{E}}, ts, V^3) \nsubseteq \mathcal{P}(V^{\mathrm{SK}})$
(iii) $agr(\mathbf{LK}^{-i}_{+\mathrm{E}-\mathrm{I}}, ts, V^3) \nsubseteq \mathcal{P}(V^{\mathrm{Sch}})$

Proof Consider a set of valuations V conformed by both the valuation $v_{1/2}$ (the one that assigns $1/2$ to every formula), and a valuation $v \in V$ such that it belongs to neither V^{SK} nor V^{Sch} nor V^{SSch}. Trivially, V is included in V^3. Also, every rule of $\mathbf{LK}^{-i}$, $\mathbf{LK}^{-i}_{+\mathrm{E}}$ and $\mathbf{LK}^{-i}_{+\mathrm{E}-\mathrm{I}}$ is globally ts-valid in V since their instances all have non-empty premises, which are all ts-invalid in V (see Fact 4.1.7). Therefore, V belongs to $agr(\mathbf{LK}^{-i}_{+\mathrm{E}}, ts, V^3)$, to $agr(\mathbf{LK}^{-i}, ts, V^3)$ and to $agr(\mathbf{LK}^{-i}_{+\mathrm{E}-\mathrm{I}}, ts, V^3)$. But clearly V is not in the above mentioned powersets. □

Moreover, the absolute global ts-range in V^3 of $\mathbf{LK}^{-i}_{+\mathrm{E}}$ is strictly included both in the absolute global ts-ranges in V^3 of $\mathbf{LK}^{-i}$ and $\mathbf{LK}^{-i}_{+\mathrm{E}-\mathrm{I}}$, which are incomparable, as stated in Fact 4.2.14.

Fact 4.2.14

(i) $agr(\mathbf{LK}^{-i}_{+\mathrm{E}}, ts, V^3) \subset agr(\mathbf{LK}^{-i}, ts, V^3)$
(ii) $agr(\mathbf{LK}^{-i}_{+\mathrm{E}}, ts, V^3) \subset agr(\mathbf{LK}^{-i}_{+\mathrm{E}-\mathrm{I}}, ts, V^3)$
(iii) $agr(\mathbf{LK}^{-i}, ts, V^3) \nsubseteq agr(\mathbf{LK}^{-i}_{+\mathrm{E}-\mathrm{I}}, ts, V^3)$
(iv) $agr(\mathbf{LK}^{-i}_{+\mathrm{E}-\mathrm{I}}, ts, V^3) \nsubseteq agr(\mathbf{LK}^{-i}, ts, V^3)$

Proof For the the first two claims, the inclusion follows from the fact that both $\mathbf{LK}^{-i}$ and $\mathbf{LK}^{-i}_{+\mathrm{E}-\mathrm{I}}$ are subsystems of $\mathbf{LK}^{-i}_{+\mathrm{E}}$. For all the '$\nsubseteq$' claims, consider valuations $v_1 \in V^{SSch}$ and $v_2 \in V^{Sch}$, such that, on the one hand, $v_1(p) = 1/2$ and $v_1(\neg p) = 1$, and on the other hand, $v_2(p) = 0$ and $v_2(\neg p) = 1/2$.

The former valuation, v_1, is a local ts-counterexample of the metainference

$$\frac{\Rightarrow \neg p}{p \Rightarrow}$$

[5] Unfortunately, Fact 4.2.13 shows that the claim made by Da Ré et al. (2021, p. 1544), regarding $\mathcal{P}(V^{\mathrm{SK}})$ being the $agr(\mathbf{LK}^{-i}_{+\mathrm{E}}, ts, V^3)$, is false.

which is an instance of the rule L¬E of both $\mathbf{LK}^{-i}_{+\mathrm{E}}$ and $\mathbf{LK}^{-i}_{+\mathrm{E}-\mathrm{I}}$. As a consequence, the singleton $\{v_1\}$ is neither in $agr(\mathbf{LK}^{-i}_{+\mathrm{E}}, ts, V^3)$ nor in $agr(\mathbf{LK}^{-i}_{+\mathrm{E}-\mathrm{I}}, ts, V^3)$. Since v_1 is a Strong Schütte valuation, by Fact 4.2.8 we get that it is in the local ts-range in V^3 of $\mathbf{LK}^{-i}$. By Fact 4.2.12 we then know that $\{v_1\} \in agr(\mathbf{LK}^{-i}, ts, V^3)$. Therefore, $agr(\mathbf{LK}^{-i}, ts, V^3) \nsubseteq agr(\mathbf{LK}^{-i}_{+\mathrm{E}}, ts, V^3)$ and $agr(\mathbf{LK}^{-i}, ts, V^3) \nsubseteq agr(\mathbf{LK}^{-i}_{+\mathrm{E}-\mathrm{I}}, ts, V^3)$. The former completes the proof for the first claim and the latter gives the third claim.

The other valuation, v_2, is a local ts-counterexample of the metainference

$$\frac{p \Rightarrow}{\Rightarrow \neg p}$$

which is an instance of the rule R¬ of both $\mathbf{LK}^{-i}$ and $\mathbf{LK}^{-i}_{+\mathrm{E}}$. Thus, the singleton $\{v_2\}$ is neither in $agr(\mathbf{LK}^{-i}, ts, V^3)$ nor in $agr(\mathbf{LK}^{-i}_{+\mathrm{E}}, ts, V^3)$. Since v_2 is a Schütte valuation, by Fact 4.2.8 we get that it is in the local ts-range in V^3 of $\mathbf{LK}^{-i}_{+\mathrm{E}-\mathrm{I}}$. Fact 4.2.12 tells us that $\{v_2\} \in agr(\mathbf{LK}^{-i}_{+\mathrm{E}-\mathrm{I}}, ts, V^3)$. Therefore, $agr(\mathbf{LK}^{-i}_{+\mathrm{E}-\mathrm{I}}, ts, V^3) \nsubseteq agr(\mathbf{LK}^{-i}, ts, V^3)$ and $agr(\mathbf{LK}^{-i}_{+\mathrm{E}-\mathrm{I}}, ts, V^3) \nsubseteq agr(\mathbf{LK}^{-i}_{+\mathrm{E}}, ts, V^3)$. The former completes the proof for the second claim and the latter gives the fourth claim. □

Finally, we want to complete what was presented at the end of the Sect. 3.3.2.3, and show here the potential mismatch between local and absolute global validity (for a given sequent calculus and its corresponding local and absolute global ranges), but for the non-reflexive consequence relations we present in this chapter.

Let $\mathbf{S_2}$ be a sequent calculus containing all the structural rules except for Id, and the following rules for disjunction:

$$\frac{\Rightarrow \varphi}{\Rightarrow \varphi \vee \psi}\ (1a) \qquad \frac{\Rightarrow \psi}{\Rightarrow \varphi \vee \psi}\ (1b)$$

$$\frac{\varphi \vee \psi \Rightarrow}{\varphi \Rightarrow}\ (2a) \qquad \frac{\varphi \vee \psi \Rightarrow}{\psi \Rightarrow}\ (2b)$$

$$\frac{\varphi \Rightarrow \qquad \psi \Rightarrow}{\varphi \vee \psi \Rightarrow}\ (3)$$

$$\frac{\Rightarrow \varphi \vee \psi}{\varphi \Rightarrow \varphi, \psi}\ (4a) \qquad \frac{\Rightarrow \varphi \vee \psi}{\psi \Rightarrow \varphi, \psi}\ (4b)$$

The valuations belonging to the local ts-range in V^3 of $\mathbf{S_2}$ are all and only those satisfying the following conditions:

- If $v(\varphi) = 1$ or $v(\psi) = 1$, then $v(\varphi \vee \psi) = 1$ [rules (1a) and (1b)]
- If $v(\varphi \vee \psi) = 0$, then $v(\varphi) = v(\psi) = 0$ [rules (2a) and (2b)]
- If $v(\varphi) = 0$ and $v(\psi) = 0$, then $v(\varphi \vee \psi) = 0$ [rule (3)]
- If $v(\varphi \vee \psi) = 1$, then $v(\varphi) = 1$ or $v(\psi) = 1$ or $v(\varphi) = v(\psi) = 0$ [rules (4a) and (4b)]

The valuations in V^3 that satisfy these conditions are exactly the valuations that respect the Strong Kleene table for disjunction. Consider now the following rule:

$$\frac{\Rightarrow \varphi \vee \psi}{\Rightarrow \varphi, \psi} *$$

which is locally ts-valid in V^{SK}. Therefore, it is locally ts-valid in the local ts-range of $\mathbf{S_2}$.

But if we take the set of valuations $\{v_1, v_2\}$, with v_1 and v_2 defined as follows:

$$v_1(\varphi) = 1 \text{ for every } \varphi \qquad v_2(\varphi) = \begin{cases} 1 & \text{if } \varphi \text{ is of the form } \delta \vee \gamma \\ 0 & \text{otherwise} \end{cases}$$

then, the set $\{v_1, v_2\}$ belongs to the absolute global ts-range of $\mathbf{S_2}$, since rules (1a)–(4b) are globally ts-valid in it. However, the rule $(*)$ is not globally ts-valid in $\{v_1, v_2\}$. For example, the following instance

$$\frac{\Rightarrow p \vee q}{\Rightarrow p, q}$$

is such that the premise is ts-valid in $\{v_1, v_2\}$, but the conclusion is not. Therefore, the rule $(*)$ is not absolutely globally ts-valid in the absolute global ts-range of $\mathbf{S_2}$. This entails the aforementioned mismatch between local and absolutely global validity when we consider non-reflexive sequent calculi.

4.2.3 Metainferential Soundness and Completeness

In this section we show metainferential Soundness and Completeness for $\mathbf{LK}^{-i}$, $\mathbf{LK}^{-i}_{+\mathrm{E}}$ and $\mathbf{LK}^{-i}_{+\mathrm{E}-\mathrm{I}}$ with respect to ts-validity in V^{SSch} (Hösli & Jäger, 1994), V^{SK}, and V^{Sch}, respectively.

The strategy for the proofs is just the same as that in Sect. 3.3.3, with slight adjustments (the details can also be found in (Hösli & Jäger, 1994)).

Theorem 4.2.15 *A rule $S \Rrightarrow r$ is derivable in* $\mathbf{LK}^{-i}$ *if and only if it is locally ts-valid in V^{SSch}.*

Proof Soundness is, also here, an immediate consequence of Fact 4.2.8, so we leave it to the reader and move to Completeness directly. First, in the proof of Theorem 3.3.29 we need to modify the closure conditions displayed at Stage 0 in the tree construction. Since there are no axioms in $\mathbf{LK}^{-i}$, a tree for $\Gamma_0 \Rightarrow \Delta_0$ is closed only when its leafs are obtained applying Weakening or Contraction to sequents of S. Second, since Cut is now among the rules in the system, we need to add right after Stage 0, a further step that takes care of the cuts so that all possible cut formulas are treated (notice that this allows for infinite branches). For this we need a fixed enumeration of the formulas in the language $\xi_1, \ldots, \xi_i, \ldots$:

Stage $\mathbf{n+1=2m}$. If $\Gamma_n \Rightarrow \Delta_n$ is not obtained by applications of Weakening or Contractions to sequents of S, then extend the tree with $\xi_m, \Gamma_n \Rightarrow \Delta_n$ and $\Gamma_n \Rightarrow \Delta_n, \xi_m$.

The other construction steps only concern the operational rules and the structural rules of **LK** except Identity, so what is called Stage $n+1$ in Theorem 3.3.29 remains the same. We just call it Stage $n+1 = 2m+1$, since it alternates with the previously stated step for Cut. And again, is is easy to check that, if the tree is closed, then we can construct a derivation of r from S in $\mathbf{LK}^{-i}$.

Hence, when r is not derivable from S in $\mathbf{LK}^{-i}$, there is an open branch in the reduction tree constructed according to the above instruction and, as it is easy to see given Stage $n+1 = 2m$, it will be infinite. Here comes the next modification in the proof of Theorem 3.3.29: this time the counterexample that we can get when the tree for r is open will be a Strong Schütte valuation that ts-satisfies all sequents in S but does not ts-satisfy r. Consider the open branch $\Gamma_0 \Rightarrow \Delta_0, \Gamma_1 \Rightarrow \Delta_1 \ldots$ and take $\Gamma_\omega = \bigcup_{i\in\omega}(\Gamma_i)$ and $\Delta_\omega = \bigcup_{i\in\omega}(\Delta_i)$. We construct a valuation v such that:

$$v(\varphi) = \begin{cases} 1 & \text{if } \varphi \in \Gamma_\omega \text{ and } \varphi \notin \Delta_\omega \\ 0 & \text{if } \varphi \in \Delta_\omega \text{ and } \varphi \notin \Gamma_\omega \\ 1/2 & \text{if } \varphi \in \Gamma_\omega \text{ and } \varphi \in \Delta_\omega \end{cases}$$

Notice that contrary to the valuation defined in Theorem 3.3.29, here it can happen that φ is both in Γ_n and in Δ_n, since Weakening/Contraction on Identity is not a closing condition for a branch. As Theorem 3.3.29, you may notice that v is defined on formulas and not, as usual, on propositional variables. The reason for this is lack of compositionality: in the case of Strong Schütte tables, we do not have a guarantee that assigning 1/2 to all variables such that they are both in the premises and in the conclusions, the resulting formulas will also have value 1/2. Moreover, the proof that $v \in V^{\mathrm{SSch}}$ consists in considering each possible non-atomic formulas and is analogous to the proof of the corresponding fact in Theorem 3.3.29. The fact that v is a ts-counterexample to $\Gamma_0 \Rightarrow \Delta_0$ follows from the definition of v and from the fact that $\Gamma_0 \subseteq \Gamma_\omega$ and $\Delta_0 \subseteq \Delta_\omega$. At the same time the sequents in S are ts-satisfied by v, because if it did not ts-satisfy some sequent $\Sigma \Rightarrow \Pi$ in S, by definition of v we would have that $\Sigma \subseteq \Gamma_\omega$ and $\Pi \subseteq \Delta_\omega$. If that were the case, $\Gamma_0 \Rightarrow \Delta_0$ would be derivable from $\Sigma \Rightarrow \Pi$, contradicting the assumption that there is an open branch in the tree. □

Moving on to the case of $\mathbf{LK}^{-i}_{+\mathrm{E}}$, we have the following:

Theorem 4.2.16 *A rule $S \Rightarrow r$ is derivable in $\mathbf{LK}^{-i}_{+\mathrm{E}}$ if an only if it is locally ts-valid in V^{SK}.*

Proof Soundness is again left to the reader, since it is an easy consequence of Fact 4.2.7. The Completeness proof follows again a similar strategy to the previous theorem. We already assumed an enumeration of the formulas in the language to take care of cut formulas and here we also need it to progressively consider all the ways in which formulas in the tree could be obtained applying the elimination rules. Since elimination rules are just the same as those in $\mathbf{LK}^{-c}_{+\mathrm{E}}$, we have the same additional construction step that one in the proof of Theorem 3.3.31 called Stage $n + 1 = 2m$. The only difference for our case is that we also have another step before that which takes care of Cut and is now named Stage $n + 1 = 3m$, while the Stage for the elimination rules is $n + 1 = 3m + 1$ and that one for introduction rules is now called Stage $n + 1 = 3m + 2$.

Again, we can easily prove that if the reduction tree for r is closed, then there is a derivation from S to r in $\mathbf{LK}^{-i}_{+\mathrm{E}}$. Hence, if there is not such derivation there will be an (infinite) open branch $\Gamma_0 \Rightarrow \Delta_0, \Gamma_1 \Rightarrow \Delta_1 \ldots$. Once more we take $\Gamma_\omega = \bigcup_{i\in\omega}(\Gamma_i)$ and $\Delta_\omega = \bigcup_{i\in\omega}(\Delta_i)$ and construct v as in Theorem 4.2.15. The proof of v being such that is ts-satisfies the sequents in S while not ts-satisfying r is analogous to the proof or Theorem 4.2.15. As for the proof of v being an SK valuation, it is just the same as that in Theorem 3.3.31. □

In Sect. 4.1 we also introduced the sequent calculus $\mathbf{LK}^{-i}_{+\mathrm{E-I}}$, which only includes, apart from the structural rules except Id, the elimination rules. We have a result of Soundness and Completeness for this case too, whose proof is just analogous to the strategies displayed above, where the construction of the reduction tree will now only count the stages taking care of closure conditions, cut formulas and elimination rules. Hence, we only state the result here and leave the proof to the reader as an Exercise 65.

Theorem 4.2.17 *A rule $S \Rrightarrow r$ is derivable in $\mathbf{LK}^{-i}_{+\mathrm{E-I}}$ if an only if it is locally ts-valid in V^{Sch}.*

Exercises

Exercise 64 Prove Soundness for one of the Theorems above.

Exercise 65 Prove Theorem 4.2.17

4.3 $\mathbb{TS}$ and Other Logics

This section serves as a junction point since it presents many connections between some of the logics studied so far in the book. In particular, these connections concern translations between logics and their dualities. Before introducing the concepts and results that illustrate these links, a brief but important reminder is in order.

In the previous chapter logics were introduced as sets of valid inferences and locally valid metainferences relatively to some consequence relation (see Definition 3.3.6). In this section, we will use this definition in an extended sense in order to account for the connections between non-reflexive and non-transitive logics. The new notions to be included in this extended sense will be formally introduced later. However, we can already anticipate that, from this point on, given a valuation schema V and a standard xy, a logic $\mathbb{L}$ will be understood as the set of all and only xy-valid (positive) inferences, xy-valid anti-inferences (Definition 4.3.5), xy-valid negative inferences (Definition 4.3.11), xy-antivalid inferences (Definition 4.3.17), locally xy-valid metainferences with positive or negative inferences (Definition 4.3.12) and locally xy-valid meta-anti-inferences (Definition 4.3.7). In particular, in this section we will consider the following logics: $\mathbb{CL}$ is the logic associated to the ss (or tt, st, ts) standard in V^{B}, $\mathbb{ST}$ and $\mathbb{TS}$ are the logics associated to the st and the ts standard in V^{SK} respectively, while $\mathbb{K}3$ and $\mathbb{LP}$ are the logics associated to the ss and tt standard in V^{SK}, respectively.

Although this long list may initially seem unwieldy or as an *ad hoc* redefinition of what a logic is, here and in the next section we will provide not only significant technical results which build on these notions, but also provide some intuitions motivating their introduction.

4.3.1 A Sheep in Wolf's Clothing

In Sect. 3.4 we used the metaphor by Cobreros et al. (2020) that describes $\mathbb{ST}$ as $\mathbb{LP}$ in sheep's clothing, meaning that the former's inferences can be seen, modulo translation, *just as* $\mathbb{LP}$'s inferences. Here, we want to use the reverse metaphor for the non-reflexive logic too, the sheep in wolf's clothing, to show the deep connections it holds with $\mathbb{K}3$.

We follow the strategy from Sect. 3.4, and we use the same definition of *formula translation* of a sequent $\Gamma \Rightarrow \Delta$, $\tau(\Gamma \Rightarrow \Delta)$, and of a rule of the form $S \Rrightarrow r$, $\tau(S \Rrightarrow s) = \tau(S) \Rightarrow \tau(r)$. These notions are what we need to prove the main result previously announced:

Theorem 4.3.1 *Let $S \Rrightarrow r$ be a metainference whose inferences are all sequents. $S \Rrightarrow r$ is in $\mathbb{TS}$ if and only if $\tau(S \Rrightarrow r)$ is in $\mathbb{K}3$.*

Proof From left to right: assume that $S \Rrightarrow r$ is locally ts-valid. Now consider a valuation $v \in V^{\mathrm{SK}}$ such that $u(\tau(s)) = 1$ for all $s \in S$. Take any $s \in S$ of the form $\Gamma \Rightarrow \Delta$. Since $v(\tau(s)) = 1$, we cannot have that both $\Gamma = \Delta = \emptyset$ because $v(\tau(s)) \neq \bot$. So $\tau(s)$ must be either $\neg \bigwedge \Gamma$, $\bigvee \Delta$ or $\bigwedge \Gamma \rightarrow \bigvee \Delta$. Since $v(\tau(s)) = 1$, by the Strong Kleene tables we have that:

if $\tau(s) = \neg \bigwedge \Gamma$, $v(\gamma) = 0$ for some $\gamma \in \Gamma$
if $\tau(s) = \bigvee \Delta$, $v(\delta) = 1$ for some $\delta \in \Delta$
if $\tau(s) = \bigwedge \Gamma \rightarrow \bigvee \Delta$, $v(\gamma) = 0$ for some $\gamma \in \Gamma$ or $v(\delta) = 1$ for some $\delta \in \Delta$

Either way, we have that s is ts-satisfied. Since s was arbitrary, the same holds for any $s \in S$ assuming that $v(\tau(s)) = 1$. Since by our first assumption we have that $S \Rrightarrow r$ is locally ts-valid, we have that also r is ts-satisfied by v. r is of the form $\Sigma \Rightarrow \Pi$, so either $v(\sigma) = 0$ for some $\sigma \in \Sigma$, or $v(\pi) = 1$ for some $\pi \in \Pi$. So by the Strong Kleene truth tables, $v(\bigwedge \Sigma) = 0$ or $v(\bigvee \Pi) = 1$. Hence $v(\tau(r)) = 1$. So we have that under the assumption that $S \Rrightarrow r$ is locally ts-valid, if $v(\tau(s)) = 1$ for all $s \in S$, then $v(\tau(r)) = 1$, i.e. $t(S \Rrightarrow r)$ is ss-valid.

Now, from right lo left: assume that $\tau(S \Rrightarrow r)$ is ss-valid. Suppose that for all $s \in S$, s is ts satisfied by a valuation $v \in V^{\text{SK}}$. Again using the Strong Kleene truth tables we show that $v(\tau(s)) = 1$ for all $s \in S$. By our first assumption, we have that also $v(\tau(r)) = 1$ so we find that r is ts-satisfied by v. Hence $S \Rrightarrow r$ is locally ts-valid. □

Just as in the case of $\mathbb{LP}$ and $\mathbb{ST}$, this result tells us that there is a deep connection between the conditional of $\mathbb{K3}$ and the $\mathbb{TS}$ consequence relation. We can illuminate it further with Fact 4.3.2:

Fact 4.3.2 Let $v \in V^{\text{SK}}$; $\varphi \Rightarrow \psi$ is ts-satisfied by v if and only if $v(\varphi \to \psi) = 1$

The proof is left as an exercise (Exercise 66).

What we can gather from this is that what we know about conditionals in $\mathbb{K3}$ is enough to tell us something important about inferences in $\mathbb{TS}$, namely that none of them is valid. There is always a V^{SK} valuation making a conditional, or any other formula, not exactly true (not 1)—in fact, $\mathbb{K3}$ has no tautologies.

Let us anticipate that the result established in Theorem 4.3.1 serves us to further illuminate an interesting relation between $\mathbb{TS}$ and another logic in the framework that we are considering, which is $\mathbb{LP}$. However, this result (Fact 4.3.9) needs the notion of negation duality that we will introduce in the next section and therefore we present it later.

To conclude this section, let us point out that the results presented here can be interpreted as another demonstration of the intriguing nature of this non-reflexive logic, despite its initial appearance. Ultimately, an inferentially empty logic is no more astonishing than a system devoid of theorems.

Exercises

Exercise 66 Prove Fact 4.3.2.

4.3.2 Duality Between $\mathbb{ST}$ *and* $\mathbb{TS}$

The literature on substructural logics points to an existing duality between $\mathbb{ST}$ and $\mathbb{TS}$, and, in particular, a duality between Id and Cut. For instance, claims that support the idea of a duality holding between these two approaches to paradoxes can be found, more or less explicitly, in (Frankowski, 2004; French, 2016; Hösli & Jäger, 1994; Ripley, 2013).

Some authors proposed several ways to frame the kind of duality existing between $\mathbb{ST}$ and $\mathbb{TS}$. Among them, Cobreros et al. (2021) and Da Ré et al. (2020) adopt two different points of view: the former remaining at the inferential level, the latter moving up to metainferences. Blomet & Égré (2024) also elaborates a notion of duality building on that by Cobreros et al. (2021).

The common feature of these proposals lies in the formulation of a notion of duality alternative to the usual. So, let us first consider what are the limitations of the standard notion of duality when accounting for the intuitive duality between $\mathbb{TS}$ and $\mathbb{ST}$ and then present some results that serve to justify this intuition. Next, we introduce, in this order, the proposed solutions by Da Ré et al. (2020), Cobreros et al. (2021) and Blomet & Égré (2024).

The usual notion of duality, that Da Ré et al. (2020) called 'negation duality',[6] requires the definition of the 'negation dual' of an inference. Let $\neg\Sigma = \{\neg\sigma | \sigma \in \Sigma\}$:

Definition 4.3.3 The *negation dual* of an inference token $\Gamma \Rightarrow \Delta$ is the inference token $\neg\Delta \Rightarrow \neg\Gamma$. An inference schema is the negation dual of another inference schema iff each inference token of one of them has an inference dual that is a token of the other.

Hence, we have the following:

Definition 4.3.4 A logic $\mathbb{L}$ is negation dual to a logic $\mathbb{L}'$ if and only if, if an inference schema is in $\mathbb{L}$ then its negation dual is in $\mathbb{L}'$ and vice versa.

With this definition at hand, we have that $\mathbb{CL}$ is self-dual, i.e. the negation dual of itself, while $\mathbb{LP}$ and $\mathbb{K3}$ are one the negation dual of the other. In particular, we get that Ex Falso ($\varphi \wedge \neg\varphi \Rightarrow$) and the Law of Excluded Middle ($\Rightarrow \varphi \vee \neg\varphi$) are negation duals only *modulo logical equivalence* in these logics (or any logic satisfying De Morgan laws). By this we mean that the negation dual of the former, i.e. $\Rightarrow \neg(\varphi \wedge \neg\varphi)$, is not $\Rightarrow \varphi \vee \neg\varphi$ but it is logically equivalent to it. Similarly, the negation dual of the latter, i.e. $\neg(\varphi \vee \neg\varphi) \Rightarrow$, is not $\varphi \wedge \neg\varphi \Rightarrow$ but it is logically equivalent to it.

The duality between $\mathbb{LP}$ and $\mathbb{K3}$ and the translation of $\mathbb{ST}$ into $\mathbb{LP}$ (see Sect. 3.4)—as well as that of $\mathbb{TS}$ into $\mathbb{K3}$ that we just showed in the previous section—provide good reasons for believing that there should also be a duality between $\mathbb{TS}$ and $\mathbb{ST}$, and between Id and Cut (at least modulo logical equivalence), as there is for Ex Falso and the Law of Excluded Middle.

This claim can be further supported by two other results. The first connects $\mathbb{TS}$ to $\mathbb{CL}$ in a surprising, and—as it will be evident—specular manner with respect to the equivalence relation holding between classically valid inferences and $st-$valid inferences that was shown in the previous chapter. The second result is the one that we anticipated at the end of the previous section, relating $\mathbb{TS}$ also to $\mathbb{LP}$ (analogous to that proved by Cobreros et al. (2020) relating $\mathbb{ST}$ to $\mathbb{K3}$). To state these results, we need some new notions: anti-inferences and, for the latter, meta-anti-inferences. An intuitive understanding of these notions comes from the idea (see e.g. (Cobreros

[6] Blomet & Égré (2024) and Cobreros et al. (2021) call it 'operational duality'. Here we'll keep calling it as in (Da Ré et al., 2020).

et al., 2020) and (Scambler, 2020a)) that at the inferential level, logics can be characterized at least by two sets: the inferences they validate on the one hand, and the anti-inferences they validate on the other (we will return to this idea in Sect. 4.4). This can be informally spelled out as follows. There are two limit cases when we look at inferences in a logic: those inferences that are satisfied by all valuations—hence valid inferences—and those inferences that are never satisfied by any valuation. This last case is captured by the notion of anti-inference. More formally:

Definition 4.3.5 An *anti-inference* with premises Γ and conclusions Δ is such that a valuation v xy-satisfies it if and only if v does not xy-satisfy $\Gamma \Rightarrow \Delta$. We write this as $\Gamma \not\Leftrightarrow \Delta$.[7] An anti-inference is xy-valid if and only if all valuations xy-satisfy it, namely if and only if no valuation xy-satisfies $\Gamma \Rightarrow \Delta$.

The first result we can obtain from Definition 4.3.5 relates ts-valid anti-inferences and classically valid anti-inferences, similar to Theorem 3.2.8. Recall that Theorem 3.2.8, proved by Ripley (2012), stated that $\mathbb{ST}$ and $\mathbb{CL}$ validate the same inferences. Also recall that the proof of that result depended on Lemma 3.2.7 (that we also need here), stating that for any $v \in V^{\mathrm{SK}}$, if v evaluates some formulas as true or false, there is a $v' \in V^{\mathrm{B}}$ that coincides with v in evaluating each of those formulas.

The dual result connecting $\mathbb{TS}$ to $\mathbb{CL}$, which will again be important in Sect. 4.4, shows that $\mathbb{TS}$ is just like classical logic if one looks at anti-inferences, namely:

Fact 4.3.6 An anti-inference $\Gamma \not\Leftrightarrow \Delta$ is ts-valid in V^{SK} if and only if $\Gamma \not\Leftrightarrow \Delta$ is valid in V^{B}

Proof We prove both directions of the biconditional by contraposition. For the left-to-right direction, assume that there is $v \in V^{\mathrm{B}}$ satisfying $\Gamma \Rightarrow \Delta$. Hence either $v(\gamma) = 0$ for some $\gamma \in \Gamma$, or $v(\delta) = 1$ for some $\delta \in \Delta$. Since $V^{\mathrm{B}} \subseteq V^{\mathrm{SK}}$, v is also such that it ts-satisfies $\Gamma \Rightarrow \Delta$ in V^{SK}. For the right-to-left direction, assume that there is $v \in V^{\mathrm{SK}}$ such that is ts-satisfies $\Gamma \Rightarrow \Delta$, namely it is such that $v(\gamma) = 0$ for some $\gamma \in \Gamma$, or $v(\delta) = 1$ for some $\delta \in \Delta$. By Lemma 3.2.7, there is also a valuation $v' \in V^{\mathrm{B}}$ assigning the same value to the formulas in Γ and Δ, hence classically satisfying $\Gamma \Rightarrow \Delta$. □

This theorem can be seen as supporting the idea that $\mathbb{ST}$ and $\mathbb{TS}$ are duals since, with Theorem 3.2.8, it establishes that while $st-$validity captures one side of $\mathbb{CL}$, namely the set of valid inferences, $ts-$validity captures, so to say, the other side of the coin, namely the set of valid anti-inferences.

This idea can also be reproduced at the level of metainferences, with the second result that we announced. We need the notion of meta-anti-inferences, which follows then quite naturally from that of anti-inferences:

[7] An example of a classically valid anti-inference is $p \vee \neg p \not\Leftrightarrow p \wedge \neg p$. Notice that these classes, namely valid inferences and valid anti-inferences, are not (even in the classical case) complementary, since an inference can be neither.

Definition 4.3.7 A *meta-anti-inference* is a metainference whose inferences are anti-inferences, so something of the form $\Gamma \nRightarrow \Delta \Rrightarrow \Gamma' \nRightarrow \Delta'$, which is xy-satisfied by a valuation v iff v does not xy-satisfy $\Gamma \nRightarrow \Delta$ or v xy-satisfies $\Gamma' \nRightarrow \Delta'$.[8]

From this definition we get that a valuation v xy-satisfies $\Gamma \nRightarrow \Delta \Rrightarrow \Gamma' \nRightarrow \Delta'$ if and only if it also xy-satisfies $\Gamma' \Rightarrow \Delta' \Rrightarrow \Gamma \Rightarrow \Delta$. Hence, we can state the following:

Fact 4.3.8 A meta-anti-inference $\Gamma \nRightarrow \Delta \Rrightarrow \Gamma' \nRightarrow \Delta'$ is locally xy-valid if and only if the metainference $\Gamma' \Rightarrow \Delta' \Rrightarrow \Gamma \Rightarrow \Delta$ is locally xy-valid.

Connecting to the previous section, Cobreros et al. (2020) extend the definition of translation of a metainference to meta-anti-inferences, in the following way: given a meta-anti-inference $\Gamma \nRightarrow \Delta \Rrightarrow \Gamma' \nRightarrow \Delta'$, its translation $\tau(\Gamma \nRightarrow \Delta \Rrightarrow \Gamma' \nRightarrow \Delta')$ is defined as $\bigwedge \Gamma \wedge \neg \bigvee \Delta \Rightarrow \bigwedge \Gamma' \wedge \neg \bigvee \Delta'$. We can finally state the result connecting $\mathbb{TS}$ meta-anti-inferences and $\mathbb{LP}$ inferences:

Fact 4.3.9 A meta-anti-inference $\Gamma \nRightarrow \Delta \Rrightarrow \Gamma' \nRightarrow \Delta'$ is locally ts-valid in V^{SK} if and only if its translation is locally tt-valid in V^{SK}.

Proof By Fact 4.3.8 we have that $\Gamma \nRightarrow \Delta \Rrightarrow \Gamma' \nRightarrow \Delta'$ is locally ts-valid in V^{SK} if and only if $\Gamma' \Rightarrow \Delta' \Rrightarrow \Gamma \Rightarrow \Delta$ is locally ts-valid in V^{SK}. By Theorem 4.3.1, we have that $\Gamma' \Rightarrow \Delta' \Rrightarrow \Gamma \Rightarrow \Delta$ is locally ts-valid in V^{SK} if and only if $\tau(\Gamma' \Rightarrow \Delta' \Rrightarrow \Gamma \Rightarrow \Delta)$, i.e. $\bigwedge \Gamma' \supset \bigvee \Delta' \Rightarrow \bigwedge \Gamma \supset \bigvee \Delta$, is ss-valid in V^{SK}. Since $\mathbb{K}3$ is negation dual to $\mathbb{LP}$, $\bigwedge \Gamma' \supset \bigvee \Delta' \Rightarrow \bigwedge \Gamma \supset \bigvee \Delta$, is ss-valid in V^{SK} if and only if $\neg(\bigwedge \Gamma \supset \bigvee \Delta) \Rightarrow \neg(\bigwedge \Gamma' \supset \bigvee \Delta')$, i.e. $\bigwedge \Gamma \wedge \neg \bigvee \Delta \Rightarrow \bigwedge \Gamma' \wedge \neg \bigvee \Delta'$, is tt-valid in V^{SK}. □

Given this and the tight relation between $\mathbb{ST}$ and $\mathbb{LP}$ showed in Chap. 3, it is, once more, very natural to expect $\mathbb{TS}$ and $\mathbb{ST}$ to be dual one to the other. However, the notion of negation duality doesn't provide us with a satisfactory account of this relation between $\mathbb{TS}$ and $\mathbb{ST}$, and in particular between Cut and Id. It is easy to check that, according to this definition, $\mathbb{TS}$ and $\mathbb{ST}$ are both self-negation duals. Let us check the case of $\mathbb{TS}$ (the one of $\mathbb{ST}$ is left to the reader in Exercise 67):

Fact 4.3.10 $\mathbb{TS}$ is self-negation dual.

Proof Obviously, since $\mathbb{TS}$ is empty neither an inference $\Gamma \Rightarrow \Delta$ nor its negation dual $\neg\Delta \Rightarrow \neg\Gamma$ are in it, thus the result readily follows. □

To overcome this undesirable result, Da Ré et al. (2020) introduce an alternative notion of negation duality that moves, as one might expect, the level of analysis

[8] Notice that this definition and the following result are restricted to the simpler case of single premise meta-anti-inferences, as in (Cobreros et al., 2020), but this is not essential and they can be easily extended to meta-anti-inferences with multiple premises (so something of the form $\Gamma_1 \nRightarrow \Delta_1, \ldots \Gamma_n \nRightarrow \Delta_n \Rrightarrow \Gamma' \nRightarrow \Delta'$).

from the inference level to that of metainferences. To do so, they define a notion of negation that applies to inferences themselves. This comes with a peculiar understanding of inferences, that can now be seen as triples of two multisets of formulas and a polarity. Just as formulas, they consider inference tokens to be either positive or negative: the former are denoted as $\Gamma \Rightarrow^{+} \Delta$; the latter are the (involutive) 'negation'[9] of a positive inference and are denoted as $\Gamma \Rightarrow^{-} \Delta$. The satisfaction conditions for positive inferences are the same as those for regular inferences, while in the case of a negative inference, we have that a valuation satisfies it if and only if it satisfies the premises at the right of the sequent and the conclusions at the left. More formally, this is equivalent to the following:

Definition 4.3.11 A valuation v xy-satisfies a *negative inference* $\Gamma \Rightarrow^{-} \Delta$ if and only if, for every $\gamma \in \Gamma$ and $\delta \in \Delta$, v xy-satisfies $\Rightarrow^{+} \gamma$ and $\delta \Rightarrow^{+}$.

As usual, a negative inference schema is valid only if, for all valuations, all its negative inference tokens are satisfied.

Notice that in a logical consequence with the same standard for premises and conclusions, as those characterizing $\mathbb{CL}$, $\mathbb{K3}$ and $\mathbb{LP}$, the condition expressed by Definition 4.3.11 is the same as demanding v to be a countermodel to the positive inference. This implies that the satisfaction conditions for negative inferences and anti-inferences, used earlier for Facts 4.3.6 and 4.3.9, collapse in these cases. This, however, is not the case for consequence relations with different standards for premises and conclusions, as in the case of $\mathbb{ST}$ and $\mathbb{TS}$, and it can be seen just applying the definition of negative inference. In the case of $\mathbb{ST}$, by Definition 4.3.11, we have that a valuation $v \in V^{SK}$ st-satisfies $\Gamma \Rightarrow^{-} \Delta$ iff $v(\gamma) \in \{1, 1/2\}$ for all $\gamma \in \Gamma$ and $v(\delta) \in \{1/2, 0\}$ for all $\delta \in \Delta$, whereas by Definition 4.3.5 it st-satisfies $\Gamma \not\Rightarrow \Delta$ iff $v(\gamma) = 1$ for all $\gamma \in \Gamma$ and $v(\delta) = 0$ for all $\delta \in \Delta$. Similarly for $\mathbb{TS}$, a valuation $v \in V^{SK}$ ts-satisfies $\Gamma \Rightarrow^{-} \Delta$ iff $v(\gamma) = 1$ for all $\gamma \in \Gamma$ and $v(\delta) = 0$ for all $\delta \in \Delta$, whereas it ts- satisfies $\Gamma \not\Rightarrow \Delta$ iff $v(\gamma) \in \{1, 1/2\}$ for all $\gamma \in \Gamma$ and $v(\delta) \in \{1/2, 0\}$ for all $\delta \in \Delta$.

Notice that the definition of satisfaction of negative inferences makes $\mathbb{ST}$ and $\mathbb{TS}$ behave in a sort of 'paraconsistent' and 'paracomplete' manner, respectively (in a way, this is not surprising given the paraconsistency of $\mathbb{LP}$ and the paracompleteness of $\mathbb{K3}$ and the relation that holds between these and $\mathbb{ST}$ and $\mathbb{TS}$, respectively). By this we mean that in V^{SK} a valuation could st-satisfy both a positive inference and its negative counterpart, while a valuation may not ts-satisfy neither a positive inference nor its negative counterpart (find such a valuation in Exercise 68).[10]

[9] This kind of negation is not part of the language, hence the scare quotes.

[10] The paraconsistent and paracomplete character of negative inferences in $\mathbb{ST}$ and $\mathbb{TS}$, respectively, makes the difference between negative inferences and anti-inferences even clearer: contrary to negative inferences, anti-inferences are satisfied if and only if the corresponding (positive) inference is not.

Having a notion of negation that works for inferences, negation duality is now moved to the level of metainferences. To do so, we also need to consider a framework that not only allows for multiple premises in metainferences but also for multiple conclusions (in fact, even negation duality requires this symmetry between premises and conclusions).

Let us then consider the following definition of metainference:

Definition 4.3.12 A metainference token is a pair of sets of inference tokens (that can be positive or negative) appearing as premises and conclusions,

$$\frac{\Gamma_1 \Rightarrow^{+/-} \Delta_1 \quad \ldots \quad \Gamma_n \Rightarrow^{+/-} \Delta_n}{\Sigma_1 \Rightarrow^{+/-} \Pi_1 \quad \ldots \quad \Sigma_m \Rightarrow^{+/-} \Pi_m}$$

A valuation v xy-satisfies a metainference token if and only if either it does not xy-satisfy one of the premises or it xy-satisfies one of the conclusions.[11]

Hence a metainference token is locally xy-valid iff all valuations xy-satisfy it, and a metainference schema is xy-valid iff all its instances are xy-valid.

We can now define the metainferential dual of a metainference:

Definition 4.3.13 The *metainferential dual* of the metainference token

$$\frac{\Gamma_1 \Rightarrow^{+/-} \Delta_1 \quad \ldots \quad \Gamma_n \Rightarrow^{+/-} \Delta_n}{\Sigma_1 \Rightarrow^{+/-} \Pi_1 \quad \ldots \quad \Sigma_m \Rightarrow^{+/-} \Pi_m}$$

is the metainference token

$$\frac{\Sigma_1 \Rightarrow^{-/+} \Pi_1 \quad \ldots \quad \Sigma_m \Rightarrow^{-/+} \Pi_m}{\Gamma_1 \Rightarrow^{-/+} \Delta_1 \quad \ldots \quad \Gamma_n \Rightarrow^{-/+} \Delta_n}$$

This definition extends to schemas as usual. As expected, this definition is analogous to negation duality at the level of metainferences: we invert premises and conclusions (that this time are not formulas but inferences) and negate them (so a positive premise becomes a negative conclusion and vice versa, and a negative premise becomes a positive conclusion and vice versa).

Definition 4.3.14 A logic $\mathbb{L}$ is metainferential dual to a logic $\mathbb{L}'$ if and only if, if a metainference schema is in $\mathbb{L}$ then its metainferential dual is in $\mathbb{L}'$ and vice versa.

With this new notion of duality at hand, we can prove the following:

Fact 4.3.15 $\mathbb{TS}$ and $\mathbb{ST}$ are metainferentially dual.

Proof To prove this, it is sufficient to show that a positive inference $\Gamma \Rightarrow^{+} \Delta$ is not $(ts)st$-satisfied by a valuation v in V^{SK} iff v $(st)ts$-satisfies $\Gamma \Rightarrow^{-} \Delta$ in V^{SK}, and that it does not $(ts)st$-satisfy a negative inference $\Gamma \Rightarrow^{-} \Delta$ in V^{SK} iff it $(st)ts$-satisfies $\Gamma \Rightarrow^{+} \Delta$ in V^{SK}. This is left as an exercise to the reader (Exercise 69).

[11] Notice that just as in the case of multiple-conclusion inferences, in metainferences with multiple conclusions these are read disjunctively.

This is enough to prove the metainferential duality between $\mathbb{ST}$ and $\mathbb{TS}$ for the following reason: a metainference token is in $\mathbb{ST}$ ($\mathbb{TS}$) if and only if for every valuation $v \in V^{SK}$, either it does not $(ts)st$-satisfy one of the positive or negative inferences appearing as premises, or it $(ts)st$-satisfies one of the positive or negative inferences appearing as conclusions. The metainferential dual takes premises to be conclusions and vice versa and also inverts the positive and negative of an inference, hence a metainference token is $ts(st)$-satisfied by some valuation in V^{SK} if and only if its metainferential dual is $st(ts)$-satisfied. □

We can now say something about the duality of Cut and Id. As in the case of Ex Falso and the Law of Excluded Middle, they are not exactly the (metainferential) dual of one another, but they are so *modulo logical equivalence* (see Footnote 12 for an explanation of this). We can see this with an example:

Consider this instance of Cut with empty conclusions (that is in $\mathbb{TS}$):

$$\text{(i)}\ \frac{\Rightarrow^{+} p \quad p \Rightarrow^{+}}{}$$

and its metainferential dual:

$$\text{(ii)}\ \frac{}{\Rightarrow^{-} p \quad p \Rightarrow^{-}}$$

(ii), which is in $\mathbb{ST}$, is not an instance of Id, but it is equivalent to this:[12]

$$\text{(iii)}\ \frac{}{p \Rightarrow^{+} p}$$

We can say the same about the metainferential dual of (iii), that is:

$$\text{(iv)}\ \frac{p \Rightarrow^{-} p}{}$$

(iv) is not strictly speaking (i), but it is logically equivalent to it.

Therefore, this new notion of duality between logics delivers the desired duality between $\mathbb{TS}$ and $\mathbb{ST}$, and, as expected, we find it by lifting our look from the inferential level to that of metainferences, just as we have to do to find out other interesting facts about these logics. In this way, we can also account for the duality (*modulo logical equivalence*) of Cut and Id.

Notice that under this definition $\mathbb{CL}$ is metainferentially self-dual, just as it was under the previous definition of negation duality. However, while metainferential duality is by itself an appropriate notion of duality for logics that are intuitively dual at the metainferential level, it cannot be seen exactly as an extension of the notion of negation duality. In fact, under this notion of duality $\mathbb{K}3$ and $\mathbb{LP}$ cease to be dual of one another.

Fact 4.3.16 $\mathbb{K}3$ and $\mathbb{LP}$ are not metainferentially dual.

[12] To see why (ii) and (iii) are equivalent in the same logic (here in particular in $\mathbb{ST}$), recall the definition of negative inference: a valuation v does not satisfy (ii) if and only if it does not satisfy $\Rightarrow^{+} p$ and it does not satisfy $p \Rightarrow^{+}$. In the same logic this valuation is also a counterexample to the instance of Id in (iii). The other direction of the equivalence is similar.

Proof Here we show a counterexample to the fact that for any metainference that in $\mathbb{K}3$, its dual is in $\mathbb{LP}$. A counterexample to the other direction of the duality (i.e. that for any metainference that is in $\mathbb{LP}$, its dual is in $\mathbb{K}3$) is left to the reader (Exercise 70).

Consider the metainference schema:

$$(a)\,\frac{\varphi \wedge \neg\varphi \Rightarrow^{-}}{\psi \wedge \neg\psi \Rightarrow^{+}}$$

Since all instances of $\psi \wedge \neg\psi \Rightarrow^{+}$ are always ss-satisfied in V^{SK}, this metainference schema is in $\mathbb{K}3$. Now consider the metainference dual of (a):

$$(\overline{a})\,\frac{\psi \wedge \neg\psi \Rightarrow^{-}}{\varphi \wedge \neg\varphi \Rightarrow^{+}}$$

and take an instance of $(\overline{a})$, e.g. the metainference token:

$$(\overline{a})\,\frac{p \wedge \neg p \Rightarrow^{-}}{q \wedge \neg q \Rightarrow^{+}}$$

Now consider a valuation v in V^{SK} assigning 1/2 to all variables. This valuation is such that it tt-satisfies $p \wedge \neg p \Rightarrow^{-}$ but it does not tt-satisfy $q \wedge \neg q \Rightarrow^{+}$. So $(\overline{a})$ is not in $\mathbb{LP}$.

□

An alternative notion to that of negation duality which accounts for the duality between $\mathbb{TS}$ and $\mathbb{ST}$ can be found in (Cobreros et al., 2021). To introduce it, we need the notion of *anti-validity*[13] (also called *inversevalidity* in (Fiore et al., 2022)). While validity accounts for inferences understood as pairs of (multi)sets of sentences where the *acceptance* of one is supposed to ground the *acceptance* of the other, anti-validity captures the idea that the *rejection* of the first set is supposed to ground the *rejection* of the second one.

Definition 4.3.17 An inference token $\Gamma \Rightarrow \Delta$ is *xy-anti-valid* in a logic $\mathbb{L}$ with standard x for the premises and y for the conclusions, i.e. $=\!\mid_{\mathbb{L}} \Gamma \Rightarrow \Delta$, if and only if for all valuations v, if $v(\gamma) \notin x$ for all $\gamma \in \Gamma$ then $v(\delta) \notin y$ for some $\delta \in \Delta$.

An inference schema is xy-anti-valid iff all its instances are xy-anti-valid.[14]

[13] This terminology partly overlaps with the term 'antitautology' that we used above but they should not be confused (see Definition 3.2.9). Notice that if the standards of satisfaction for validity in a logic are the same, then antitautologies are a limit case of both valid anti-inferences and anti-valid inferences, i.e. formulas on the right-hand-side of both when the left side can be empty—but this is not the case for logics with mixed standards.

[14] The notions of anti-validity of an inference and that of validity of an anti-inference are not to be conflated. Recall that an anti-inference $\gamma \not\Leftrightarrow \delta$ is xy-valid iff for any valuation v, $v(\gamma) \in x$ *AND* $v(\delta) \notin y$, while an inference $\gamma \Rightarrow \delta$ is xy-antivalid, *i.e.* $=\!\mid_{xy} \gamma \Rightarrow \delta$, iff for any valuation v, $v(\gamma) \in x$ *OR* $v(\delta) \notin y$. Hence a valid anti-inference correlates with an anti-valid inference but not necessarily the converse.

With this new notion of anti-validity, we can then define when two logics are *structural duals*:

Definition 4.3.18 A logic $\mathbb{L}$ is *structural dual* to a logic $\mathbb{L}'$ if and only if, if an inference schema $\Gamma \Rightarrow \Delta$ is such that $\vDash_{\mathbb{L}} \Gamma \Rightarrow \Delta$, then $\dashv\!|_{\mathbb{L}'} \Delta \Rightarrow \Gamma$ and vice versa.

We can easily see that $\mathbb{CL}$, $\mathbb{LP}$ and $\mathbb{K3}$ are structural self-duals and that $\mathbb{TS}$ and $\mathbb{ST}$ are each other's structural dual:

Fact 4.3.19

$$\vDash_{\mathbb{CL}} \Gamma \Rightarrow \Delta \text{ iff } \dashv\!|_{\mathbb{CL}} \Delta \Rightarrow \Gamma$$

$$\vDash_{\mathbb{LP}} \Gamma \Rightarrow \Delta \text{ iff } \dashv\!|_{\mathbb{LP}} \Delta \Rightarrow \Gamma$$

$$\vDash_{\mathbb{K3}} \Gamma \Rightarrow \Delta \text{ iff } \dashv\!|_{\mathbb{K3}} \Delta \Rightarrow \Gamma$$

$$\vDash_{\mathbb{ST}} \Gamma \Rightarrow \Delta \text{ iff } \dashv\!|_{\mathbb{TS}} \Delta \Rightarrow \Gamma$$

$$\vDash_{\mathbb{TS}} \Gamma \Rightarrow \Delta \text{ iff } \dashv\!|_{\mathbb{ST}} \Delta \Rightarrow \Gamma$$

So the notion of structural duality, remaining at the level of inferences, offers an alternative framing of the duality between $\mathbb{ST}$ and $\mathbb{TS}$, but falls short again to account for the case of $\mathbb{LP}$ and $\mathbb{K3}$.

Instead, this is achieved by Blomet & Égré (2024), who put forth another notion of duality which is the result of combining a refined version of negation duality with structural duality (by Cobreros et al. (2021)), that simultaneously counts $\mathbb{LP}$ and $\mathbb{K3}$ as duals as well as $\mathbb{ST}$ and $\mathbb{TS}$.

First, the 'refined' notion of negation duality, which can be found in (Schröder, 2017) and goes back to (Kleene, 1952), can be defined as follows:

Definition 4.3.20 Let $\sim : \mathcal{L} \to \mathcal{L}$ be a mapping such that:

- $\sim p = p$, if $p \in Var$,
- $\sim\top = \bot$
- $\sim\bot = \top$
- $\sim\neg\varphi = \neg\sim\varphi$
- $\sim(\varphi \wedge \psi) = \sim\varphi \vee \sim\psi$
- $\sim(\varphi \vee \psi) = \sim\varphi \wedge \sim\psi$

and let $\sim\Sigma = \{\sim\sigma : \sigma \in \Sigma\}$.
The *refined-negation dual* of an inference schema $\Gamma \Rightarrow \Delta$, is the inference $\sim\Delta \Rightarrow \sim\Gamma$.

So, while premises and conclusion are reversed, as in the standard case, the kind of negation used here is simply a function that exchanges disjunction and conjunction (as well as $\top$ and $\bot$).

Definition 4.3.21 A logic $\mathbb{L}$ is refined-negation dual to a logic $\mathbb{L}'$ when, if an inference schema $\Gamma \Rightarrow \Delta$ is such that $\vDash_{\mathbb{L}} \Gamma \Rightarrow \Delta$, then $\vDash_{\mathbb{L}'} \sim\Delta \Rightarrow \sim\Gamma$ and vice-versa.

Blomet & Égré (2024) prove that the refined notion of negation duality once again delivers a duality between $\mathbb{K}3$ and $\mathbb{LP}$, and self-duality in the case of $\mathbb{ST}$ and $\mathbb{TS}$:

Fact 4.3.22

$$\vDash_{\mathbb{K}3} \Gamma \Rightarrow \Delta \text{ iff } \vDash_{\mathbb{LP}} \sim\Delta \Rightarrow \sim\Gamma \text{ and } \vDash_{\mathbb{LP}} \Gamma \Rightarrow \Delta \text{ iff } \vDash_{\mathbb{K}3} \sim\Delta \Rightarrow \sim\Gamma$$

$$\vDash_{\mathbb{ST}} \Gamma \Rightarrow \Delta \text{ iff } \vDash_{\mathbb{ST}} \sim\Delta \Rightarrow \sim\Gamma \text{ and } \vDash_{\mathbb{TS}} \Gamma \Rightarrow \Delta \text{ iff } \vDash_{\mathbb{TS}} \sim\Delta \Rightarrow \sim\Gamma$$

As Blomet & Égré (2024) notice, the key difference between the use of $\neg$ and $\sim$ in the definition of our dualities lies in the fact that, while the former does not allow for a straightforward extensional interdefinability (of the respective sets of valid inferences), i.e. a one-to-one correspondence, between $\mathbb{K}3$ and $\mathbb{LP}$, the latter does. To show this, it is enough to notice that not all $\mathbb{K}3$ or $\mathbb{LP}$ valid inferences are of the form $\neg\Gamma \Rightarrow \neg\Delta$. More precisely, we cannot identify the set of valid inferences in $\mathbb{K}3$ with the set of their negation duals that are valid in $\mathbb{LP}$, and vice-versa the set of valid inferences in $\mathbb{LP}$ is not equal to the set of their negation duals that are valid in $\mathbb{K}3$. To see why this is the case, recall what we already said about the duality between Law of Excluded Middle and Ex Falso which only holds *modulo logical equivalence*. Instead, as Blomet & Égré (2024) show, working with the refined notion of negation duality obtained with $\sim$, allows us to prove the extensional interdefinability between $\mathbb{K}3$ and $\mathbb{LP}$. We will not further discuss this, and we will keep focusing on the duality results.

What is now interesting to us is the fact that combining the refined notion of negation duality with that of structural duality that we have already introduced allows us to hold together all the expected (inferential) dualities. In fact, notice that using the previous facts we can show that the following holds (which corresponds to Theorem 7.2 in (Blomet & Égré, 2024)):

Fact 4.3.23

$$\vDash_{\mathbb{K}3} \Gamma \Rightarrow \Delta \text{ iff } \dashv\!\mid_{\mathbb{LP}} \sim\Gamma \Rightarrow \sim\Delta$$

$$\vDash_{\mathbb{LP}} \Gamma \Rightarrow \Delta \text{ iff } \dashv\!\mid_{\mathbb{K}3} \sim\Gamma \Rightarrow \sim\Delta$$

$$\vDash_{\mathbb{ST}} \Gamma \Rightarrow \Delta \text{ iff } \dashv\!\mid_{\mathbb{TS}} \sim\Gamma \Rightarrow \sim\Delta$$

$$\vDash_{\mathbb{TS}} \Gamma \Rightarrow \Delta \text{ iff } \dashv\!\mid_{\mathbb{ST}} \sim\Gamma \Rightarrow \sim\Delta$$

We only prove the first and third biconditionals, leaving the others as an exercise to the reader (Exercise 71).

Proof For the first biconditional: by Fact 4.3.22 $\vDash_{\mathbb{K}3} \Gamma \Rightarrow \Delta$ iff $\vDash_{\mathbb{LP}} {\sim}\Delta \Rightarrow {\sim}\Gamma$ iff (by Fact 4.3.19) $\dashv_{\mathbb{LP}} {\sim}\Gamma \Rightarrow {\sim}\Delta$.

For the third biconditional: by Fact 4.3.22 $\vDash_{\mathbb{ST}} \Gamma \Rightarrow \Delta$ iff $\vDash_{\mathbb{ST}} {\sim}\Delta \Rightarrow {\sim}\Gamma$ iff (by Fact 4.3.19) $\dashv_{\mathbb{TS}} {\sim}\Gamma \Rightarrow {\sim}\Delta$. □

We might define the notion of duality that emerges from these results as a refined-negation-structural duality:

Definition 4.3.24 A logic $\mathbb{L}$ is *refined-negation-structural-dual* to a logic $\mathbb{L}'$ if and only if, if an inference schema $\Gamma \Rightarrow \Delta$ is such that $\vDash_{\mathbb{L}} \Gamma \Rightarrow \Delta$, then $\dashv_{\mathbb{L}'} {\sim}\Gamma \Rightarrow {\sim}\Delta$ and vice-versa.

By Fact 4.3.23 and under this new notion of duality, we have that $\mathbb{K}3$ and $\mathbb{LP}$ are each other's dual and also that $\mathbb{ST}$ and $\mathbb{TS}$ are duals.

Notice that we can also prove a similar result replacing the function $\sim$ by $\neg$ and this time using the fact that $\mathbb{LP}$ and $\mathbb{K}3$ are negation duals according to Definition 4.3.4, instead of using Fact 4.3.22.

Fact 4.3.25

$$\vDash_{\mathbb{K}3} \Gamma \Rightarrow \Delta \text{ iff } \dashv_{\mathbb{LP}} \neg\Gamma \Rightarrow \neg\Delta$$

$$\vDash_{\mathbb{LP}} \Gamma \Rightarrow \Delta \text{ iff } \dashv_{\mathbb{K}3} \neg\Gamma \Rightarrow \neg\Delta$$

$$\vDash_{\mathbb{ST}} \Gamma \Rightarrow \Delta \text{ iff } \dashv_{\mathbb{TS}} \neg\Gamma \Rightarrow \neg\Delta$$

$$\vDash_{\mathbb{TS}} \Gamma \Rightarrow \Delta \text{ iff } \dashv_{\mathbb{ST}} \neg\Gamma \Rightarrow \neg\Delta$$

Hence, a notion of negation-structural duality in the sense that is expressed by Fact 4.3.25 is enough to account uniformly for the dualities between $\mathbb{LP}$ and $\mathbb{K}3$ and between $\mathbb{ST}$ and $\mathbb{TS}$, just as its refined version does.

However, these results only apply to inferences and, therefore, cannot deliver a notion of duality that also links Cut and Id as in (Da Ré et al., 2020).

Exercises

Exercise 67 Prove that $\mathbb{CL}$ and $\mathbb{ST}$ are self-negation duals and that $\mathbb{LP}$ and $\mathbb{K}3$ are the negation dual of one another.

Exercise 68 Find an example of a valuation in V^{SK} such that: it st-satisfies both a positive inference and its negative counterpart; it ts-satisfies neither a positive inference nor its negative.

Exercise 69 Prove that a positive inference $\Gamma \Rightarrow^{+} \Delta$ is not $st(ts)$-satisfied by a valuation v in V^{SK} iff it $ts(st)$-satisfies $\Gamma \Rightarrow^{-} \Delta$, and that it does not $st(ts)$-satisfy a negative inference $\Gamma \Rightarrow^{-} \Delta$ iff it $ts(ts)$-satisfies $\Gamma \Rightarrow^{+} \Delta$.

Exercise 70 Find a metainference that is in $\mathbb{LP}$ such that its metainferential dual is not in $\mathbb{K}3$.

Exercise 71 Prove that $\vDash_{\mathbb{LP}} \Gamma \Rightarrow \Delta$ iff $\dashv_{\mathbb{K3}} {\sim}\Gamma \Rightarrow {\sim}\Delta$ and that $\vDash_{\mathbb{TS}} \Gamma \Rightarrow \Delta$ iff $\dashv_{\mathbb{ST}} {\sim}\Gamma \Rightarrow {\sim}\Delta$.

4.3.3 Metainferential Mixed Logics

When considering metainferential validity in previous sections, we worked with two notions of preservation: local xy-validity involves the preservation of xy-satisfaction, and global xy-validity involves the preservation of xy-validity. But while in the first chapter we used the same standard for premises and conclusions, non-transitive and non-reflexive logics introduced the use of different standards at their inferential level; nothing prevents us then from using different standards for metainferential premises and conclusions as well. In doing so we can redefine local and global validity:

Definition 4.3.26 Let $x, y, z, w \in \{s, t\}$, and $V \subseteq V^3$. A metainference $S \Rrightarrow r$ is *globally* xy/zw*-valid in* V just in case some $s \in S$ is not xy-valid in V or r is zw-valid in V.

Definition 4.3.27 Let $x, y, z, w \in \{s, t\}$, and $V \subseteq V^3$. A metainference $S \Rrightarrow r$ is *locally* xy/zw*-valid in* V just in case, for every valuation $v \in V$, v does not xy-satisfy some $s \in S$ or it zw-satisfies r.

Thus, up to here, we were only working with notions of metainferential validity of the kind xy/xy. At the inferential level we had that satisfaction for premises had to meet a stricter standard than that for the conclusion or vice-versa, so how can this same strategy be lifted to the level of metainferences? Notice that, just as s is arguably a stronger standard than t since it is more restrictive, i.e. fewer valuations assign it, ts is a stronger standard than ss, tt and, *a fortiori*, st, since less inferences can meet it. So, for instance, the standard ts/st for metainferences is analogous to the choice of an st standard for inferences, as it moves from a stronger standard to a weaker one. Following the previous strategy, we can define mixed consequence relations xy/zw for metainferences with standards with different strengths, such as st/ss or ts/st.

In what follows, we present some results of working with mixed metainferential validity, applied to the logics $\mathbb{ST}$, $\mathbb{LP}$, $\mathbb{K3}$ and $\mathbb{TS}$ as in previous sections. We refer to the notions of metainferential validity with their names forming pairs. These results are tied to the local validity notion for metainferences.

Let us start with $\mathbb{TS}/\mathbb{ST}$, presented in (Pailos, 2019). In this context, $\mathbb{TS}$ introduces the strict element for the premises while $\mathbb{ST}$ brings with it the tolerant one. $\mathbb{TS}/\mathbb{ST}$ shares its valid metainferences with Classical Logic.[15] What is interesting here is although that neither $\mathbb{ST}$ nor $\mathbb{TS}$ are structural, but structural properties hold

[15] The proof can be found in (Pailos, 2019).

for $\mathbb{TS}/\mathbb{ST}$, just as in Classical Logic. However, $\mathbb{TS}/\mathbb{ST}$ can be safely extended with some vocabulary that could lead to triviality in Classical Logic.[16]

For the case of $\mathbb{ST}/\mathbb{TS}$, the empty-feature that $\mathbb{TS}$ presents at the inferential level is also present at the metainferential level of $\mathbb{ST}/\mathbb{TS}$: there are no metainferences in $\mathbb{ST}/\mathbb{TS}$. This comes from the more demanding role that the standard for conclusions has over the standard for premises, as was the case for the evaluation of metainferences in $\mathbb{TS}$. This makes $\mathbb{ST}/\mathbb{TS}$ substructural, not only due to the failure of Identity, as in $\mathbb{TS}$, but also due to the failure of every other structural property.

$\mathbb{LP}$ and $\mathbb{K}3$, in opposition to $\mathbb{ST}$ and $\mathbb{TS}$, are structural logics, i.e. every structural property holds when we consider metainferential validity as preservation of inferential standards—the logics $\mathbb{LP}/\mathbb{LP}$ and $\mathbb{K}3/\mathbb{K}3$. But if we drop this, we can define $\mathbb{LP}/\mathbb{K}3$ and $\mathbb{K}3/\mathbb{LP}$, both with mixed impure metainferential standards, though pure inferential standards (since $\mathbb{LP}$ builds on the tt-validity notion and $\mathbb{K}3$ does it on the notion of ss-validity). However, Cut, Weakening and Contraction do not hold neither in $\mathbb{LP}/\mathbb{K}3$ nor in $\mathbb{K}3/\mathbb{LP}$. They are not empty, since, for example, every instance of Identity holds. This shows a way of obtaining substructural logics from some usual structural logics.

When combining the pure and impure inferential standards, we obtain both structural and substructural logics. For the structural ones, it is enough to have ts as the standard for premises or st as the standard for conclusions. As a special case, $\mathbb{TS}/\mathbb{ST}$ satisfies both conditions. Though they are all structural, only $\mathbb{TS}/\mathbb{ST}$ coincides with Classical Logic, since in the other four logics that we could build with those conditions some classical inferences fail. The substructural ones share the above structural failures and can be obtained by having st as the standard for premises or ts as the standard for conclusions. $\mathbb{ST}/\mathbb{TS}$ satisfies both conditions, though it is even weaker than any of the other substructural logics since it is fully empty.

The proposal of mixed metainferential logics allows for a different approach to logic. We have defined metainferences as relations between inferences, now this can be pushed by defining higher-order relations, for instance between metainferences. But this would be redundant if we work only with pure metainferential validity, with pure meta-metainferential validity, and so on. With the inclusion of impure metainferential validity, every level can be mixed in one way or another, allowing for different hierarchical systems to be defined, with non-obvious structural properties. Delving into this take us apart from the contents of this book, but one result is worth mentioning: one of these hierarchies[17] follows the spirit of $\mathbb{TS}/\mathbb{ST}$ by having a strict standard for the corresponding premises and a tolerant one for the conclusions, which results in a system that contains a logic that matches Classical Logic at

[16] This feature is used as an argument in favour of $\mathbb{TS}/\mathbb{ST}$ and other logics above $\mathbb{ST}$ in (Barrio et al., 2021b).

[17] $\mathbb{ST}^{\omega}$ as defined by Barrio et al. (2020).

every metainferential level[18] and that can also be extended with a truth predicate without the usual problems. The concept of hierarchies gave rise to some debates that still continue, including but not restricted to (Ripley, 2022), (McAllister, 2022), and (Porter, 2022) in (Barrio and Egré, 2023), or (Pailos, 2020), (Scambler, 2020a, 2020b), and (Barrio et al., 2021a).

4.4 Non-reflexive Counterparts of any Tarskian Logic

The previous chapter summarized the strategy used by Szmuc (2024) to obtain a non-transitive counterpart $\mathbb{L}^{nt}$ of any Tarskian logic $\mathbb{L}$, via infectious extensions of an algebra. More precisely, the so obtained counterparts are not only non-transitive, but they share all valid inferences of their Tarskian sibling.

The motivating question of this result was whether, just as we can find a logic which cannot be taken to be $\mathbb{CL}$ even though it validates the same inferences, the same phenomenon can arise for any Tarskian logic. The answer was positive and the results showed that $\mathbb{ST}$ is just one of the non-transitive counterparts of $\mathbb{CL}$ with this feature, as $\mathbb{wST}$ is so as well.

In (Borzi & Zirattu, n.d.) the same question is posed, dually, for the non-reflexive case considered in the present chapter. Just as the non-reflexive $\mathbb{TS}$ shares the same valid anti-inferences as $\mathbb{CL}$ (recall Fact 4.3.6) but differs from it regarding its inferences, is there some analogous phenomenon for any Tarskian logic? Although this issue might seem less natural or compelling than that concerning different logics sharing the same valid *inferences*—rather than *anti*-inferences—there are good reasons to think that it is interesting on its own, and not just as a dualized version. The idea that a logic is not only a set of valid inferences (or meta-inferences) is discussed in (Scambler, 2020a) and used in some works in the literature (e.g. (Cobreros et al., 2020)).

The strategy to construct the non-reflexive counterpart, call it $\mathbb{L}^{nr}$, of a Tarskian logic $\mathbb{L}$, will be dual to that in (Szmuc, 2024). We can build on the definitions and preliminary notions already introduced in the previous chapter and only add what we need for our case.

The new notion that we need is that of logical q-matrices, due to Malinowski (1990a, 1990b), which can be understood as structures that relate to logics where reflexivity might be invalid:

[18] Although with the caveat of needing potentially infinite levels, that is, being at least the first limit ordinal of its system. Pailos takes the first step from finitary to infinitary systems of hierarchies in (Pailos, 2020) with the logic $\mathbf{CM}_\omega$, conceived as the union of every logic in the system up to level n. However, it was observed by Scambler (2020b) that the failure of classicality extends from the arbitrarily finite to any transfinite generalization. McAllister (2022) then showed that both at the finite or transfinite level, $\mathbb{CL}$ as $\mathbb{ST}$ cannot be uniquely characterized within the classical set theory and that alternatives like a paraconsistent set theory seem to recover the uniqueness and characterizability of $\mathbb{CL}$ only to produce highly non-classical meta-results.

Definition 4.4.1 A *logical q-matrix* $\mathfrak{M}$ for a language $\mathcal{L}$ is a structure $\langle \mathcal{A}, D^p, D^c \rangle$ where $\mathcal{A}$ is an algebra whose universe is A, and D^p, D^c are non-empty subsets of A, such that $D^c \subseteq D^p$.

It is easy to check that q-matrices give us a generalization of the regular logical matrices described in Definition 3.5.2: the latter can be represented as an instance of the former where $D^p = D^c$. And just as a logical matrix for $\mathcal{L}$ brings about a Tarskian consequence relation, a logical q-matrix $\mathfrak{M}$ characterizes validity of inferences in a q-consequence relation $\vDash_{\mathfrak{M}}$ as follows:

$$\vDash_{\mathfrak{M}} \Gamma \Rightarrow \varphi \iff \text{For all } v_{\mathfrak{M}}\text{: if } v_{\mathfrak{M}}(\gamma) \in D_p \text{ for every } \gamma \in \Gamma, \text{ then } v_{\mathfrak{M}}(\varphi) \in D_c$$

When we have a class $\mathbf{\mathfrak{M}}$ of logical q-matrices for a language $\mathcal{L}$, the q-consequence relation $\vDash_{\mathbf{\mathfrak{M}}}$ is understood as $\bigcap\{\vDash_{\mathfrak{M}} \mid \mathfrak{M} \in \mathbf{\mathfrak{M}}\}$. In the case of logics characterized by a q-consequence relation (also termed q-logics), both Transitivity and Monotonicity hold, as in Tarskian consequence relations, but this isn't necessarily the case for Reflexivity.

Notice that this way of characterizing logical consequence with ($q-$)matrices, only captures the notion of (in)validity of regular inferences in a logic, which is not exactly the same as characterizing the notion of valid anti-inferences. In fact, recalling Definition 4.3.5, we have that a logical (q-)matrix $\mathfrak{M}$ for $\mathcal{L}$ characterizes validity of anti-inferences in a (q-)consequence relation as follows (which extends to classes of (q-) matrices in the expected way):

$$\vDash_{\mathfrak{M}} \Gamma \not\Rightarrow \varphi \iff \text{For all } v_{\mathfrak{M}} : v_{\mathfrak{M}}(\gamma) \in D(D_p)$$
$$\text{for every } \gamma \in \Gamma, \text{ and } v_{\mathfrak{M}}(\varphi) \notin D(D_c)$$

For this reason, in contrast with Sect. 3.5 we cannot appeal to Wójcicki's theorem and show that *any* Tarskian logic has a non-reflexive counterpart, but we can only prove that there is such counterpart for any Tarskian logic whose validity of anti-inferences is captured by a class of (standard) logical matrices, namely such that $\Gamma \not\Rightarrow \varphi \in \mathbb{L}$ if and only if $\vDash_{\mathfrak{M}} \Gamma \not\Rightarrow \varphi$. With this restriction in mind, the following can be proved (Borzi & Zirattu, n.d.):

Theorem 4.4.2 *For every consequence relation* $\vDash_{\mathfrak{M}}$ *characterized by the logical matrix* $\mathfrak{M} = \langle \mathcal{A}, D \rangle$, *there is a corresponding logical q-matrix* $\mathfrak{M}[e] = \langle \mathcal{A}[e], D^p \cup \{e\}, D^c \rangle$ *such that* $\vDash_{\mathfrak{M}[e]}$ *has the same valid anti-inferences as* $\vDash_{\mathfrak{M}}$.

Proof We demonstrate that $\vDash_{\mathfrak{M}[e]} \Gamma \not\Rightarrow \varphi$ implies $\vDash_{\mathfrak{M}} \Gamma \not\Rightarrow \varphi$ by contraposition. Assume that $\nvDash_{\mathfrak{M}} \Gamma \not\Rightarrow \varphi$. Thus, there's an $\mathfrak{M}$-valuation v such that $v \vDash_{\mathfrak{M}} \Gamma \Rightarrow \varphi$, that is: $v(\gamma) \notin D^p$ for some $\gamma \in \Gamma$ or $v(\varphi) \in D^c$. But it is straightforward to notice that every $\mathfrak{M}$-valuation is also a $\mathfrak{M}[e]$-valuation (since the former is a sub-matrix of the latter), and so v is such that $v \vDash_{\mathfrak{M}[e]} \Gamma \Rightarrow \varphi$. Therefore, $\nvDash_{\mathfrak{M}[e]} \Gamma \not\Rightarrow \varphi$.

To prove that $\vDash_{\mathfrak{M}} \Gamma \not\Rightarrow \varphi$ implies $\vDash_{\mathfrak{M}[e]} \Gamma \not\Rightarrow \varphi$ we proceed in the same manner. Suppose that $\nvDash_{\mathfrak{M}[e]} \Gamma \not\Rightarrow \varphi$. That means there is an $\mathfrak{M}[e]$-valuation such that $v \vDash_{\mathfrak{M}[e]} \Gamma \Rightarrow \varphi$, that is: $v(\gamma) \notin D^p \cup \{e\}$ for some $\gamma \in \Gamma$ or $v(\varphi) \in D^c$. Furthermore, by the infectious character of e (Fact 3.5.8), we can infer that for all propositional variables p appearing in $\Gamma \cup \{\varphi\}$, $v(p) \neq e$. Thus, v coincides with an $\mathfrak{M}$-valuation v' on the set of sentences $\Gamma \cup \{\varphi\}$. Hence, $\nvDash_{\mathfrak{M}} \Gamma \not\Rightarrow \varphi$. □

The same result easily extends to logics characterized by a collection of matrices: in that case, there is a collection of logical q-matrices, $\mathfrak{M}$, where for each $\mathfrak{M}_i \in \mathfrak{M}$, we have that $\mathfrak{M}_i[e_i]$ shares the same valid anti-inferences.

Corollary 4.4.3 *For every consequence relation* $\vDash_{\mathfrak{M}}$ *characterized by a class* $\mathfrak{M}$ *of logical matrices* $\{\langle \mathcal{A}_1, D_1\rangle, \ldots, \langle \mathcal{A}_n, D_n\rangle\}$*, there is a corresponding class* $\mathfrak{M}[e]$ *of logical q-matrices* $\{\langle \mathcal{A}_1[e_1], D_1^p \cup \{e_1\}, D_1^c\rangle, \ldots, \langle \mathcal{A}_n[e_n], D_n^p \cup \{e_n\}, D_n^c\rangle\}$ *such that* $\vDash_{\mathfrak{M}[e]}$ *has the same valid anti-inferences as* $\vDash_{\mathfrak{M}}$.

From these, the main result of this section can be proved (Borzi & Zirattu, n.d.):

Theorem 4.4.4 *Every Tarskian logic* $\mathbb{L}$ *whose valid anti-inferences are characterized by a class of logical matrices* $\mathfrak{M}$ *has a non-reflexive counterpart* $\mathbb{L}^{nr}$ *such that* $\mathbb{L}^{nr}$ *has the same valid anti-inferences as* $\mathbb{L}$.

Proof Let $\mathbb{L}$ be any Tarskian logic whose valid anti-inferences are characterized by a class of logical matrices $\mathfrak{M}$. By Theorem 4.4.2 and Corollary 4.4.3, there is also a class of logical q-matrices $\mathfrak{M}[e]$ extending each $\mathfrak{M}_i \in \mathfrak{M}$ with an infectious element, without affecting the set of valid anti-inferences.

Consider now the axiom Id:

$$\text{Id}\ \frac{}{\varphi \Rightarrow \varphi}$$

There is an $\mathfrak{M}_i[e_i]$-valuation v such that $v(\varphi) = e_i$, and thus, $v(\varphi) \in D_i^p \cup \{e_i\}$ but $v(\varphi) \notin D_i^c$. Hence, v constitutes an $\mathfrak{M}_i[e_i]$-counterexample to the Identity axiom. From this, it follows that the collection $\mathfrak{M}$ of matrices provides a semantics for a non-reflexive counterpart $\mathbb{L}^{nr}$ of the Tarskian logic $\mathbb{L}$, in the sense that it leaves unaltered the set of valid anti-inferences and inconsistent formulas, but it is not reflexive because the axiom Id is not valid in it. □

Therefore, the answer to our initial question is also affirmative in this case. Similarly to what was stated for $\mathbb{ST}$, the results above demonstrate that not only $\mathbb{TS}$, but also w$\mathbb{TS}$, shares the same valid anti-inferences as $\mathbb{CL}$, despite differing from it due to its non-reflexivity.

4.5 Three-sided Sequent Calculi

Here we provide an alternative (proof-theoretic) characterization of some of the logical systems introduced in this and in the previous chapter, *via* the so-called three-sided sequent calculi. As the name indicates, they differ from the standard Gentzen-style sequents, which can also be called two-sided, in the number of 'sides' where a rule can be applied. In what follows, we will set forth three-sided calculi for the logics based on the strong and weak Kleene truth-tables.

Let us begin with a three-sided calculus that can be sound and complete with respect to either $\mathbb{ST}$, $\mathbb{TS}$, $\mathbb{LP}$ or $\mathbb{K3}$—depending on how one uses it to induce the logic. It can be found in (Ripley, 2012).

Definition 4.5.1 (Three-sided Sequent) A *three-sided sequent* is an expression of the form $\Gamma \mid \Sigma \mid \Delta$, where Γ, Σ, Δ are finite, possibly empty set of formulas.

Definition 4.5.2 A valuation v of V^{SK} *satisfies a three-sided sequent* $\Gamma \mid \Sigma \mid \Delta$ if and only if:

- For some $\gamma \in \Gamma$, $v(\gamma) = 0$, or
- For some $\sigma \in \Sigma$, $v(\sigma) = 1/2$, or
- For some $\delta \in \Delta$, $v(\delta) = 1$.

Intuitively, $\Gamma \mid \Sigma \mid \Delta$ can be given a disjunctive reading: in each three-sided sequent at least one of the formulas in Γ takes value 0, or at least one of the formulas in Σ takes value $1/2$, or at least one of the formulas in Δ takes value 1.

Definition 4.5.3 A three-sided sequent $\Gamma \mid \Sigma \mid \Delta$ is valid if and only if it is satisfied by every valuation v in V^{SK}. Conversely, we say that a valuation v of V^{SK} is a *counterexample* to a sequent if and only if it does not satisfy it.

Definition 4.5.4 Let Γ, Δ, Σ and Π be finite sets of formulas, φ and ψ be formulas of $\mathcal{L}$ of any complexity, and φ^{At} be any atomic formula of $\mathcal{L}$. The three-sided sequent calculus $3\mathcal{SK}$ for strong Kleene logics is defined by the following rules:

Structural rules

$$\frac{}{\varphi^{At} \mid \varphi^{At} \mid \varphi^{At}}\ \text{Id}$$

$$\frac{\Gamma \mid \Sigma \mid \Delta}{\Gamma, \Pi \mid \Sigma \mid \Delta}\ 0\text{W} \qquad \frac{\Gamma \mid \Sigma \mid \Delta}{\Gamma \mid \Sigma \mid \Delta, \Pi}\ 1\text{W} \qquad \frac{\Gamma \mid \Sigma \mid \Delta}{\Gamma \mid \Sigma, \Pi \mid \Delta}\ 1/2\text{W}$$

$$\frac{\Gamma, \varphi \mid \Sigma \mid \Delta \qquad \Gamma \mid \Sigma \mid \Delta, \varphi}{\Gamma \mid \Sigma \mid \Delta}\ \text{Cut}(0, 1) \qquad \frac{\Gamma, \varphi \mid \Sigma \mid \Delta \qquad \Gamma \mid \Sigma, \varphi \mid \Delta}{\Gamma \mid \Sigma \mid \Delta}\ \text{Cut}(0, 1/2)$$

$$\frac{\Gamma \mid \Sigma, \varphi \mid \Delta \qquad \Gamma \mid \Sigma \mid \Delta, \varphi}{\Gamma \mid \Sigma \mid \Delta}\ \text{Cut}(1/2, 1)$$

Operational rules

$$\frac{\Gamma \mid \Sigma \mid \Delta, \varphi}{\Gamma, \neg\varphi \mid \Sigma \mid \Delta}\,0\neg \qquad \frac{\Gamma, \varphi \mid \Sigma \mid \Delta}{\Gamma \mid \Sigma \mid \Delta, \neg\varphi}\,1\neg \qquad \frac{\Gamma \mid \Sigma, \varphi \mid \Delta}{\Gamma \mid \Sigma, \neg\varphi \mid \Delta}\,{}^{1}\!/\!_{2}\neg$$

$$\frac{\Gamma, \varphi, \psi \mid \Sigma \mid \Delta}{\Gamma, \varphi \wedge \psi \mid \Sigma \mid \Delta}\,0\wedge \qquad \frac{\Gamma \mid \Sigma \mid \Delta, \varphi \qquad \Gamma \mid \Sigma \mid \Delta, \psi}{\Gamma \mid \Sigma \mid \Delta, \varphi \wedge \psi}\,1\wedge$$

$$\frac{\Gamma \mid \Sigma, \varphi, \psi \mid \Delta \qquad \Gamma \mid \Sigma, \varphi \mid \Delta, \varphi \qquad \Gamma \mid \Sigma, \psi \mid \Delta, \psi}{\Gamma \mid \Sigma, \varphi \wedge \psi \mid \Delta}\,{}^{1}\!/\!_{2}\wedge$$

$$\frac{\Gamma, \varphi \mid \Sigma \mid \Delta \qquad \Gamma, \psi \mid \Sigma \mid \Delta}{\Gamma, \varphi \vee \psi \mid \Sigma \mid \Delta}\,0\vee \qquad \frac{\Gamma \mid \Sigma \mid \Delta, \varphi, \psi}{\Gamma \mid \Sigma \mid \Delta, \varphi \vee \psi}\,1\vee$$

$$\frac{\Gamma \mid \Sigma, \varphi, \psi \mid \Delta \qquad \Gamma, \varphi \mid \Sigma, \varphi \mid \Delta \qquad \Gamma, \psi \mid \Sigma, \psi \mid \Delta}{\Gamma \mid \Sigma, \varphi \vee \psi \mid \Delta}\,{}^{1}\!/\!_{2}\vee$$

Definition 4.5.5 Let Γ, Δ be sets of formulas. Then:

- $\Gamma \vdash_{\mathbb{ST}} \Delta$ if and only if the sequent $\Gamma \mid \Gamma, \Delta \mid \Delta$ is derivable in $3\mathcal{SK}$,
- $\Gamma \vdash_{\mathbb{TS}} \Delta$ if and only if the sequent $\Gamma \mid \varnothing \mid \Delta$ is derivable in $3\mathcal{SK}$,
- $\Gamma \vdash_{\mathbb{LP}} \Delta$ if and only if the sequent $\Gamma \mid \Delta \mid \Delta$ is derivable in $3\mathcal{SK}$,
- $\Gamma \vdash_{\mathbb{K}3} \Delta$ if and only if the sequent $\Gamma \mid \Gamma \mid \Delta$ is derivable in $3\mathcal{SK}$.

The Soundness and Completeness proofs of $3\mathcal{SK}$ can be found in (Ripley, 2012).

Moving on, we now present a three sided sequent calculus that is sound and complete with respect to $\mathbb{wST}$, $\mathbb{wTS}$, $\mathbb{PWK}$ and $\mathbb{K}_3^{\mathbb{w}}$. It can be found in the unpublished (Bonzio & Pra Baldi, n.d.), but it can also be obtained following the method presented by Avron and Konikowska (2005).

Definition 4.5.6 A valuation v of V^{WK} satisfies a three-sided sequent $\Gamma \mid \Sigma \mid \Delta$ if and only if:

- For some $\gamma \in \Gamma$, $v(\gamma) = 0$, or
- For some $\sigma \in \Sigma$, $v(\sigma) = 1/2$, or
- For some $\delta \in \Delta$, $v(\delta) = 1$.

Definition 4.5.7 A three-sided sequent $\Gamma \mid \Sigma \mid \Delta$ is valid if and only if it is satisfied by every valuation v in V^{WK}. Conversely, we say that a valuation v of V^{WK} is a counterexample to a sequent if and only if it does not satisfy it.

Definition 4.5.8 Let Γ, Δ, Σ and Π be finite sets of formulas, φ and ψ be formulas of $\mathcal{L}$ of any complexity, and φ^{At} be any atomic formula of $\mathcal{L}$. The three-sided sequent calculus $3\mathcal{WK}$ for weak Kleene logics is defined by the following rules:

Structural rules

$$\frac{}{\varphi^{At} \mid \varphi^{At} \mid \varphi^{At}}\ \text{Id}$$

$$\frac{\Gamma \mid \Sigma \mid \Delta}{\Gamma, \Pi \mid \Sigma \mid \Delta}\ 0\text{W} \qquad \frac{\Gamma \mid \Sigma \mid \Delta}{\Gamma \mid \Sigma, \Pi \mid \Delta}\ {}^{1}\!/_{2}\text{W} \qquad \frac{\Gamma \mid \Sigma \mid \Delta}{\Gamma \mid \Sigma \mid \Delta, \Pi}\ 1\text{W}$$

$$\frac{\Gamma, \varphi \mid \Sigma \mid \Delta \qquad \Gamma \mid \Sigma, \varphi \mid \Delta}{\Gamma \mid \Sigma \mid \Delta}\ \text{Cut}(0, {}^{1}\!/_{2}) \qquad \frac{\Gamma, \varphi \mid \Sigma \mid \Delta \qquad \Gamma \mid \Sigma \mid \Delta, \varphi}{\Gamma \mid \Sigma \mid \Delta}\ \text{Cut}(0, 1)$$

$$\frac{\Gamma \mid \Sigma, \varphi \mid \Delta \qquad \Gamma \mid \Sigma \mid \Delta, \varphi}{\Gamma \mid \Sigma \mid \Delta}\ \text{Cut}({}^{1}\!/_{2}, 1)$$

Operational rules

$$\frac{\Gamma \mid \Sigma \mid \Delta, \varphi}{\Gamma, \neg\varphi \mid \Sigma \mid \Delta}\ 0\neg \qquad \frac{\Gamma \mid \Sigma, \varphi \mid \Delta}{\Gamma \mid \Sigma, \neg\varphi \mid \Delta}\ {}^{1}\!/_{2}\neg \qquad \frac{\Gamma, \varphi \mid \Sigma \mid \Delta}{\Gamma \mid \Sigma \mid \Delta, \neg\varphi}\ 1\neg$$

$$\frac{\Gamma, \varphi, \psi \mid \Sigma \mid \Delta \qquad \Gamma, \varphi \mid \Sigma \mid \Delta, \varphi \qquad \Gamma, \psi \mid \Sigma \mid \Delta, \psi}{\Gamma, \varphi \wedge \psi \mid \Sigma \mid \Delta}\ 0\wedge$$

$$\frac{\Gamma \mid \Sigma, \varphi, \psi \mid \Delta}{\Gamma \mid \Sigma, \varphi \wedge \psi \mid \Delta}\ {}^{1}\!/_{2}\wedge \qquad \frac{\Gamma \mid \Sigma \mid \Delta, \varphi \qquad \Gamma \mid \Sigma \mid \Delta, \psi}{\Gamma \mid \Sigma \mid \Delta, \varphi \wedge \psi}\ 1\wedge$$

$$\frac{\Gamma, \varphi \mid \Sigma \mid \Delta \qquad \Gamma, \psi \mid \Sigma \mid \Delta}{\Gamma, \varphi \vee \psi \mid \Sigma \mid \Delta}\ 0\vee \qquad \frac{\Gamma \mid \Sigma, \varphi, \psi \mid \Delta}{\Gamma \mid \Sigma, \varphi \vee \psi \mid \Delta}\ {}^{1}\!/_{2}\vee$$

$$\frac{\Gamma \mid \Sigma \mid \Delta, \varphi, \psi \qquad \Gamma, \varphi \mid \Sigma \mid \Delta, \varphi \qquad \Gamma, \psi \mid \Sigma \mid \Delta, \psi}{\Gamma \mid \Sigma \mid \Delta, \varphi \wedge \psi}\ 1\vee$$

Definition 4.5.9 Let Γ, Δ be sets of formulas. Then:

- $\Gamma \vdash_{\mathsf{wST}} \Delta$ if and only if the sequent $\Gamma \mid \Gamma, \Delta \mid \Delta$ is derivable in $3\mathcal{WK}$,
- $\Gamma \vdash_{\mathsf{wTS}} \Delta$ if and only if the sequent $\Gamma \mid \varnothing \mid \Delta$ is derivable in $3\mathcal{WK}$,
- $\Gamma \vdash_{\mathbb{PWK}} \Delta$ if and only if the sequent $\Gamma \mid \Delta \mid \Delta$ is derivable in $3\mathcal{WK}$,
- $\Gamma \vdash_{\mathbb{K}_3^w} \Delta$ if and only if the sequent $\Gamma \mid \Gamma \mid \Delta$ is derivable in $3\mathcal{WK}$.

The Soundness and Completeness proofs of $3\mathcal{WK}$ can be found in (Bonzio & Pra Baldi, n.d.).

Bibliography

Avron, A., & Konikowska, B. (2005). Multi-valued calculi for logics based on non-determinism. *Logic Journal of the IGPL, 13*(4), 365–387.

Barrio, E., & Egré, P. (2023). Substructural logics and metainferences. *Journal of Philosophical Logic, 51*(Special Issue 6).

Barrio, E. A., Pailos, F., & Szmuc, D. (2020). A hierarchy of classical and paraconsistent logics. *Journal of Philosophical Logic, 49*(1), 93–120.

Barrio, E. A., Pailos, F., & Szmuc, D. (2021a). (Meta) inferential levels of entailment beyond the tarskian paradigm. *Synthese*, 198(22), 5265–5289.

Barrio, E. A., Pailos, F., & Toranzo Calderón, J. S. (2021b). Anti-exceptionalism, truth and the BA-plan. *Synthese, 199*(5), 12561–12586.

Blomet, Q., & Égré, P. (2024). ST and TS as product and sum. *Journal of Philosophical Logic, 53*, 1673–1700.

Bonzio, S., & Pra Baldi, M. (n.d.). Undefinability of standard sequent calculi for paraconsistent three-valued logics. Unpublished. https://doi.org/10.48550/arXiv.1612.00963.

Borzi, A., & Zirattu, M. (n.d.) The dark side of a structural logic. Manuscript in Preparation.

Cobreros, P., Egré, P., Ripley, E., & van Rooij, R. (2020). Inferences and metainferences in ST. *Journal of Philosophical Logic, 49*(6), 1057–1077.

Cobreros, P., La Rosa, E., & Tranchini, L. (2021). (I can't get no) antisatisfaction. *Synthese, 198*(9), 8251–8265.

Da Ré, B., & Pailos, F. (2022). Sequent-calculi for metainferential logics. *Studia Logica, 110*(2), 319–353.

Da Ré, B., Pailos, F., Szmuc, D., & Teijeiro, P. (2020). Metainferential duality. *Journal of Applied Non-Classical Logics, 30*(4),312–334.

Da Ré, B., Szmuc, D., & Teijeiro, P. (2021). Derivability and metainferential validity. *Journal of Philosophical Logic, 51*(6), 1521–1547.

Fiore, C., Pailos, F., & Rubin, M. (2022). Inferential constants. *Journal of Philosophical Logic, 52*(3), 767–796.

Frankowski, S. (2004). p-consequence versus q-consequence operations. *Bulletin of the Section of Logic, 33*(4), 197–207.

French, R. (2016). Structural reflexivity and the paradoxes of self-reference. *Ergo, 3*(5), 113–131.

Hacking, I. (1979). What is logic? *Journal of Philosophy, 76*(6), 285–319.

Hösli, B., Jäger, G. (1994). About some symmetries of negation. *The journal of symbolic logic, 59*(2), 473–485.

Kleene, S. C. (1952). *Introduction to metamathematics*. North-Holland.

Malinowski, G. (1990a). q-consequence operation. *Reports on Mathematical Logic, 24*, 49–59.

Malinowski, G. (1990b). Towards the concept of logical many-valuedness. *Acta Universitatis Lodziensis. Folia Philosophica, 7*, 97–103.

McAllister, I. (2022). Classical logic is not uniquely characterizable. *Journal of Philosophical Logic, 51*(6), 1345–1365.

Pailos, F. M. (2019). A family of metainferential logics. *Journal of Applied Non-Classical Logics, 29*(1), 97–120.

Pailos, F. (2020). A fully classical truth theory characterized by substructural means. *The Review of Symbolic Logic, 13*(2), 249–268.

Porter, B. (2022). Supervaluations and the strict-tolerant hierarchy. *Journal of Philosophical Logic, 51*(6), 1367–1386.

Ripley, E. (2012). Conservatively extending classical logic with transparent truth. *Review of Symbolic Logic, 5*(2), 354–378.

Ripley, E. (2013). Paradoxes and failures of cut. *Australasian Journal of Philosophy, 91*(1), 139–164.

Ripley, E. (2022). One step is enough. *Journal of Philosophical Logic, 51*(6), 1233–1259.

Scambler, C. (2020a). Classical logic and the strict tolerant hierarchy. *Journal of Philosophical Logic, 49*(2), 351–370.
Scambler, C. (2020b). Transfinite meta-inferences. *Journal of Philosophical Logic, 49*(6), 1079–1089.
Schröder, E. (2017). *Der Operationskreis des Logikkalkuls*. Leopold Classic Library.
Szmuc, D. E. (2024). Non-transitive counterparts of every tarskian logic. *Analysis, 84*(2), 320–326.
Wintein, S. (2016). On all strong kleene generalizations of classical logic. *Studia Logica, 104*(3), 503–545.

Chapter 5
Non-Contractive and Non-Monotonic Logics

Mariela Rubin, Miguel Álvarez Lisboa, and Yll Buzoku

Abstract In this chapter, we will inquire into some non-contractive and or non-monotonic logics that can be found in the literature. Our starting point, as in the previous chapters, will be subtracting some structural rules from **LK**, in this case Contraction and/or Weakening. But unlike previous chapters, here we will move away from the Strong Kleene family of logics and present some other logics of different nature. We begin by introducing Classical Linear Logic, $\mathbb{CLL}$, a logic that is sometimes presented as Classical Logic with devices to control Weakening and Contraction along the proofs. This logic will occupy most of the chapter. We will start by showing some syntactic features of its language, then we will introduce phase semantics, and finally study some of its metainferential properties. Then we will move to logics where either Weakening or Contraction fails (but not both). We proceed in the following order: first, we begin with some logics that drop Contraction but retain Weakening. In particular, we will study a way to characterize the family of finitely many-valued logics of Łukasiewicz by adding some axioms and bounded Contraction to Affine Linear Logic (a system that is obtained from the calculus **CLL** plus Weakening). We will also survey a non-contractive presentation of **LK**, a logic that has the same admissible rules as **LK**. And finally, we will make the opposite move, and we will present $\mathbb{LKNW}$, a logic that is obtained from taking the additive fragment of **CLL** and adding Contraction.

M. Rubin (✉)
UBA/IIF-SADAF-CONICET, Bulnes, Buenos Aires, Argentina

M. Álvarez Lisboa
IIF-SADAF-CONICET, Bulnes, Buenos Aires, Argentina
e-mail: miguel.alvarez@um.uchile.cl

Y. Buzoku
University College London, London, UK
e-mail: y.buzoku@ucl.ac.uk

P. Teijeiro, E. A. Barrio (eds.), *Metainferences in Substructural Logics*, Synthese Library 528, https://doi.org/10.1007/978-3-032-05449-4_5

5.1 Introduction

In the previous chapters we saw what happens when we take Cut or Identity from **LK**. In this chapter, we will inquire what happens when we take away Contraction and/or Weakening from the calculus. Unlike $\mathbb{ST}$ and $\mathbb{TS}$, which are fairly simple logics, at least in the sense that they have calculi that look very similar to **LK** and have quite simple semantics, when we mess around with Weakening and Contraction things become more complicated. For a start, the language becomes richer. That is, every dyadic connective and every constant splits in two. At the same time, the semantics becomes way more intricate.[1]

In this chapter we will present a family of logics. We begin by introducing Classical Linear Logic ($\mathbb{CLL}$), a version of Classical Logic in which Weakening and Contraction are handled 'in a controlled way'. This system shall entertain us for the largest part of this chapter, for we want to provide a detailed account of its semantics, proof theory and metainferential properties. Then we will move to logics where either Weakening or Contraction fails. We do so in the following order. We begin with some logics that drop Contraction but keep Weakening. Here we present Affine Linear Logic, a system that obtains from the calculus **CLL** (to be introduced in Sect. 5.2) by adding Weakening, and show how to extend it to get the family of finitely many-valued logics of Łukasiewicz. We also consider a non-contractive logic that can be seen as a non-contractive presentation of **LK**, in the sense that it has the same admissible rules as **LK** when the vocabulary is the customary one. Then, we make the opposite move: to drop Weakening but keep Contraction. Hence we present $\mathbb{LKNW}$, a logic that obtains from the additive fragment of **CLL** by adding Contraction.

5.2 Classical Linear Logic

In this section we will present what is usually referred to as Classical Linear Logic ($\mathbb{CLL}$), originally introduced by Girard (1987). One may think of this logic (or family of logics) as what one obtains when the structural rules of Weakening and Contraction are substracted or constrained in **LK**. The most notable effect of these changes is that the usual connectives—conjunction, disjunction, implication and the truth and falsity constants—split in two. We shall call them the 'multiplicative' and the 'additive' fragments of the logic (cf. Sect. 2.2). One also adds two extra operators, called the 'exponentials', that allow one to recover the structural rules of Weakening and Contraction in *a controlled way*.

What we call $\mathbb{CLL}$ refers to the propositional fragment of standard (Classical) Linear Logic. The notation contains the following connectives:

[1] As it will become clear at the end of the chapter the lack of Weakening is probably more responsible for this than the lack of Contraction.

$$\neg; \rightarrow, \circ, +, 1, \bot; \supset, \wedge, \vee, \top, 0; !, ?$$

Here is a list of what each symbol represents and the names one might find in the literature to refer to them:

- *The multiplicative fragment:* also called the group-theoretical fragment. It includes $\neg$ ('linear negation'), $\rightarrow$ ('linear implication'), $\circ$ ('tensor', 'times', 'product' or 'multiplicative conjunction'), $+$ ('par' or 'multiplicative disjunction'), $\bot$ ('bottom' or 'false') and 1 ('unit').
- *The additive fragment:* also called the lattice-theoretical fragment. It includes $\supset$ ('additive implication'), $\wedge$ ('and' or 'additive conjunction'), $\vee$ ('plus' or 'additive disjunction'), 0 ('zero') and $\top$ ('top' or 'true').
- *The exponentials or modalities:* ! ('bang' or 'of course!') and ? ('question mark' or 'why not?').

The reader might notice that there is no additive negation; the reason for this will become clear in Sect. 5.2.4.3. The exponentials are unary operators and shall be used in the prefix position. So $!!\varphi$, $?!\varphi$, $\neg!\varphi$ are all well-formed formulas. Aside from this, well-formed formulas of $\mathcal{L}$ are defined in the usual way.

As for sequents, we follow the previous convention of taking them to be pairs of multisets. Customarily, only the multiplicative implication is presented in the calculus, yet we will work with both additive and multiplicative implications here.

Let us make a short comment on the notation before we proceed. Linear Logic has been widely studied not only by philosophers, but also by mathematicians and computer scientists. As a consequence, there is not a standard notation for it. In Table 5.1 we depict various different notations that are used in the literature.[2] Ours is heavily based on Restall (2000), with a couple of exceptions. The reason for these divergences is that we wanted to be as homogeneous as possible with the rest of the book, while avoiding possible confusions if one compares our work with the rest of the literature. This is why we refrained from the use of certain symbols that are already used by others in a different sense.

We begin with **CLL**, a calculus for $\mathbb{CLL}$. In Sect. 5.2.2 we will review some inferential properties of the system. Then we dig into the semantics of $\mathbb{CLL}$ in Sect. 5.2.3, where we will present Phase semantics. Finally, in Sect. 5.2.4 we will turn to some of the metainferential properties of the logic.

5.2.1 Sequent Calculus for $\mathbb{CLL}$

In this section we present $\mathbb{CLL}$ and some of its properties. We shall begin with its sequent calculus, **CLL**, because it is the most natural way of presenting Linear Logic. As we will see in Sect. 5.2.3, semantics for this logic can be a little more

[2] This table was taken from Paoli (2002) and extended with our own notation.

Table 5.1 Comparative table of notations

Our notation	Restall (2000)	Paoli (2002)	Girard (1987)	Troelstra (1992)
$\neg$	$\sim$	$\neg$	$\perp$	$\sim$
$\circ$	$\circ$	$\otimes$	$\otimes$	$\star$
$+$	$+$	$\oplus$	$⅋$	$+$
$\wedge$	$\wedge$	$\wedge$	$\&$	$\sqcap$
$\vee$	$\vee$	$\vee$	$\oplus$	$\sqcup$
$\rightarrow$	$\rightarrow$	$\rightarrow$	$\multimap$	$\multimap$
!	!	!	!	!
?	?	?	?	?
1	1	1	1	1
$\perp$	0	0	$\perp$	0
$\top$	$\top$	$\top$	$\top$	$\top$
0	$\perp$	$\perp$	0	$\perp$

convoluted than the ones already found in the book for other logics. In contrast, the proof-theoretic properties of the calculus are comparatively easier to get. For instance, it is claimed to have the cleanest possible proof of Cut-elimination—the reader can evaluate this on their own. Here is the calculus, **CLL** for the logic $\mathbb{CLL}$:

Structural Rules

$$\text{Id}\ \frac{}{\varphi \Rightarrow \varphi} \qquad \text{Cut}\ \frac{\Gamma \Rightarrow \varphi, \Delta \qquad \Sigma, \varphi \Rightarrow \Pi}{\Gamma, \Sigma \Rightarrow \Delta, \Pi}$$

Multiplicative Rules

$$\text{L}\neg\ \frac{\Gamma \Rightarrow \varphi, \Delta}{\Gamma, \neg\varphi \Rightarrow \Delta} \qquad \text{R}\neg\ \frac{\Gamma, \varphi \Rightarrow \Delta}{\Gamma \Rightarrow \neg\varphi, \Delta}$$

$$\text{L}\rightarrow\ \frac{\Gamma \Rightarrow \varphi, \Delta \qquad \Sigma, \psi \Rightarrow \Pi}{\Gamma, \Sigma, \varphi \rightarrow \psi \Rightarrow \Delta, \Pi} \qquad \text{R}\rightarrow\ \frac{\Gamma, \varphi \Rightarrow \psi, \Delta}{\Gamma \Rightarrow \varphi \rightarrow \psi, \Delta}$$

$$\text{L}\circ\ \frac{\Gamma, \varphi, \psi \Rightarrow \Delta}{\Gamma, \varphi \circ \psi \Rightarrow \Delta} \qquad \text{R}\circ\ \frac{\Gamma \Rightarrow \varphi, \Delta \qquad \Sigma \Rightarrow \psi, \Pi}{\Gamma, \Sigma \Rightarrow \varphi \circ \psi, \Delta, \Pi}$$

$$\text{L+}\ \frac{\Gamma, \varphi \Rightarrow \Delta \qquad \Sigma, \psi \Rightarrow \Pi}{\Gamma, \Sigma, \varphi + \psi \Rightarrow \Delta, \Pi} \qquad \text{R+}\ \frac{\Gamma \Rightarrow \varphi, \psi, \Delta}{\Gamma \Rightarrow \varphi + \psi, \Delta}$$

$$\text{L1}\ \frac{\Gamma \Rightarrow \Delta}{\Gamma, 1 \Rightarrow \Delta} \qquad \text{R1}\ \frac{}{\Rightarrow 1}$$

$$\text{L}\perp\ \frac{}{\perp \Rightarrow} \qquad \text{R}\perp\ \frac{\Gamma \Rightarrow \Delta}{\Gamma \Rightarrow \perp, \Delta}$$

Additive Rules

$$\text{L}\supset \frac{\Gamma \Rightarrow \varphi, \Delta \qquad \Gamma, \psi \Rightarrow \Delta}{\Gamma, \varphi \supset \psi \Rightarrow \Delta} \qquad \text{R}\supset \frac{\Gamma \Rightarrow \psi, \Delta}{\Gamma \Rightarrow \varphi \supset \psi, \Delta}$$

$$\text{R}\supset \frac{\Gamma, \varphi \Rightarrow \Delta}{\Gamma \Rightarrow \varphi \supset \psi, \Delta}$$

$$\text{L}\wedge \frac{\Gamma, \varphi \Rightarrow \Delta}{\Gamma, \varphi \wedge \psi \Rightarrow \Delta} \qquad \text{R}\wedge \frac{\Gamma \Rightarrow \varphi, \Delta \qquad \Gamma \Rightarrow \psi, \Delta}{\Gamma \Rightarrow \varphi \wedge \psi, \Delta}$$

$$\text{L}\wedge \frac{\Gamma, \psi \Rightarrow \Delta}{\Gamma, \varphi \wedge \psi \Rightarrow \Delta}$$

$$\text{L}\vee \frac{\Gamma, \varphi \Rightarrow \Delta \qquad \Gamma, \psi \Rightarrow \Delta}{\Gamma, \varphi \vee \psi \Rightarrow \Delta} \qquad \text{R}\vee \frac{\Gamma \Rightarrow \varphi, \Delta}{\Gamma \Rightarrow \varphi \vee \psi, \Delta}$$

$$\text{R}\vee \frac{\Gamma \Rightarrow \psi, \Delta}{\Gamma \Rightarrow \varphi \vee \psi, \Delta}$$

$$\text{L}0 \frac{}{\Gamma, 0 \Rightarrow \Delta} \qquad \text{R}\top \frac{}{\Gamma \Rightarrow \top, \Delta}$$

Exponentials

$$\text{L!} \frac{\Gamma, \varphi \Rightarrow \Delta}{\Gamma, !\varphi \Rightarrow \Delta} \qquad \text{L?} \frac{!\Gamma, \varphi \Rightarrow ?\Delta}{!\Gamma, ?\varphi \Rightarrow ?\Delta}$$

$$\text{R!} \frac{!\Gamma \Rightarrow \varphi, ?\Delta}{!\Gamma \Rightarrow !\varphi, ?\Delta} \qquad \text{R?} \frac{\Gamma \Rightarrow \varphi, \Delta}{\Gamma \Rightarrow ?\varphi, \Delta}$$

$$\text{W!} \frac{\Gamma \Rightarrow \Delta}{\Gamma, !\varphi \Rightarrow \Delta} \qquad \text{W?} \frac{\Gamma \Rightarrow \Delta}{\Gamma \Rightarrow ?\varphi, \Delta}$$

$$\text{C!} \frac{\Gamma, !\varphi, !\varphi \Rightarrow \Delta}{\Gamma, !\varphi \Rightarrow \Delta} \qquad \text{C?} \frac{\Gamma \Rightarrow ?\varphi, ?\varphi, \Delta}{\Gamma \Rightarrow ?\varphi, \Delta}$$

As we said before, $\mathbb{CLL}$ has more connectives than other logical systems studied in this volume and so we proceed with a short commentary on their peculiarities.

5.2.2 *Linear Inferential Logic*

We begin by the usual connectives found in other logical systems: conjunction, disjunction, conditional and negation. Then we will move to exponentials and learn what it means for these connectives to recover Contraction and Weakening.

5.2.2.1 The Usual Connectives

Let's start with some inferential properties of linear conjunctions and disjunctions. As we saw in Sect. 2.2, in **LK** one can define conjunction and disjunction by means of any pair of rules (be it both left and right multiplicative/additive rules, or a combination thereof) because Weakening and Contraction make any pair of such rules interderivable. Yet in **CLL**, as neither Weakening nor Contraction are freely available, these connectives split. This allows one to see some interesting relations between the two kinds of disjunctions and conjunctions, and how sensitive are connectives in general regarding rule formation.

These four connectives validate some key distributive properties:

- $\varphi \circ (\psi \vee \delta) \Rightarrow (\varphi \circ \psi) \vee (\varphi \circ \delta)$
- $\varphi + (\psi \wedge \delta) \Rightarrow (\varphi + \psi) \wedge (\varphi + \delta)$

Here are derivations that prove these facts:

$$\text{L}\circ\,\dfrac{\text{L}\vee\,\dfrac{\text{R}\vee\,\dfrac{\text{R}\circ\,\dfrac{\text{Id}\,\dfrac{}{\varphi \Rightarrow \varphi} \qquad \text{Id}\,\dfrac{}{\psi \Rightarrow \psi}}{\varphi, \psi \Rightarrow \varphi \circ \psi}}{\varphi, \psi \Rightarrow (\varphi \circ \psi) \vee (\varphi \circ \delta)} \qquad \text{R}\vee\,\dfrac{\text{R}\circ\,\dfrac{\text{Id}\,\dfrac{}{\varphi \Rightarrow \varphi} \qquad \text{Id}\,\dfrac{}{\delta \Rightarrow \delta}}{\varphi, \delta \Rightarrow \varphi \circ \delta}}{\varphi, \delta \Rightarrow (\varphi \circ \psi) \vee (\varphi \circ \delta)}}{\varphi, \psi \vee \delta \Rightarrow (\varphi \circ \psi) \vee (\varphi \circ \delta)}}{\varphi \circ (\psi \vee \delta) \Rightarrow (\varphi \circ \psi) \vee (\varphi \circ \delta)}$$

$$\text{R}\wedge\,\dfrac{\text{R}+\,\dfrac{\text{L}+\,\dfrac{\text{Id}\,\dfrac{}{\varphi \Rightarrow \varphi} \qquad \text{L}\wedge\,\dfrac{\text{Id}\,\dfrac{}{\psi \Rightarrow \psi}}{\psi \wedge \delta \Rightarrow \psi}}{\varphi + (\psi \wedge \delta) \Rightarrow \varphi, \psi}}{\varphi + (\psi \wedge \delta) \Rightarrow \varphi + \psi} \qquad \text{R}+\,\dfrac{\text{L}+\,\dfrac{\text{Id}\,\dfrac{}{\varphi \Rightarrow \varphi} \qquad \text{L}\wedge\,\dfrac{\text{Id}\,\dfrac{}{\delta \Rightarrow \delta}}{\psi \wedge \delta \Rightarrow \delta}}{\varphi + (\psi \wedge \delta) \Rightarrow \varphi, \delta}}{\varphi + (\psi \wedge \delta) \Rightarrow \varphi + \delta}}{\varphi + (\psi \wedge \delta) \Rightarrow (\varphi + \psi) \wedge (\varphi + \delta)}$$

One can also show that these connectives are linked through negation by the usual De Morgan equivalences, as the following proposition states.

Fact 5.2.1 The following sequents are derivable in **CLL**:

$$\neg(\varphi \circ \psi) \Leftrightarrow (\neg\varphi + \neg\psi)$$
$$\neg(\varphi + \psi) \Leftrightarrow (\neg\varphi \circ \neg\psi)$$
$$\neg(\varphi \vee \psi) \Leftrightarrow (\neg\varphi \wedge \neg\psi)$$
$$\neg(\varphi \wedge \psi) \Leftrightarrow (\neg\varphi \vee \neg\psi)$$

Where $\Gamma \Leftrightarrow \Delta$ is an abbreviation for the sequents $\Gamma \Rightarrow \Delta$ and $\Delta \Rightarrow \Gamma$.

If the sequent calculus has the rules of Contraction and Weakening on both sides of the turnstile, then $\wedge$ collapses with $\circ$ and $\vee$ with $+$ (as it was mentioned in Sect. 2.2). We will come back to this later (Sect. 5.2.4.3).

As with conjunctions and disjunctions, $\mathbb{CLL}$ comes with two conditionals: a multiplicative and an additive one. Both of them are definable in terms of negation and disjunction:

- $\varphi \to \psi \Leftrightarrow \neg\varphi + \psi$
- $\varphi \supset \psi \Leftrightarrow \neg\varphi \vee \psi$

Therefore, there is not much to say about them at this point; but we shall add few more things about them later on (see Sect. 5.2.4.3).

Concerning the multiplicative connectives, as the structural rules of Weakening and Contraction are not available, they will lack some of the properties usually attributed to conjunction and disjunction.

Fact 5.2.2 The following sequents are not derivable in **CLL**:

$$\begin{aligned} \varphi \circ \psi &\Rightarrow \varphi \\ \varphi \circ \psi &\Rightarrow \psi \\ \varphi &\Rightarrow \varphi + \psi \\ \psi &\Rightarrow \varphi + \psi \end{aligned}$$

Proof Given that Cut will be shown to be eliminable (see Sect. 5.2.4.2), if these sequents were to be derivable, then the last rule applied should be L$\circ$ (resp, R$+$). But the sequent to which this rule is to be applied is not derivable in general without Weakening. Therefore, the sequents are not derivable. □

Yet, they still retain other usual properties, such as associativity and commutativity.

Finally, in **CLL** we find four constants. 1 and $\bot$ are the multiplicative constants, and $\top$ and 0 are the additive ones. As the proposition below states, $\top$ is the neutral element of additive conjunction, and absorbs additive disjunction; whereas 0 is the neutral element of additive disjunction and absorbs additive conjunction.

Fact 5.2.3 The following sequents are derivable in **CLL**:

$$\begin{aligned} \varphi \vee \top &\Rightarrow \top \\ \top &\Rightarrow \varphi \vee \top \\ \varphi \wedge 0 &\Rightarrow 0 \\ 0 &\Rightarrow \varphi \wedge 0 \end{aligned}$$

Exercises

Exercise 72 Prove that the multiplicative conjunction and disjunction are associative and commutative:

- $\varphi \circ (\psi \circ \delta) \Leftrightarrow (\varphi \circ \psi) \circ \delta$
- $\varphi + (\psi + \delta) \Leftrightarrow (\varphi + \psi) + \delta$
- $\varphi \circ \psi \Leftrightarrow \psi \circ \varphi$
- $\varphi + \psi \Leftrightarrow \psi + \varphi$

Exercise 73 What about the distributive properties $\varphi \vee (\psi \circ \delta) \Rightarrow (\varphi \vee \psi) \circ (\varphi \vee \delta)$ and $\varphi \wedge (\psi + \delta) \Rightarrow (\varphi \wedge \psi) + (\varphi \wedge \delta)$? Are they valid or not? The reader is invited to try to derive these sequents for themselves and see what the problem is with them.

Exercise 74 Find the derivations that prove the Facts 5.2.1 and 5.2.3.

Exercise 75 It is instructive to try the derivation of $\varphi + 1 \Rightarrow 1$ and $\varphi \circ \bot \Rightarrow \bot$. Why are they not derivable? Would it be possible assuming a structural rule like Weakening or Contraction?

Exercise 76 Give proofs to the following sequents above mentioned:

- $\varphi \rightarrow \psi \Leftrightarrow \neg\varphi + \psi$
- $\varphi \supset \psi \Leftrightarrow \neg\varphi \vee \psi$

Exercise 77 Linear negation provides means not only to define some connectives in terms of the others, but also to define new ones. Find Left and Right rules for a new connective $\varphi \| \psi$, equivalent to $\neg(\varphi \wedge \psi)$. That is, you should be able to use these rules, together with the available rules of **CLL** to prove the equivalence:

$$\varphi \| \psi \Leftrightarrow \neg(\varphi \wedge \psi)$$

5.2.2.2 The Exponentials

It is sometimes said that Linear Logic is motivated by a more careful handling of the structural properties in its relation with the connectives. This 'careful handling' is permitted by the introduction of the exponentials.[3]

If we look at their rules, we see that the main difference between $!\varphi$ and $?\varphi$ is that the former allows the use of Weakening and Contraction on the antecedent of the sequent and the latter in the succedent. This is also reflected on their rules: L! and R? are pretty straightforward rules, but their counterparts are a bit more demanding. In both cases, we introduce the exponential just by prefixing it to a preexisting formula, but all the formulas in the contexts need to be already with their corresponding exponential: ! on the left, ? on the right.

One can prove that in fact these two operations are dual, in the sense that the following result holds.

Fact 5.2.4 $\neg\,!\varphi$ is equivalent to $?\neg\varphi$; and conversely, $\neg\,?\varphi$ is equivalent to $!\neg\varphi$.

Proof We have to prove the following two equivalences:

- $\neg\,!\varphi \Leftrightarrow\, ?\neg\varphi$

[3] Maybe it is worth commenting that although Girard proposed the reading of $!\varphi$ as 'of course that φ!' and $?\varphi$ as 'why not φ?', this is not meant to be a formalization of these interjections in natural language, but only a colorful way of reading these symbols.

- $\neg ?\varphi \Leftrightarrow !\neg\varphi$

We provide proofs for the sequents that correspond to the first equivalence; the other two are analogous.

$$\dfrac{\dfrac{\dfrac{\dfrac{\dfrac{}{\varphi \Rightarrow \varphi}\,\text{Id}}{\Rightarrow \varphi, \neg\varphi}\,\text{R}\neg}{\Rightarrow \varphi, ?\neg\varphi}\,\text{R}?}{\Rightarrow !\varphi, ?\neg\varphi}\,\text{R}!}{\neg !\varphi \Rightarrow ?\neg\varphi}\,\text{L}\neg \qquad \dfrac{\dfrac{\dfrac{\dfrac{\dfrac{}{\varphi \Rightarrow \varphi}\,\text{Id}}{\varphi, \neg\varphi \Rightarrow}\,\text{L}\neg}{!\varphi, \neg\varphi \Rightarrow}\,\text{L}!}{!\varphi, ?\neg\varphi \Rightarrow}\,\text{L}?}{?\neg\varphi \Rightarrow \neg !\varphi}\,\text{R}\neg$$

□

Note that in the first derivation, we need to introduce $?\neg\varphi$ before applying the rule R!, because of the requirement on contexts; the same applies in the second proof for $!\varphi$.

With the power of the exponentials one can see that allowing Weakening and Contraction makes it possible to prove that the additive conjunction (disjunction) is definable in terms of multiplicative conjunction (disjunction), and vice-versa. The exponentials in the following equivalences show how and where the structural rules are needed in this result.

Fact 5.2.5 $!(\varphi \wedge \psi)$ is equivalent to $!\varphi \circ !\psi$. $?(\varphi \vee \psi)$ is equivalent to $?\varphi + ?\psi$.

Other relevant equivalences and entailments that involve exponentials are depicted in Exercise 79.

Given all that has been said one can show that **LK** is, in fact, embeddable into **CLL**.

Definition 5.2.6 (Girard (1987)) The *classical embedding translation functions* τ^+ and τ^- from the language of **LK** to the language of **CLL** are defined as follows:

- $\tau(p)^+ = \tau(p)^- = p$
- $\tau^+(\neg\varphi) = \neg\tau^-(\varphi)$
- $\tau^-(\neg\varphi) = \neg\tau^+(\varphi)$
- $\tau^+(\varphi \wedge \psi) = ?\tau^+(\varphi) \wedge ?\tau^+(\psi)$
- $\tau^-(\varphi \wedge \psi) = \tau^-(\varphi) \wedge \tau^-(\psi)$
- $\tau^+(\varphi \vee \psi) = \tau^+(\varphi) + \tau^+(\psi)$
- $\tau^-(\varphi \vee \psi) = !\tau^-(\varphi) + !\tau^-(\psi)$
- $\tau^+(\varphi \rightarrow \psi) = \neg\tau^-(\varphi) \vee \tau^+(\psi)$
- $\tau^-(\varphi \rightarrow \psi) = !\neg\tau^+(\varphi) \vee !\tau^-(\psi)$

Definition 5.2.7 The translation of a multiset is defined as follows: $\tau^+(\gamma_1, \ldots, \gamma_n)$ where $n > 0$ (abbreviated $\tau^+(\Gamma)$) is defined as $\tau^+(\gamma_1), \ldots, \tau^+(\gamma_n)$. $\tau^-(\gamma_1, \ldots, \gamma_m)$ where $m > 0$ (abbreviated $\tau^-(\Gamma)$) is defined as $\tau^-(\gamma_1),\ldots,\tau^-(\gamma_m)$. $\tau^+(\emptyset) = \tau^-(\emptyset) = \emptyset$

Theorem 5.2.8 *The sequent* $\Gamma \Rightarrow \Delta$ *is provable in* **LK** *if and only if the sequent* $!\tau^-(\Gamma) \Rightarrow ?\tau^+(\Delta)$ *is provable in* **CLL**.

For the proof of the Theorem we refer the reader to Paoli (2013). A nice feature of these translation functions, among many others, is that they preserve the structure of the proof. Yet it is important to remark that they apply only on Cut-free proofs. Luckily Cut is eliminable both **LK** and **CLL**.

Exercises

Exercise 78 Prove Fact 5.2.5. Follow the strategy of the Proof of Theorem 5.2.4.

Exercise 79 Prove that:

$$\begin{aligned}
!\varphi &\Leftrightarrow 1 \wedge (\varphi \wedge (!\varphi \circ !\varphi)) \\
!(\varphi \wedge \psi) &\Leftrightarrow !\varphi \circ !\psi \\
!(!\varphi \wedge !\psi) &\Rightarrow !(\varphi \wedge \psi) \\
1 &\Leftrightarrow !\top \\
0 &\Leftrightarrow ?\bot \\
?\varphi + ?\psi &\Leftrightarrow ?(\varphi \vee \psi) \\
!\varphi \rightarrow \psi &\Leftrightarrow !\varphi \rightarrow !\psi
\end{aligned}$$

5.2.3 *Semantics for* $\mathbb{CLL}$

Classical Linear Logic, like other logics, has many different approaches to how one may give it a semantics. More famously though, it is known for having what some might consider unwieldy semantic theories. One may attribute the unwieldiness of these semantics to the nature of the logic of study; a logic which has no structural rules, the cornerstone of classical reasoning, surely demands "out of the box" semantic structures. In this chapter, we will stick to looking at a particular semantics for Linear Logic called *phase semantics*. In fact, phase semantics can be considered a provability semantics; that is to say, it is a semantics which models the provability of some formula. Phase semantics has its origins in Girard's seminal paper (1987) though the version of the semantics we will be looking at will be somewhat different. It is left as an exercise to the reader to convince themselves that the semantics we present here is equivalent to those found in Girard (1987).

5.2.3.1 Mathematical Preliminaries

We begin this section with some mathematical preliminaries that will equip us well to understand the semantical structures we will be studying:

Definition 5.2.9 (Monoid) Given a set M with a binary operation $\star : M \times M \rightarrow M$, the pair $(M, \star)$ is called a monoid when the binary operation satisfies the following two properties:

- Associativity, that is, for all $a, b, c \in M$, we have that the equality $a \star (b \star c) = (a \star b) \star c$ holds.
- Identity, that is, there exists an $e \in M$ such that for all $a \in M$ we have that the equalities $a \star e = e \star a = a$ all hold.

Should the monoid $(M, \star)$ also satisfy the following axiom it may be said to be commutative:

- For all $a, b \in M$, the equality $a \star b = b \star a$ holds.

As usual, where the operation is known, the monoid is simply represented by the underlying set.

Some examples of monoids include:

- $(\mathbb{Z}, \cdot)$—the integers under multiplication.
- $(\mathcal{P}(A), \cup)$—where $\mathcal{P}(A)$ is the power set of the set A and $\cup$ is set-theoretic union.
- $(\mathcal{M}, \star)$—where $\mathcal{M}$ is the set of multisets of Linear Logic formulas, with $\star$ being the multiset sum (the union of multisets including multiplicities).

Monoids, like other common algebraic structures also include substructures, such as submonoids:

Definition 5.2.10 (Submonoid) A submonoid $(N, \star)$ of the monoid $(M, \star)$ is such that $N \subseteq M$, the identity of M is in N and for all $a, b \in N$ we have that $a \star b \in N$.

In the remainder of this section, we will use multiplicative notation with e taking the role of the identity element, and for any two elements m, n of a monoid, their product, $m \star n$, will be written as mn. Using the notion of a monoid, we can now introduce the idea of a phase space.

Definition 5.2.11 (Phase Space) Let M be a commutative monoid. A phase space is a pair $(M, \bot_M)$ where $\bot_M \subseteq M$. Elements of M are called phases, with elements of $\bot_M$ being called antiphases. Where it is clear $\bot_M$ may just be written as $\bot$.

For any subset $N \subseteq M$ we can define the notion of its dual $N^{\bot}$.

Definition 5.2.12 (Duality) Given a phase space $(M, \bot)$, and a subset $N \subseteq M$, we define the dual of N as $N^{\bot} = \{m \in M \mid \forall n \in N,\ mn \in \bot_M\}$.

Finally, we introduce the notion of a fact as so:

Definition 5.2.13 (Fact) Given a phase space M, and a subset of it N, the subset is said to be a fact when $N = N^{\bot\bot}$.

Definition 5.2.14 (Validity) A subset N of a phase space M is said to be valid if the identity element, e, of M, is in N.

We now prove some useful properties about facts that allow us to more easily construct some examples later.

Proposition 5.2.15 *Let M be a phase space. Then the following points hold:*

(*a*) *For any* $N \subseteq M$*, we have that* $N \subseteq N^{\perp\perp}$.
(*b*) *For any* $N, P \subseteq M$*, we have that* $N \subseteq P$ *implies* $P^{\perp} \subseteq N^{\perp}$.
(*c*) *For any* $N \subseteq M$*, we have that* N *is a fact if and only if* N *is of the form* $P^{\perp}$ *for some* $P \subseteq M$.

Proof

(*a*) We want to show for any $a \in N$ that $a \in N^{\perp\perp}$. We know by definition that $N^{\perp\perp} = \{m \in M \mid \forall n \in N^{\perp}, mn \in \perp_M\}$ and so we must show that for any element $a \in N$, when multiplied by an element $b \in N^{\perp}$, we obtain an element $ab \in \perp_M$. But we know that elements of $N^{\perp}$ are precisely those elements of M that, when multiplied by an element of N, deliver an element of $\perp_M$. Therefore, given any $b \in N^{\perp}$, it holds that $ba \in \perp_M$, which by commutativity allows us to conclude that $ab \in \perp_M$. Thus, $a \in N^{\perp\perp}$ and consequently, $N \subseteq N^{\perp\perp}$.
(*b*) Assume the hypothesis and let $p \in P^{\perp}$. We want to show $p \in N^{\perp}$, that is, for any $n \in N$ we have $pn \in \perp_M$. Since $p \in P^{\perp}$ then by definition we have that $pq \in \perp_M$ for all $q \in P$. Furthermore, as $N \subseteq P$, then if $q \in N$ we have that $pq \in \perp_M$, and thus $p \in N^{\perp}$, as required.
(*c*) Here we consider two cases:

- Going left to right: Since N is a fact, by definition $N = N^{\perp\perp}$. Thus, by setting $P = N^{\perp}$, we are done.
- Going right to left: Assume $N = P^{\perp}$. We want to show that $N = N^{\perp\perp}$. By (a), we have that $P^{\perp} \subseteq P^{\perp\perp\perp}$, so all we need to show is that $P^{\perp\perp\perp} \subseteq P^{\perp}$. We do this by noting that $P \subseteq P^{\perp\perp}$, also because of (a). By (b), we can therefore assert $P^{\perp\perp\perp} \subseteq P^{\perp}$, as required.

□

Before we move on to developing the semantics, we should look at an example of how to find the facts of a phase space.

Example 5.2.16 Let $M = \mathbb{Z}$ be the integers and multiplication, as the first of the examples above. We see that this satisfies the definition of a commutative monoid as multiplication is associative, multiplication is commutative and 1 is an identity element. To turn this into a phase space, we need to fix a subset of antiphases. For example, let us fix $\perp_M = \{0\}$. For this phase space $(\mathbb{Z}, \perp_M)$, let us try to characterize what are its facts. Given a subset $N \subseteq \mathbb{Z}$, we wish to compute $N^{\perp}$, that is, to find the set $\{m \in \mathbb{Z} \mid \forall n \in N, mn = 0\}$.

Let us consider some examples for N. We start with $N = \mathbb{Z}$. In this instance, we see that $\mathbb{Z}^{\perp} = \{m \in \mathbb{Z} \mid \forall n \in \mathbb{Z}, mn = 0\}$. The only value of m which satisfies the equation $mn = 0$ for all n is $m = 0$. Thus $\mathbb{Z}^{\perp} = \{0\}$.

As we would like to see if $\mathbb{Z}$ is a fact (and it should be), we have to check that $\{0\}^{\perp} = \mathbb{Z}$. Expanding $\{0\}^{\perp} = \{m \in \mathbb{Z} \mid \forall n \in \{0\}, mn = 0\}$ we see that, in fact, we have to find all m such that $m0 = 0$. This, incidentally, is all $m \in \mathbb{Z}$. So indeed, it is the fact that $\mathbb{Z}$ is a fact. Note that, since $\mathbb{Z}^{\perp} = \{0\}$ and $\{0\}^{\perp} = \mathbb{Z}$, then it is the case that $\{0\}$ is also a fact.

Can we find any other facts? Suppose N is an arbitrary, non-empty, strict subset of $\mathbb{Z}$, not equal to $\{0\}$. Let us compute the dual $N^{\perp} = \{m \in \mathbb{Z} \mid \forall n \in N,\ mn = 0\}$. For the same reasons as before, we have that $N^{\perp} = \{0\}$. Thus, $N^{\perp\perp} = \mathbb{Z} \neq N$. This means that no strict, non-empty, subsets of $\mathbb{Z}$ are facts, other than $\{0\}$.

Finally, that leaves us with considering $\varnothing$. In this case, we see that $\varnothing$ is also not a fact as $\varnothing^{\perp} = \{m \in \mathbb{Z} \mid \forall n \in \varnothing,\ mn = 0\} = \mathbb{Z}$. Thus, it is not the case that $\varnothing^{\perp\perp} = \varnothing$. Therefore, we can conclude that the only facts of our phase space $(\mathbb{Z}, \perp_M)$ are indeed $\mathbb{Z}$ and $\perp_M$.

It is also worth noting that it immediately follows that every subset containing 1 is valid.

Exercises

Exercise 80 Convince yourselves that the examples in 5.2.3.1 are all examples of monoids. What are the identity elements in the above examples? Are all the example monoids commutative?

Exercise 81 Find some examples of submonoids of the previously mentioned monoids in 5.2.3.1.

Exercise 82 Given a phase space $(M, \perp_M)$, show that both M and $\perp_M$ are always facts.

Exercise 83 Given a phase space $(M, \perp_M)$ and an arbitrary subset $N \subseteq M$ show that $N^{\perp\perp}$ is the smallest fact containing N.

Exercise 84 Given a phase space $(M, \perp_M)$ with identity element e, show that $\{e\}^{\perp} = \perp_M$.

Exercise 85 Given a phase space $(M, \perp_M)$ with identity element e, show that $\varnothing^{\perp} = M$.

Exercise 86 Given a phase space $(M, \perp_M)$ and facts $N, P \subseteq M$ show that $N \cap P$ is a fact. Show that $N \cup P$ is not necessarily a fact.

Exercise 87 Given a phase space $(M, \perp_M)$ and facts $N, P \subseteq M$ show that $(N \cup P)^{\perp} = N^{\perp} \cap P^{\perp}$.

Exercise 88 Given a phase space $(M, \perp_M)$ and facts $N, P \subseteq M$ show that $(N \cap P)^{\perp} = (N^{\perp} \cup P^{\perp})^{\perp\perp}$.

Exercise 89 How does the Example 5.2.16 change if we change the operation of the monoid to addition instead of multiplication? What are all the duals then? Which subsets are facts?

5.2.3.2 Obtaining a Phase Semantics for $\mathbb{CLL}$

Our goal is now to use these tools to give a semantics for Linear Logic. We do this in a series of steps, analyzing fragments of $\mathbb{CLL}$ to slowly build up the semantics for the whole logic. We start by first obtaining a semantics for the purely multiplicative fragment of $\mathbb{CLL}$, called 'Multiplicative Linear Logic', $\mathbb{MLL}$, and prove Soundness and Completeness for it. Using this as a foundation, we then consider, first the additive connectives, and then the exponentials, to obtain an adequate semantics for the whole logic.

Let us start with something simple: Given any formula of Linear Logic φ, we already know that $\neg\neg\varphi \Leftrightarrow \varphi$. Recall that facts of a phase space $(M, \bot_M)$ are defined to be those subsets of M for which duality acts as an involution. This raises the question: Is it possible to interpret formulas of Linear Logic as facts of some phase space? As will be shown in the rest of this section, the answer will turn out to be yes. With this in mind, we need to somehow obtain this interpretation. At a first glance this might seem difficult, since we only have a vague notion of a possible correspondence between formulas and facts and their duals. But recall that $\neg\varphi \Leftrightarrow \varphi \to \bot$. It turns out that this is the key to developing this correspondence.

In what follows, we will write the fact associated with the formula φ as $[\![\varphi]\!]$. Given a formula φ, we know how to construct the dual of its interpretation, namely $[\![\varphi]\!]^{\bot} = \{m \in M \mid \forall a \in [\![\varphi]\!], ma \in \bot_M\}$ and so we have that $[\![\varphi]\!] = [\![\varphi]\!]^{\bot\bot}$, since $[\![\varphi]\!]$ is a fact. But we also know that $\varphi \Leftrightarrow \neg\neg\varphi$, and thus we have $[\![\neg\neg\varphi]\!] = [\![\varphi]\!]^{\bot\bot}$. Thus, it might be the case that $[\![\neg\varphi]\!] = [\![\varphi]\!]^{\bot}$. Let us assume this and see if this assumption, along with the setup we have so far, works out. Since we know that $[\![\neg\varphi]\!] = [\![\varphi \to \bot]\!]$, this tells us that $[\![\varphi \to \bot]\!] = \{m \in M \mid \forall a \in [\![\varphi]\!], ma \in \bot_M\}$ holds. Furthermore, we know that $\bot_M$ is itself a fact, so we can set $[\![\bot]\!] = \bot_M$. If we therefore define a map $\multimap$ from pairs of facts of M to facts of M, we can thus write $[\![\varphi \to \bot]\!] = [\![\varphi]\!] \multimap \bot_M$, which suggests that for any implication $\varphi \to \psi$, we may be able to write $[\![\varphi \to \psi]\!] = \{m \in M \mid \forall a \in [\![\varphi]\!], ma \in [\![\psi]\!]\} = [\![\varphi]\!] \multimap [\![\psi]\!]$. An immediate consequence of this definition is the following lemma:

Lemma 5.2.17 *For any phase space M and phases $G, H \subseteq M$, we have that $p \in G \multimap H$ iff $\forall q \in G, pq \in H$*

Using this idea—to define all the connectives as operations on facts of some phase space—we will obtain a semantics for Linear Logic. So far, we have perhaps convinced ourselves that we have a way of sensibly expressing $\to$ and $\bot$. This has been, until now, but a convincing argument. We will, however, later show that these expressions, along with similar expressions for the other connectives, indeed form a sound and complete semantics for Linear Logic. But first, let us keep trying to obtain such expressions for the other connectives.

In Proposition 5.2.1 we saw that the two multiplicative connectives $\circ$ and $+$ are dual to one another in a very precise way, namely that $\neg(\varphi \circ \psi) \Leftrightarrow (\neg\varphi + \neg\psi)$ and $\neg(\varphi + \psi) \Leftrightarrow (\neg\varphi \circ \neg\psi)$. Thus, if we had a way of representing either of these connectives in terms of $\to$, we would be able to obtain the desired operations on facts, which would give us a semantics for $\mathbb{MLL}$. Indeed, as we have seen, $\varphi \to \psi \Leftrightarrow \neg\varphi + \psi$. Thus we have that:

- $\varphi \circ \psi \Leftrightarrow \neg(\varphi \to \neg\psi)$
- $\varphi + \psi \Leftrightarrow \neg\varphi \to \psi$

Using these De Morgan equivalences, we can give immediate interpretations to the connectives $\circ$ and $+$. Such an approach has the benefit of satisfying the De Morgan properties by definition. Doing as such so gives us the following:

- $[\![\varphi \circ \psi]\!] = [\![\neg(\varphi \to \neg\psi)]\!] = [\![\varphi \to \neg\psi]\!]^\perp$
- $[\![\varphi + \psi]\!] = [\![\neg\varphi \to \psi]\!]$

We are still in want of an expression for 1. As we will show in Exercise 90, it is the case that $\neg 1 \Leftrightarrow \perp$ and $1 \Leftrightarrow 1 + \perp$. Using these equivalences, we derive the following:

$$\begin{aligned} [\![1]\!] &= [\![1 + \perp]\!] \\ &= [\![\neg 1 \to \perp]\!] \\ &= [\![\perp \to \perp]\!] \\ &= [\![\perp]\!] \multimap [\![\perp]\!] \\ &= [\![\perp]\!]^\perp \\ &= {\perp_M}^\perp \end{aligned}$$

In Exercise 84 we showed that, given a phase space M with identity e, it holds that $\{e\}^\perp = \perp_M$. Negating both sides gives us $\{e\}^{\perp\perp} = {\perp_M}^\perp$. We therefore obtain an expression for $[\![1]\!]$, namely that it is equal to the fact $\{e\}^{\perp\perp}$.

Thus, we have obtained expressions for each connective of $\mathbb{MLL}$. Given a phase space $(M, \perp_M)$ the expressions are as follows:

- $[\![\perp]\!] = \perp_M$
- $[\![\neg\varphi]\!] = [\![\varphi]\!]^\perp = \{m \in M \mid \forall a \in [\![\varphi]\!],\ ma \in \perp_M\}$
- $[\![\varphi \to \psi]\!] = [\![\varphi]\!] \multimap [\![\psi]\!] = \{m \in M \mid \forall a \in [\![\varphi]\!],\ ma \in [\![\psi]\!]\}$
- $[\![\varphi \circ \psi]\!] = ([\![\varphi]\!] \multimap [\![\psi]\!]^\perp)^\perp$
- $[\![\varphi + \psi]\!] = [\![\varphi]\!]^\perp \multimap [\![\psi]\!]$
- $[\![1]\!] = \{e\}^{\perp\perp}$, where e is the identity element of M.

As stated earlier, these are simply expressions which seem to be behaving correctly. We have made a lot of claims and used logical equivalences to obtain these expressions which gives rise to the following question: Are these expressions in any way meaningful, or are we hoping for more than what we have really found?

As the next step in our exposition, we will show that the above expressions do faithfully represent some aspect of $\mathbb{MLL}$. But before we proceed, we will make a short digression that will simplify things quite some in the coming sections.

Exercises

Exercise 90 Show that the following equivalences hold:

- $\neg 1 \Leftrightarrow \bot$
- $\varphi \Leftrightarrow \varphi + \bot$

Exercise 91 Show that the following equivalences hold:

- $[\![\varphi \circ \psi]\!] = [\![\psi \circ \varphi]\!]$
- $[\![\varphi + \psi]\!] = [\![\psi + \varphi]\!]$

5.2.3.3 An Important Digression: A Single-sided Sequent System

An apt reader might have noticed that, in the sequent calculus **CLL**, the rules seem to display a kind of symmetry. That is to say, the 'form' of every left rule (except those governing implications and negation) corresponds to the 'form' of some right rule in a very natural way. Take L$\vee$ and R$\wedge$ as an example: they mirror each other, with similar conditions in the premises and the conclusions but in opposite 'sides' of the sequents. Using the rules for linear negation, we can always move every formula in the antecedent of a sequent to a formula in the succedent (and vice-versa). Thus, we may take every left rule in the system **CLL** as a composition of right rules and negation, in order to obtain a single-sided sequent calculus for $\mathbb{CLL}$. Let us see what this system looks like.

$$\text{Id}\ \frac{}{\Rightarrow \varphi, \neg\varphi} \qquad \text{Cut}\ \frac{\Rightarrow \Gamma, \varphi \qquad \Rightarrow \neg\varphi, \Delta}{\Rightarrow \Gamma, \Delta}$$

$$\circ\ \frac{\Rightarrow \varphi, \Gamma \qquad \Rightarrow \psi, \Delta}{\Rightarrow \varphi \circ \psi, \Gamma, \Delta} \qquad \wedge\ \frac{\Rightarrow \varphi, \Gamma \qquad \Rightarrow \psi, \Gamma}{\Rightarrow \varphi \wedge \psi, \Gamma}$$

$$+\ \frac{\Rightarrow \varphi, \psi, \Gamma}{\Rightarrow \varphi + \psi, \Gamma} \qquad \vee\ (i = 0, 1)\ \frac{\Rightarrow \varphi_i, \Gamma}{\Rightarrow \varphi_0 \vee \varphi_1, \Gamma}$$

$$1\ \frac{}{\Rightarrow 1} \qquad \top\ \frac{}{\Rightarrow \top, \Gamma}$$

$$\bot\ \frac{\Rightarrow \Gamma}{\Rightarrow \bot, \Gamma} \qquad \text{No 0 rule}$$

$$\text{Promotion}\ \frac{\Rightarrow \varphi, ?\Gamma}{\Rightarrow !\varphi, ?\Gamma} \qquad \text{Weakening}\ \frac{\Rightarrow \Gamma}{\Rightarrow ?\varphi, \Gamma}$$

$$\text{Dereliction}\ \frac{\Rightarrow \varphi, \Gamma}{\Rightarrow ?\varphi, \Gamma} \qquad \text{Contraction}\ \frac{\Rightarrow ?\varphi, ?\varphi, \Gamma}{\Rightarrow ?\varphi, \Gamma}$$

We will continue to use this sequent system for the remainder of this section on semantics. It should be noted that we take propositional letters and their duals to be the atoms, and the constants and propositions only formulas composed of such atoms. In other words, the linear negation of a proposition is now defined as its De Morgan dual.

It is important to note that Linear Logic is frequently presented in this form (such as in the work of Girard, 1987, 1995) and so familiarity with this system is not undesirable. Furthermore, it allows us to focus only on the important cases in what follows and proceed with showing that our obtained clauses, along with clauses we will later obtain, do indeed 'faithfully' model $\mathbb{CLL}$.

The exercises that follow may help to bridge the gap between the two sequent calculi.

Exercises

Exercise 92 In the single-sided calculus, we do not include any rules for $\rightarrow$ and $\supset$. Why might this be the case?

Exercise 93 Convince yourself that every rule in this single-sided sequent calculus is derivable in the two-sided sequent calculus **CLL**.

Exercise 94 Show that every rule from the two-sided sequent calculus **CLL** can be translated into a derivation in the single-sided calculus. *Hint: Start by converting the sequents in each rule into single-sided sequents. You may need to use the de-Morgan equivalences.*

Exercise 95 Show that the $+$ rule of the single sided sequent calculus is invertible, i.e. that the rule

$$\frac{\Rightarrow \varphi + \psi, \Gamma}{\Rightarrow \varphi, \psi, \Gamma}$$

is admissible in our single sided sequent calculus.

5.2.3.4 Soundness and Completeness for $\mathbb{MLL}$

So far, we have not made precise what we mean by a 'faithful' representation or model of $\mathbb{MLL}$. Phase semantics is an example of what is sometimes called provability semantics (see Girard, 1995). That is to say, the semantics models the provability of a formula. A formula ξ of $\mathbb{MLL}$ is said to be provable, precisely when there is a proof of the sequent $\Rightarrow \xi$. Furthermore, a multiset Γ is said to be provable when the *par* ($+$) of its elements is. Thus, the interpretation of a multiset should be clear, in that for $\Gamma = \{\gamma_1, \ldots, \gamma_n\}$ we have that $[\![\Gamma]\!] = [\![\gamma_1 + \cdots + \gamma_n]\!]$. Let us now state the two theorems of Soundness and Completeness:

Theorem 5.2.18 (Soundness Theorem for $\mathbb{MLL}$) *Let Γ be a multiset of $\mathbb{MLL}$ formulas. If Γ is provable, its interpretation $[\![\Gamma]\!]$ is valid (Definition 5.2.14) in any phase space.*

Theorem 5.2.19 (Completeness Theorem for $\mathbb{MLL}$) *Let Γ be a multiset of $\mathbb{MLL}$ formulas. If, the interpretation $[\![\Gamma]\!]$ is valid in any phase space, then Γ is provable.*

Now that we have the single-sided sequent calculus, it suffices to show Soundness and Completeness with respect to *that* proof system.[4] We proceed as follows:

Proof of Theorem 5.2.18 To prove Soundness, we will do an induction over the height of a proof of an arbitrary sequent $\Rightarrow \Gamma$. The notion of height we will use is the same as that in Chap. 2 for **LK**. We show that whenever Γ is provable, then it is valid in any phase space. Since we do not ever fix any particular phase space in our argument, this amounts to showing that for each rule of our single-sided sequent calculus, in every case where the hypothesis is valid, the conclusion is also valid. We will do this by doing a case analysis on a proof of the sequent $\Rightarrow \Gamma$.

We now consider the cases, starting with the base cases of our induction:

Case (a). $\Gamma = \neg\varphi, \varphi$ for some arbitrary formula φ. Thus we have that $\Rightarrow \neg\varphi, \varphi$. We know this is equivalent to $\neg\varphi + \varphi$, so, by $[\![\neg\varphi + \varphi]\!] = [\![\varphi \rightarrow \varphi]\!] = [\![\varphi]\!] \multimap [\![\varphi]\!] = \{m \mid \forall x \in [\![\varphi]\!], mx \in [\![\varphi]\!]\}$ it is clear that the identity element $e \in [\![\neg\varphi + \varphi]\!]$, as required.

Case (b). $\Gamma = 1$. We have already shown that the identity element of a phase space $e \in [\![1]\!]$.

Next we move on to the inductive cases:

Case (c). *The proof of* $\Rightarrow \Gamma$ *ends with an instance of the Cut rule.* In this case we take by the inductive hypothesis that the interpretations of the premises of the rule are valid and want to show that the conclusion is also valid.

We partition Γ as $\Delta \cup \Theta$. Let us say we have $\Rightarrow \varphi, \Delta$ and $\Rightarrow \neg\varphi, \Theta$ as premises. Our goal is to show that $e \in [\![\Delta]\!]^{\perp} \multimap [\![\Theta]\!]$. Since $[\![\varphi + \Delta]\!] = [\![\Delta + \varphi]\!]$ and $[\![\varphi + \Delta]\!]$ is valid, we have that $[\![\Delta]\!]^{\perp} \multimap [\![\varphi]\!]$ is also valid. Our other hypothesis gives that $[\![\varphi]\!]^{\perp} \multimap [\![\Theta]\!]$ is also valid, and from here, by Lemma 5.2.17, we can conclude that $[\![\Delta]\!]^{\perp} \multimap [\![\Theta]\!]$ is valid as well.

Case (d). *The proof of* $\Rightarrow \Gamma$ *ends with an instance of the* $+$ *rule.* In this case, it follows immediately by our understanding of how we read a multiset of formulas.

Case (e). *The proof of* $\Rightarrow \Gamma$ *ends with an instance of the* $\circ$ *rule.* In this case, we have two hypotheses, again $\Rightarrow \varphi, \Delta$ and $\Rightarrow \psi, \Theta$ where $\Gamma = \Delta, \Theta, \varphi \circ \psi$. Supposing that $e \in [\![\varphi + \Delta]\!]$ and $e \in [\![\psi + \Theta]\!]$, we want to show that $e \in [\![(\varphi \circ \psi) + \Delta + \Theta]\!]$. We leave this case as an exercise, though we give the following hint: you will probably want to take an approach similar to that in the case of the Cut rule (case (c) above). That is to say, convert the formulas into a form that uses $\rightarrow$ and then using the definition of $\rightarrow$ see how elements 'move around'.

□

Proving Completeness is somewhat more interesting. Since our statement of Completeness gives us that $[\![\Gamma]\!]$ is valid in any phase space by hypothesis, we will construct a particular phase space and use that to show that $\Rightarrow \Gamma$ is indeed provable.

[4] The work in this section is heavily based on the work of Girard (1987) and Girard (1995).

We do so by working in the phase space $(\mathcal{M}, \star)$ which is the monoid of multisets of formulas of Linear Logic under the operation of 'multiset sum', which we will call 'the union'. The identity element, e, of this monoid is the empty multiset, $\varnothing$. We take as our set of antiphases $\bot_M$ the set of sequents provable in the single-sided sequent calculus. It might seem odd at first to take the set of provable multisets of formulas as the set of antiphases, but observe that the set of provable multisets contains exactly the multisets Γ which prove the sequent $\Rightarrow \Gamma, \bot$.

Before we move any further, we present a new object that will be the key tool in our study of the proof of Completeness and give some technical lemmas about it:

Definition 5.2.20 (Provability Set) The provability set of φ is the set $\mathfrak{P}(\varphi) = \{\Delta \in \mathcal{M} \mid \ \Rightarrow \varphi, \Delta \text{ is provable}\}$, that is, the set of all multisets of formulas with which a particular formula φ is provable.

It should be clear that this set is a subset of the phase space $\mathcal{M}$. A nice example of such an object is $\mathfrak{P}(\bot) = \bot_{\mathcal{M}}$. We can also extend the definition of the provability set of a formula to that of a multiset of formulas in the obvious way: Given the multiset $\Gamma = \{\gamma_1, \dots \gamma_n\}$, the provability set $\mathfrak{P}(\Gamma)$ is defined as the provability set $\mathfrak{P}(\gamma_1 + \dots + \gamma_n)$.

Lemma 5.2.21 $\mathfrak{P}(\neg\varphi)^\perp = \mathfrak{P}(\varphi)$.

Proof We take this proof in two steps.

- We start by showing $\mathfrak{P}(\varphi) \subseteq \mathfrak{P}(\neg\varphi)^\perp$. Let $\Gamma \in \mathfrak{P}(\varphi)$ and $\Delta \in \mathfrak{P}(\neg\varphi)$. Thus it is the case that $\Rightarrow \Gamma, \varphi$ and $\Rightarrow \Delta, \neg\varphi$. By the Cut rule of the single sided sequent calculus, $\Rightarrow \Gamma, \Delta$ also holds. But this, by definition, means that $\Gamma, \Delta \in \mathfrak{P}(\bot) = \bot_{\mathcal{M}}$. Since Δ is an arbitrary element of $\mathfrak{P}(\neg\varphi)$, we therefore have that $\Gamma, \Delta \in \bot_{\mathcal{M}}$, for all $\Delta \in \mathfrak{P}(\neg\varphi)$. Thus, by Lemma 5.2.17, we conclude $\Gamma \in (\mathfrak{P}(\neg\varphi) \multimap \bot_{\mathcal{M}})$, which is equivalent to $\Gamma \in \mathfrak{P}(\neg\varphi)^\perp$, as required.
- We now show that $\mathfrak{P}(\neg\varphi)^\perp \subseteq \mathfrak{P}(\varphi)$. Let $\Gamma \in \mathfrak{P}(\neg\varphi)^\perp$ and observe that by the Id rule of the sequent calculus we have that $\varphi \in \mathfrak{P}(\neg\varphi)$. Since $\Gamma \in \mathfrak{P}(\neg\varphi)^\perp$, we thus have for all $\psi \in \mathfrak{P}(\neg\varphi)$ that $\Gamma, \psi \in \bot_{\mathcal{M}}$. Crucially, this means that $\Gamma, \varphi \in \bot_{\mathcal{M}}$. This implies that $\Rightarrow \Gamma, \varphi$ is provable, and therefore, that $\Gamma \in \mathfrak{P}(\varphi)$, as required.

□

Corollary 5.2.22 $\mathfrak{P}(\varphi)$ *is a fact of* $\mathcal{M}$.

Proof Our goal is to show that $\mathfrak{P}(\varphi)$ is a fact. By the third point of Proposition 5.2.15, it suffices to find some subset $P \subseteq \mathcal{M}$ such that $P^\perp = \mathfrak{P}(\varphi)$. This follows immediately by Lemma 5.2.21. □

We are now ready to finally proceed with proving Completeness. Our main goal, from which Completeness follows, will be to show that $[\![\varphi]\!] = \mathfrak{P}(\varphi)$ in our particular phase space. Since we have by the above corollary that $\mathfrak{P}(\varphi)$ is a fact, it is legitimate to use this fact as the interpretation of atoms; that is to say $[\![p]\!] = \mathfrak{P}(p)$. Doing so allows us to give an inductive argument for proving Completeness.

Proof of Theorem 5.2.19 We start by showing that for any arbitrary formula ξ we have that $[\![\xi]\!] = \mathfrak{P}(\xi)$. In this case, we work by induction on the complexity of the formula.

Case (a). Let ξ be an atom p. In this case, by definition, we have that $[\![p]\!] = \mathfrak{P}(p)$.

Case (b). Let $\xi = \neg p$, for some atom p.[5] We want to show that $[\![\neg p]\!] = \mathfrak{P}(\neg p)$. To this end, we start by noting that since $[\![p]\!] = \mathfrak{P}(p)$ then $[\![p]\!]^{\perp} = \mathfrak{P}(p)^{\perp}$. Furthermore, we know, by Lemma 5.2.21, that $\mathfrak{P}(p)^{\perp} = \mathfrak{P}(\neg p)$. Thus, $[\![p]\!]^{\perp} = \mathfrak{P}(\neg p)$. Finally, since $[\![\neg p]\!] = [\![p]\!]^{\perp}$, we have that $[\![\neg p]\!] = \mathfrak{P}(\neg p)$, as required.

Case (c). Let $\xi = \perp$. In this case, we have that $\mathfrak{P}(\perp) = \perp_{\mathcal{M}} = [\![\perp]\!]$.

Case (d). Let $\xi = 1$. In this case, we have that $\mathfrak{P}(1) = \mathfrak{P}(\neg 1)^{\perp} = \mathfrak{P}(\perp)^{\perp} = {\perp_{\mathcal{M}}}^{\perp} = [\![1]\!]$.

Case (e). Let $\xi = \varphi \circ \psi$. The inductive hypothesis here gives us that $\mathfrak{P}(\varphi) = [\![\varphi]\!]$ and $\mathfrak{P}(\psi) = [\![\psi]\!]$. Our goal will be to show that $(\mathfrak{P}(\varphi) \multimap \mathfrak{P}(\psi)^{\perp})^{\perp} = \mathfrak{P}(\varphi \circ \psi)$. By negating both sides, we see that it is equivalent to show instead that $\mathfrak{P}(\varphi) \multimap \mathfrak{P}(\psi)^{\perp} = \mathfrak{P}(\neg\varphi + \neg\psi)$.

We start by showing that $\mathfrak{P}(\varphi) \multimap \mathfrak{P}(\psi)^{\perp} \subseteq \mathfrak{P}(\neg\varphi + \neg\psi)$. Let $\Gamma \in \mathfrak{P}(\varphi) \multimap \mathfrak{P}(\psi)^{\perp}$. This means that $\Rightarrow \Gamma, \Delta, \neg\psi$ is provable, for all Δ such that $\Rightarrow \Delta, \varphi$ is provable. Our goal is to show that $\Rightarrow \Gamma, \neg\varphi + \neg\psi$ is provable. Pick $\Delta = \neg\varphi$. This satisfies our hypothesis that $\Rightarrow \Delta, \varphi$ is provable by the Id rule and so we have that $\Rightarrow \Gamma, \neg\varphi, \neg\psi$ is provable. Thus, by the $+$-rule, we obtain that $\Rightarrow \Gamma, \neg\varphi + \neg\psi$ is provable, as required.

Now we want to show that $\mathfrak{P}(\neg\varphi + \neg\psi) \subseteq \mathfrak{P}(\varphi) \multimap \mathfrak{P}(\psi)^{\perp}$. Let $\Gamma \in \mathfrak{P}(\neg\varphi + \neg\psi)$. We want to show that $\Gamma \in \mathfrak{P}(\varphi) \multimap \mathfrak{P}(\psi)^{\perp}$, which is to say that $\Rightarrow \Gamma, \Delta, \neg\psi$ is provable, for all Δ such that $\Rightarrow \Delta, \varphi$ is provable. We also have that $\Rightarrow \Gamma, \neg\varphi + \neg\psi$ is provable, so by Exercise 95, applying the inverse $+$-rule, we have that $\Rightarrow \Gamma, \neg\varphi, \neg\psi$ is also provable. Therefore, by the following derivation:

$$\text{Cut}\ \frac{\Rightarrow \Gamma, \neg\varphi, \neg\psi \qquad \Rightarrow \Delta, \varphi}{\Rightarrow \Gamma, \Delta, \neg\psi}$$

since both branches are provable by hypothesis, we have that so is $\Rightarrow \Gamma, \Delta, \neg\psi$ and thus $\Gamma \in \mathfrak{P}(\varphi) \multimap \mathfrak{P}(\psi)^{\perp}$, as required.

Case (f). Let $\xi = \varphi + \psi$. This case is similar to the previous case and as such, we leave it as an exercise!

Now consider a sequent $\Rightarrow \Gamma$ where $\Gamma = \{\gamma_1, \ldots, \gamma_n\}$ for arbitrary n, where $[\![\Gamma]\!]$ is valid. To show it is provable, we have to show that $\varnothing \in \mathfrak{P}(\Gamma)$. Since we know

[5] Some might wonder why it suffices to only consider the negation of atoms. It follows from the fact that one may propagate the negation of any formula to the atoms using the De Morgan equivalences. Thus, when we consider a negated formula, we can transform it into a formula where one of the other connectives is the top-level connective of the formula and then proceed according to that case.

that $[\![\Gamma]\!] = [\![\gamma_1 + \cdots + \gamma_n]\!]$ and by our inductive argument that $[\![\gamma_1 + \cdots + \gamma_n]\!] = \mathfrak{P}(\gamma_1 + \cdots + \gamma_n) = \mathfrak{P}(\Gamma)$, we thus have that $[\![\Gamma]\!] = \mathfrak{P}(\Gamma)$. Since $[\![\Gamma]\!]$ is valid, and $\varnothing$ is the identity element of our phase space, we have that $\varnothing \in [\![\Gamma]\!]$. Thus, we have that $\varnothing \in \mathfrak{P}(\Gamma)$, as required. □

5.2.3.5 Extending to the Whole Logic

The additive connectives of Linear Logic are fair and nicely expressible in the phase semantics. Though we do not have an expression as 'on the nose' as we did for $\rightarrow$ and $\perp$, what we do have is not so different as to be difficult to use. Recall that, in Exercises 87 and 88, you were asked to show that, given a phase space $(M, \perp_M)$ and facts $N, P \subseteq M$, we have both $(N \cup P)^{\perp} = N^{\perp} \cap P^{\perp}$ and $(N \cap P)^{\perp} = (N^{\perp} \cup P^{\perp})^{\perp\perp}$. A keen observer might remark that this looks awfully similar to another De Morgan duality and such an observer would not be wrong. However, this relation only gets us part of the way through our understanding of the additives, as we will see.

We know that the additive connectives exhibit a particular De Morgan duality, namely that $\neg(\varphi \wedge \psi) \Leftrightarrow \neg\varphi \vee \neg\psi$ and $\neg(\varphi \vee \psi) \Leftrightarrow \neg\varphi \wedge \neg\psi$. Whilst it may be tempting to try and simply match symbols between this relation and De Morgan relations between facts, due to the fact that the De Morgan relations exhibit an exact symmetry between $\cap$ and $\cup$, it is unclear which additive connective one may associate to which operation. Thus we are still left with a question: which of these set theoretic operations corresponds to which logical operation?

It may perhaps be sensible to assume that the additive conjunction ($\wedge$) corresponds to the intersection and the additive disjunction ($\vee$) corresponds to the union, so let us try to convince ourselves of this. One clear way of doing so would be to try and see how these operations interact with the additive units $\top$ and 0. We know that:

- $\top \wedge \varphi \Leftrightarrow \varphi$
- $0 \vee \varphi \Leftrightarrow \varphi$

Therefore, if we can get a clause for either of these two units, we can figure out what the semantic clauses should be for all connectives. We leave the reader to convince themselves of this argument and instead, in contrast to before, state the clauses for the connectives, giving a series of exercises through which you should be able to justify why the clauses for the additives take the forms that they do.

We now state clauses for the additives. Given a phase space $(M, \perp_M)$, we define the following:

- $[\![\top]\!] = M$
- $[\![\varphi \wedge \psi]\!] = [\![\varphi]\!] \cap [\![\psi]\!]$
- $[\![0]\!] = \varnothing^{\perp\perp}$
- $[\![\varphi \vee \psi]\!] = ([\![\varphi]\!] \cup [\![\psi]\!])^{\perp\perp}$

There's only one question left: what of the exponentials? To be able to express the exponentials in the phase semantics we will introduce a new technical tool, that we will call 'the set of identity idempotents' (though this name is not standard). This set, that we write as $\mathfrak{S}$, will be a particular subset of the phase space of study. Where doubt may arise about what this phase space is, we may write $\mathfrak{S}_M$, where M represents the phase space in question. Let us state the definition:

Definition 5.2.23 Given a phase space $(M, \perp_M)$, the set of identity idempotents of M is defined as $\mathfrak{S}_M = \{m \in M \mid mm = m\} \cap \{e\}^{\perp\perp}$

A few observations about this definition are in place. Firstly, the set $\{e\}^{\perp\perp}$ should be familiar to us now, for it is the definition of $[\![1]\!]$. Thus this set is really only considering the idempotent elements of this particular fact. In general, the set of identity idempotents will not be a fact and this does impact the form of the clauses we give for the exponentials. So we can make:

- $[\![!\,\varphi]\!] = ([\![\varphi]\!] \cap \mathfrak{S})^{\perp\perp}$
- $[\![?\,\varphi]\!] = ([\![\varphi]\!]^{\perp} \cap \mathfrak{S})^{\perp}$

A natural question to ask ourselves is: what is the story that gives us these clauses for the exponentials? One answer is in unpacking the meaning of the set $\mathfrak{S}$. We see that the set $\{m \in M \mid mm = m\}$ considers elements m of the phase space M that behave as if they are contractable; that is to say, multiplicity is irrelevant for these elements. Similarly, by considering $[\![1]\!]$ also, we consider elements of M which behave like 1; that is, elements whose presence in an expression is superfluous and that we can weaken away. Thus, when we consider $[\![\varphi]\!] \cap \mathfrak{S}$, we are considering the set of elements of M that correspond to the formula φ which have these additional structural properties, such a formula taken as hypothesis. If one looks at the two-sided sequent calculus, one immediately sees that the connective ! has these properties when taken as an hypothesis. Therefore, one may be content with trying to work with the assignment $[\![!\,\varphi]\!] = [\![\varphi]\!] \cap \mathfrak{S}$. However, we must be careful, since $\mathfrak{S}$ may not in general be a fact, and we need them to be. Thus, we instead consider the smallest fact which contains $[\![\varphi]\!] \cap \mathfrak{S}$; that is, we consider $([\![\varphi]\!] \cap \mathfrak{S})^{\perp\perp}$ instead. We can then obtain an expression for $[\![?\,\varphi]\!]$ by considering De Morgan dualities.

We complete our discussion of the phase semantics for Linear Logic by finishing the proof of Soundness and Completeness for the additives and exponentials.

Exercises

Exercise 96 Consider the following questions:

- Why does it make sense to say that $[\![\top]\!] = M$?
- Given that $[\![\top]\!] = M$, give a justification for the expression $[\![\varphi \wedge \psi]\!] = [\![\varphi]\!] \cap [\![\psi]\!]$.
- Thus derive the expressions for $[\![\varphi \vee \psi]\!]$ and $[\![0]\!]$.

Exercise 97 Why do we need to do consider the double dual of the set $[\![\varphi]\!] \cup [\![\psi]\!]$ when giving an interpretation to $\varphi \vee \psi$?

5.2.3.6 Soundness and Completeness of $\mathbb{CLL}$

What follows in this section will be, in essence, a continuation of Theorems 5.2.18 and 5.2.19, but now we extend the result further to encompass all connectives of the logic. The results of Theorems 5.2.18 and 5.2.19 will still be valid so we do not need to consider those cases again. Instead, we only need to consider the cases with the rest of the connectives. So let us state the theorems for $\mathbb{CLL}$:

Theorem 5.2.24 (Soundness Theorem for $\mathbb{CLL}$) *Let Γ be a multiset of $\mathbb{CLL}$ formulas. If Γ is provable, its interpretation $[\![\Gamma]\!]$ is valid in any phase space.*

Theorem 5.2.25 (Completeness Theorem for $\mathbb{CLL}$) *Let Γ be a multiset of $\mathbb{CLL}$ formulas. If, the interpretation $[\![\Gamma]\!]$ is valid in any phase space, then Γ is provable.*

We begin with the proof of Theorem 5.2.24.

Proof of Theorem 5.2.24 As before, we consider a proof of an arbitrary sequent $\Rightarrow \Gamma$ where the proof ends using the remaining unconsidered rules of the one-sided sequent system. We proceed by induction on the height of the proof of the sequent. Recall that we interpret multisets as the *par* of all its elements and that we work in an arbitrary phase space $(M, \bot_M)$. In what follows, we are working with multisets of formulas, but at times, one may see expressions of the form $[\![\Gamma, \Delta]\!] = [\![\Delta, \Gamma]\!]$. This is, of course, the same thing but we include such steps to make explicit the formulas we wish to draw our attention to.

Case (a). *The proof ends with an instance of the $\top$ rule.* We want to show that the interpretation of this rule is valid, i.e. that if $\Gamma = \Delta, \top$ then $e \in [\![\Delta, \top]\!]$. To show this, we unpack the expression as follows: $[\![\Delta, \top]\!] = [\![\top, \Delta]\!] = [\![\top + \Delta]\!] = [\![\top]\!]^{\bot} \multimap [\![\Delta]\!] = [\![\neg\top]\!] \multimap [\![\Delta]\!] = [\![0]\!] \multimap [\![\Delta]\!] = \{m \in M \mid a \in [\![0]\!], am \in [\![\Delta]\!]\}$. Recall that $[\![0]\!] = \varnothing^{\bot\bot}$. It follows from Exercise 83 that $\varnothing^{\bot\bot}$ is the smallest fact in any phase space. Thus, we deduce that $[\![0]\!] \subseteq [\![\Delta]\!]$, therefore concluding that $1 \in [\![\top, \Delta]\!]$, as required.

We now move onto the inductive cases:

Case (b). *The proof of $\Rightarrow \Gamma$ concludes with an application of the $\wedge$ rule.* Letting $\Gamma = \Delta, \varphi \wedge \psi$ in this case, we want to show that if $e \in [\![\Delta, \varphi]\!]$ and $e \in [\![\Delta, \psi]\!]$, then $e \in [\![\Delta, \varphi \wedge \psi]\!]$. To prove this, we want show that for arbitrary facts P, Q, R of M we have that $P^{\bot} \multimap (Q \cap R) = (P^{\bot} \multimap Q) \cap (P^{\bot} \multimap R)$. This can be shown by verifying the following equivalences: $(P^{\bot} \multimap Q) \cap (P^{\bot} \multimap R) = \{m \in M \mid \forall p \in P^{\bot}, mp \in Q\} \cap \{m \in M \mid \forall p \in P^{\bot}, mp \in R\} = \{m \in M \mid \forall p \in P^{\bot}, mp \in Q \cap R\} = P^{\bot} \multimap (Q \cap R)$. Thus, we have that $[\![\Delta, \varphi \wedge \psi]\!] = [\![\Delta, \varphi]\!] \cap [\![\Delta, \psi]\!]$ and thus by the inductive hypothesis it follows immediately that $e \in [\![\Delta, \varphi \wedge \psi]\!]$.

Case (c). *The proof of $\Rightarrow \Gamma$ concludes with an application of either $\vee$ rule.* We leave these two as exercises for the reader, however the proof is very similar to that of the previous case.

Case (d). *The proof of* $\Rightarrow \Gamma$ *concludes with an application of the Weakening rule, where* $\Gamma = \Delta, ?\varphi$. In this case we have as hypothesis $e \in [\![\Delta]\!]$ and we want to show that $e \in [\![\Delta, ?\varphi]\!]$. Observe that, since we have $e \in [\![\Delta]\!]$ by hypothesis, we have that $\{e\} \subseteq [\![\Delta]\!]$. Double negating this relation gives, by point (b) of Proposition 5.2.15, that $\{e\}^{\perp\perp} \subseteq [\![\Delta]\!]^{\perp\perp}$. Since $[\![\Delta]\!]$ is also a fact, this is nothing more than $[\![1]\!] \subseteq [\![\Delta]\!]$. Thus, it follows that $[\![\varphi]\!]^{\perp} \cap \{m \in M \mid mm = m\} \cap [\![1]\!] \subseteq [\![\Delta]\!]$. As $[\![\varphi]\!]^{\perp} \cap \{m \in M \mid mm = m\} \cap [\![1]\!] = [\![\varphi]\!]^{\perp} \cap \mathfrak{S}$, by definition, we therefore have that $[\![\varphi]\!]^{\perp} \cap \mathfrak{S} \subseteq [\![\Delta]\!]$, and therefore that $[\![?\varphi]\!]^{\perp} \subseteq [\![\Delta]\!]$. Thus, we conclude that $e \in [\![?\varphi]\!]^{\perp} \multimap [\![\Delta]\!] = [\![?\varphi, \Delta]\!] = [\![\Delta, ?\varphi]\!]$, as required.

Case (e). *The proof of* $\Rightarrow \Gamma$ *concludes with an application of the Contraction rule.* In this case we have as hypothesis $e \in [\![\Delta, ?\varphi, ?\varphi]\!]$ where $\Gamma = \Delta, ?\varphi, ?\varphi$ and we want to show that $e \in [\![\Delta, ?\varphi]\!]$. We leave this case as an exercise to the reader.

Case (f). *The proof of* $\Rightarrow \Gamma$ *concludes with an application of the Dereliction rule, where* $\Gamma = \Delta, \varphi$. In this case we want to show that $e \in [\![\Delta, ?\varphi]\!]$. By hypothesis, we have that $e \in [\![\Delta, \varphi]\!] = [\![\varphi, \Delta]\!] = [\![\varphi]\!]^{\perp} \multimap [\![\Delta]\!]$. Thus, it follows by hypothesis that $[\![\varphi]\!]^{\perp} \subseteq [\![\Delta]\!]$. It therefore follows that $([\![\varphi]\!]^{\perp} \cap \mathfrak{S}) \subseteq [\![\Delta]\!]$ from which we can deduce that $e \in [\![?\varphi]\!]^{\perp} \multimap [\![\Delta]\!] = [\![?\varphi, \Delta]\!] = [\![\Delta, ?\varphi]\!]$.

Case (g). *The proof of* $\Rightarrow \Gamma$ *concludes with an application of the Promotion rule, where* $\Gamma = ?\Delta, \varphi$. In this case we have as hypothesis that $e \in [\![?\Delta, \varphi]\!]$ and we want to show that $e \in [\![?\Delta, !\varphi]\!]$. We leave this case also as an exercise to the reader.

So to finish the proof of Soundness we leave the remaining cases for the reader to prove as an exercise. □

We are now left with showing Completeness:

Proof of Theorem 5.2.25 We will make a subtle change to our setup in that we will now consider a slightly different phase space. As before we will be considering the phase space of contexts, with the monoid being the monoid of multisets of $\mathbb{CLL}$ formulas, and the set of antiphases being the set of provable multisets of formulas. However, we make a subtle adjustment to the phase space. We now identify multisets which make use of multiple instances of a $?\varphi$. That is, in what follows, the multisets $\{\Delta, ?\varphi, ?\varphi\}$ and $\{\Delta, ?\varphi\}$ are identified as equal. More broadly speaking, if multiple occurrences of a $?\varphi$ formula appears in a multiset, that multiset is identified with the same multiset with exactly one occurrence of that formula. This is done so that $\mathfrak{S}$ becomes the set of contexts of the form $?\Gamma$. Under these caveats, let us consider the rest of the cases for an arbitrary formula ξ. Note that we only need to show $[\![\xi]\!] = \mathfrak{P}(\xi)$ in the additional cases of the new connectives as the conclusion will be the same as before.

Case (a). $\xi = \top$. In this case, we want to show that monoid $\mathcal{M} = \mathfrak{P}(\top)$. This follows immediately by the definition of $\mathfrak{P}(\top)$.

Case (b). $\xi = \varphi \wedge \psi$. In this case, by the induction hypothesis, we have that $\mathfrak{P}(\varphi) = [\![\varphi]\!]$ and $\mathfrak{P}(\psi) = [\![\psi]\!]$. We want to show that $\mathfrak{P}(\varphi) \cap \mathfrak{P}(\psi) = \mathfrak{P}(\varphi \wedge \psi)$.

Going left to right, pick $\Gamma \in \mathfrak{P}(\varphi) \cap \mathfrak{P}(\psi)$. Thus $\Rightarrow \Gamma, \varphi$ and $\Rightarrow \Gamma, \psi$ are both provable. Therefore, by the $\wedge$-rule, we have that $\Rightarrow \Gamma, \varphi \wedge \psi$ is provable.

Going right to left, we have $\Gamma \in \mathfrak{P}(\varphi \wedge \psi)$. Thus, $\Rightarrow \Gamma, \varphi \wedge \psi$ is provable. Since $\wedge$-rule is invertible, we have that $\Rightarrow \Gamma, \varphi$ and $\Rightarrow \Gamma, \psi$ and thus $\Gamma \in \mathfrak{P}(\varphi)$ and $\Gamma \in \mathfrak{P}(\psi)$ which in particular means that $\Gamma \in \mathfrak{P}(\varphi) \cap \mathfrak{P}(\psi)$.

Case (c). $\xi = \varphi \vee \psi$. We leave this case as an exercise to the reader.

We are now left with the exponential cases: $\xi = ?\varphi$ and $\xi = !\varphi$. A reader might be confused why there are fewer cases here than in the proof of Soundness. But a moments thought gives that, due to the elimination of the multiplicity of $?\varphi$, the rules which handle its multiplicity now become superfluous. Thus, to finish the proof of Completeness we also leave to the reader the cases where $\xi = ?\varphi$ and $\xi = !\varphi$. □

As we have seen, the phase semantics does indeed form a sound and complete semantics with respect to $\mathbb{CLL}$. The Completeness proof that we have given can be strengthened to state the following:

Theorem 5.2.26 (Strong Completeness for $\mathbb{CLL}$) *Let Γ be a multiset of $\mathbb{CLL}$ formulas. If, the interpretation $[\![\Gamma]\!]$ is valid in any phase space, then Γ is provable with a 'cut-free' proof.*

This strengthened form of the Theorem has been discussed in detail separately by Avron (1988) and by Okada (1998) and so we will not discuss it here. Yet in a moment will prove Cut-elimination for the two-sided sequent calculus (Sect. 5.2.4.2).

Before we move on, let us take a final look at the clauses we used for the multiplicative connectives. As we have seen, proving Soundness and Completeness using the clauses we had at hand and the tools we developed in this section was at times quite technically challenging. In reality, the clauses for the multiplicatives are usually presented in a slightly different form. To present them as they are usually presented, we need yet another operation on submonoids, that of the 'product'. In the phase semantics, usually it is easier to talk in terms of products of elements than in terms of the $\multimap$ operation as we have been doing until now. We define this operation as follows:

Definition 5.2.27 Let M be a monoid and $P, Q \subseteq M$. We define $P \cdot Q$ as $\{pq \in M \mid p \in P, q \in Q\}$.

In the above definition, if P and Q are facts of some phase space, it is not necessarily the case that $P \cdot Q$ will be a fact. However, we see that this product operation treats elements in a very '$\circ$-like' way. In fact, we can give semantic clauses of the multiplicative connectives in terms of this product operation as follows:

- $[\![\varphi \circ \psi]\!] = ([\![\varphi]\!] \cdot [\![\psi]\!])^{\perp\perp}$
- $[\![\varphi + \psi]\!] = ([\![\varphi]\!]^{\perp} \cdot [\![\psi]\!]^{\perp})^{\perp}$
- $[\![\varphi \to \psi]\!] = ([\![\varphi]\!] \cdot [\![\psi]\!]^{\perp})^{\perp}$

Exercises

Exercise 98 Prove cases (c), (e), and (g) from Theorem 5.2.24.

Exercise 99 Prove case (c) and the cases when $\xi = ?\varphi$ and $\xi = !\varphi$ from Theorem 5.2.25.

Exercise 100 Show Soundness and Completeness of the above clauses for $\mathbb{MLL}$.

5.2.4 Linear Metainferential Logic

In this section we focus on four main topics: the relation between the exponentials and Weakening and Contraction, Cut-elimination, the relation between rules and inferences and some results of derivability and admisibility. We will start by resuming our presentation of Exponentials.

5.2.4.1 The Metainferential Impact of the Exponentials

As we said before, the constant ! authorizes the use of Weakening and Contraction on the left-hand side of the sequent; ?, on the right. In this subsection we shall dive deeper into the meaning of these connectives to see exactly how they manage to do it, and what is the relevance of this for further metainferential results, like Cut-elimination.

We begin with an elementary question: Why do we need two different exponentials? Why not dealing with only one, $!\varphi$, allowing the different rules in any side of the sequent? The following theorem provides an answer to these questions.

Theorem 5.2.28 *Adding to* **CLL** *the axiom*

$$?=!\,\frac{}{?\varphi \Rightarrow !\varphi}$$

Has the following effects:

(a) $?\varphi \Rightarrow \varphi$ becomes provable.
(b) $\varphi \Rightarrow !\varphi$ becomes provable.
(c) All structural rules become derivable for arbitrary formulas.

Proof First of all, note that $!\varphi \Rightarrow ?\varphi, !\varphi \Rightarrow \varphi$ and $\varphi \Rightarrow ?\varphi$ are already provable in **CLL**. Therefore, we are proving their converses. (a) and (b) are easily proved with the proposed rule, Cut and the rules L ! and R ?. (c) is a consequence of (a) and (b) by some convenient applications of Cut. As an example, we show how to derive Contraction-Left for an arbitrary φ:

$$\dfrac{\varphi \Rightarrow\, !\varphi \qquad \dfrac{\dfrac{!\varphi \Rightarrow \varphi \qquad \dfrac{!\varphi \Rightarrow \varphi \qquad \Gamma, \varphi, \varphi \Rightarrow \Delta}{\Gamma, !\varphi, \varphi \Rightarrow \Delta}\,\text{Cut}}{\Gamma, !\varphi, !\varphi \Rightarrow \Delta}\,\text{Cut}}{\Gamma, !\varphi \Rightarrow \Delta}\,\text{C!}}{\Gamma, \varphi \Rightarrow \Delta}\,\text{Cut}$$

□

This result confirms that the structural rules of Weakening and Contraction have to be treated differently when used in the left-hand side of the sequent or in the right-hand side. And the reason for this lies in their meaning, which we are now in a position to explain. In fact:

- $!\varphi$ behaves as an infinite iteration of $\circ$ over $1 \wedge \varphi$: $(1 \wedge \varphi) \circ \cdots \circ (1 \wedge \varphi)$
- $?\varphi$ behaves as an infinite iteration of $+$ over $\bot \vee \varphi$: $(\bot \vee \varphi) + \cdots + (\bot \vee \varphi)$

This is 'roughly' equivalent[6] to saying that:

- $!\varphi$ means that there are infinite copies of φ on the antecedent of the sequent;
- $?\varphi$ means that there are infinite copies of φ on the succedent of the sequent.

Under this interpretation, the derivation:

$$\text{LW}\,\dfrac{\Gamma \Rightarrow \Delta}{\Gamma, \varphi \Rightarrow \Delta}$$

May be read as: "You already had φ on the left all along", whereas:

$$\text{LC}\,\dfrac{\Gamma, \varphi, \varphi \Rightarrow \Delta}{\Gamma, \varphi \Rightarrow \Delta}$$

More or less says that: "One copy is the same as several copies of φ". In **LK** and the calculi studied in the rest of this book, this is completely harmless, because we have 'free' Weakening and Contraction rules. But, as we saw in Sect. 2.4, having these rules has an impact on the proof-theory of the logic; in particular, when we try to eliminate Cut.

Take the following Cut:

$$\text{Cut}\,\dfrac{\text{C?}\,\dfrac{\dfrac{\vdots\\ \mathcal{D}_1}{\Gamma \Rightarrow ?\varphi, ?\varphi, \Delta}}{\Gamma \Rightarrow ?\varphi, \Delta} \qquad \dfrac{\vdots\\ \mathcal{D}_2}{!\Gamma', ?\varphi \Rightarrow ?\Delta'}}{\Gamma, !\Gamma' \Rightarrow \Delta, ?\Delta'}$$

[6] See below, Sect. 5.2.4.4.

As we will see in the next subsection, the basic reduction procedure will give us:

$$
\dfrac{
\dfrac{
\begin{array}{c}\vdots\\ \mathcal{D}_1\\ \Gamma \Rightarrow ?\varphi, ?\varphi, \Delta\end{array}
\qquad
\begin{array}{c}\vdots\\ \mathcal{D}_2\\ !\Gamma' ?\varphi \Rightarrow ?\Delta'\end{array}
}{\Gamma, !\Gamma' \Rightarrow ?\varphi, \Delta, ?\Delta'}\ \text{Cut}
\qquad
\begin{array}{c}\vdots\\ \mathcal{D}_2\\ !\Gamma, ?\varphi \Rightarrow ?\Delta'\end{array}
}{
\dfrac{\Gamma, !\Gamma', !\Gamma' \Rightarrow \Delta, ?\Delta', ?\Delta'}{
\dfrac{\vdots}{\Gamma, !\Gamma' \Rightarrow \Delta, ?\Delta'}\ \text{Several C?}}
}\ \text{Cut}
$$

Yet even if the level of the Cut[7] in this derivation has decreased, we introduced a second instance of Cut. This is analogous to the case of **LK**, where some possible applications of Contraction rules forced us to consider a stronger version of Cut (the Mix rule). In the case of **CLL**, where Contraction (and Weakening) are not free, we can treat these cases with two stronger rules, one for each exponential:

$$
\text{Cut-!}\ \dfrac{\Gamma \Rightarrow !\varphi, \Delta \qquad \Pi, (!\varphi)^n \Rightarrow \Sigma}{\Gamma, \Pi \Rightarrow \Delta, \Sigma}
$$

$$
\text{Cut-?}\ \dfrac{\Gamma, ?\varphi \Rightarrow \Delta \qquad \Pi \Rightarrow (?\varphi)^n, \Sigma}{\Gamma, \Pi \Rightarrow \Delta, \Sigma}
$$

Where $(*\varphi)^n$ denotes several occurrences of $*\varphi$ with $* \in \{!, ?\}$, and $n > 1$. It might catch someone's eye that the rules are such that you can only have n occurrences of $*\varphi$ at one of the premises and only one occurrence at the other premise. For example, Cut-! allows one to have n occurrences of $!\varphi$ at the left side of the sequent. The reason for this is that we only have a rule for contracting !-formulas to the left of the sequent. And dually the same happens for Cut-? at the right side of the sequent. It is immediate to notice that the derivation of these two rules is identical to the derivation of Mix presented in Chap. 2.

5.2.4.2 Cut-elimination

We will show that Cut is redundant in **CLL**. In Sect. 2.4, Cut-elimination was proved for **LK**. The proof is based on a double induction on the grade as the main parameter and the rank as an auxiliary one. Here we will be doing something slightly different. The main induction will also be carried on the grade but instead of using the rank as the second inductive measure we will use the *level* of the Cut. It is important to point out that the main induction (the one on the grade) assumes we are dealing with derivations with a single Cut of grade n as its last step that can be reduced to

[7] One of the measures over which we will make the induction for the Cut-elimination proof (see Definition 5.2.31).

derivations of lesser grade. This in turn is proved by a second induction on the level of the terminal Cut. One can find a double induction done on the same parameters as in Chap. 2 in Roorda (1989). We thought that Roorda's presentation is the clearest possible using the measures used in the second chapter, so we chose a different proof.

As we just explained in the previous section, we will do it by proving the eliminability of Cut-! and Cut-? for the formulas with exponentials, and Cut for every other connective. We will follow Troelstra's method (Troelstra, 1992).

Definition 5.2.29 The *complexity* of a formula φ, $c(\varphi)$ is recursively defined as follows:

- If φ is either an atomic formula, $\top, 1, 0$, or $\bot$, then $c(\varphi) = 0$.
- If $\varphi = *\psi$, where $* \in \{\neg, ?, !\}$, then $c(\varphi) = c(\psi) + 1$.
- If $\varphi = \psi * \delta$, where $* \in \{\circ, \wedge, +, \vee, \rightarrow\}$, then $c(\varphi) = c(\psi) + c(\delta) + 1$.

Definition 5.2.30 Let CF($\mathcal{D}$) denote the set of formulas that are eliminated with Cut in a derivation $\mathcal{D}$. The *grade* of a deduction $\mathcal{D}$ is given by the highest complexity among the formulas in CF($\mathcal{D}$).

In other words, the grade of a derivation is the maximum complexity of the formulas that are eliminated with Cut.

Definition 5.2.31 Let $\mathcal{D}$ be a derivation and let

$$\text{Cut}\ \frac{\begin{array}{c}\mathcal{D}_1\\ \Gamma \Rightarrow \Delta, \varphi\end{array} \qquad \begin{array}{c}\mathcal{D}_2\\ \Pi, \varphi \Rightarrow \Sigma\end{array}}{\Gamma, \Pi \Rightarrow \Delta, \Sigma}$$

be an application of Cut in $\mathcal{D}$. The *level* of that Cut is $m + n$, where m is the number of sequents in $\mathcal{D}_1$ (including $\Gamma \Rightarrow \Delta, \varphi$) and n is the number of sequents in $\mathcal{D}_2$ (including $\Pi, \varphi \Rightarrow \Sigma$).

Example 5.2.32 Take the following derivation:

$$\text{R}\vee\ \frac{\text{Cut}\ \dfrac{\text{R}\circ\ \dfrac{\text{L}\wedge\ \dfrac{p \Rightarrow p}{p \wedge q \Rightarrow p} \qquad p \wedge q \Rightarrow p \wedge q}{p \wedge q, p \wedge q \Rightarrow p \circ (p \wedge q)} \qquad \text{L}\wedge\ \dfrac{p \wedge q \Rightarrow p \wedge q}{(p \wedge q) \wedge r \Rightarrow p \wedge q}}{(p \wedge q) \wedge r, p \wedge q \Rightarrow p \circ (p \wedge q)}}{(p \wedge q) \wedge r, p \wedge q \Rightarrow (p \circ (p \wedge q)) \vee s}$$

In this derivation the level of the Cut is 6.

Note that the difference between the level of the Cut and the rank is that the rank counts from the last occurrence of the Cut-formula to the moment it first appears, while here we keep on counting until the axioms. Also note that, as Cut is a two premise rule, the minimum level of a Cut is 2.

The structure of the proof is the following:

- 1. Induction on the grade.
  - 1.1 Base step: Cut is eliminable in derivations of grade 0. This means the most complex cut-formulas are atomic formulas or constants. Proved by induction on the level of an application of Cut in the derivation.
    * 1.1.1. Base step: Cut of level 2 is eliminable.
    * 1.1.2. Inductive step: Cut of level n can be substituted with a Cut of lesser level.
  - 1.2. Inductive step: Cut on derivations of grade m is eliminable, if it is eliminable on derivations of lesser grade; proved again by induction on the level of Cut:
    * 1.2.1. Base step: Cut of level 2 is eliminable.
    * 1.2.2. Inductive step: Cut of level n can be substituted with Cut of lesser level and possibly lesser grade.

We give the proof for some connectives and leave the rest as exercises to the reader (see Exercise 101).

Proof

1.1. Base step: Proof on atomic formulas or constants:

1.1.1. Base step: Cut on atomic formulas or constants of level 2 is eliminable. When the grade of a derivation is 0, it means that the Cut-formula is an atom or a constant. Which implies that it is the result of an application of an axiom or a rule for a constant. If the level of the Cut is 2, that means both formulas come from axioms, as in:

$$\text{Cut}\ \frac{1 \Rightarrow 1 \qquad \Rightarrow 1}{1 \Rightarrow 1}$$

which is simply reduced to $1 \Rightarrow 1$.

1.1.2. Inductive step: Cut on a atomic formulas or constants of level n may be substituted with Cut of a lesser level. If the level is higher than 2, we have two cases: **(a)** the Cut-formula is principal in both of the premises or **(b)** is not principal in at least one.

(a) If both formulas are principal and one comes from a rule, the conclusion is identical to the premise of the rule:

$$\text{Cut}\,\dfrac{\text{L1}\,\dfrac{\begin{matrix}\mathcal{D}\\ \vdots \\ \Gamma \Rightarrow \Delta\end{matrix}}{\Gamma, 1 \Rightarrow \Delta} \qquad \Rightarrow 1}{\Gamma \Rightarrow \Delta}$$

(b) If at least one of the Cut-formulas is not principal, we move up the Cut one level. Let's check some examples:

The + case:

$$\text{Cut}\,\dfrac{\begin{matrix}\mathcal{D}\\ \vdots \\ \Gamma, 1 \Rightarrow \Delta\end{matrix} \qquad \text{L+}\,\dfrac{\begin{matrix}\mathcal{D}_1\\ \vdots \\ \Gamma', \varphi \Rightarrow 1, \Delta'\end{matrix} \qquad \begin{matrix}\mathcal{D}_2\\ \vdots \\ \Gamma'', \psi \Rightarrow \Delta''\end{matrix}}{\Gamma', \Gamma'', \varphi + \psi \Rightarrow 1, \Delta', \Delta''}}{\Gamma, \Gamma', \Gamma'', \varphi + \psi \Rightarrow \Delta, \Delta', \Delta''}$$

Can be rearranged like this:

$$\text{L+}\,\dfrac{\text{Cut}\,\dfrac{\begin{matrix}\mathcal{D}\\ \vdots \\ \Gamma, 1 \Rightarrow \Delta\end{matrix} \qquad \begin{matrix}\mathcal{D}_1\\ \vdots \\ \Gamma', \varphi \Rightarrow 1, \Delta'\end{matrix}}{\Gamma, \Gamma', \varphi, \Rightarrow \Delta, \Delta'} \qquad \begin{matrix}\mathcal{D}_2\\ \vdots \\ \Gamma'', \psi \Rightarrow \Delta''\end{matrix}}{\Gamma, \Gamma', \Gamma'', \varphi + \psi, \Rightarrow \Delta, \Delta', \Delta''}$$

The $\vee$ case:

$$\text{Cut}\,\dfrac{\text{L}\vee\,\dfrac{\begin{matrix}\mathcal{D}\\ \vdots \\ \Gamma, \varphi, 1 \Rightarrow \Delta\end{matrix} \qquad \begin{matrix}\mathcal{D}_1\\ \vdots \\ \Gamma, \psi, 1 \Rightarrow \Delta\end{matrix}}{\Gamma, \varphi \vee \psi, 1 \Rightarrow \Delta} \qquad \begin{matrix}\mathcal{D}_2\\ \vdots \\ \Gamma' \Rightarrow 1, \Delta'\end{matrix}}{\Gamma, \Gamma', \varphi \vee \psi \Rightarrow \Delta, \Delta'}$$

Has to be rearranged like this:

$$\text{L}\vee\,\dfrac{\text{Cut}\,\dfrac{\begin{matrix}\mathcal{D}\\ \vdots \\ \Gamma, \varphi, 1 \Rightarrow \Delta\end{matrix} \qquad \begin{matrix}\mathcal{D}_2\\ \vdots \\ \Gamma' \Rightarrow 1, \Delta'\end{matrix}}{\Gamma, \Gamma', \varphi \Rightarrow \Delta, \Delta'} \qquad \text{Cut}\,\dfrac{\begin{matrix}\mathcal{D}_1\\ \vdots \\ \Gamma, \psi, 1 \Rightarrow \Delta\end{matrix} \qquad \begin{matrix}\mathcal{D}_2\\ \vdots \\ \Gamma' \Rightarrow 1, \Delta'\end{matrix}}{\Gamma, \Gamma', \psi \Rightarrow \Delta, \Delta'}}{\Gamma, \Gamma', \varphi \vee \psi \Rightarrow \Delta, \Delta'}$$

Note that we multiplied the Cuts but lowered the level. The main difference between these two cases is that for the additive connective we are contracting the 1 when applying L$\vee$: meaning that there was a 1 in both premises and after the application of the rule we have only one 1 on the succedent. So when we rearrenge the order, we have to Cut twice to have the same contexts so that we can apply L$\vee$. While in the multiplicative rules we don't have the problem of shared contexts.

1.1.3. Conclusion: Cut on atomic formulas or constants is eliminable.

1.2. Inductive step: Cut on derivations of grade m is eliminable, if it is eliminable on derivations of lesser grade. To prove this, as we said, we have to dive into the second auxiliary induction on the level of the Cut.

1.2.1. Base step: Cut of level 2 can be eliminated. If the level is 2 and the grade is higher than 0, then the only option is that one is cutting two instances of Id for complex formulas. And thus, Cut can be eliminated.

1.2.2. Inductive step: Cut of level n may be substituted with Cuts of lesser level. Again we have two cases to consider. When both formulas are principal and when at least one is not.

(a) For the fragment without exponentials we will consider just one case. Then we will review several transformations when both formulas are principal and exponentials are being Cut.

The $\circ$ case:

$$
\text{Cut}\ \dfrac{\text{R}\circ\ \dfrac{\begin{matrix}\vdots\\ \mathcal{D}_1\\ \Gamma \Rightarrow \varphi, \Delta\end{matrix} \qquad \begin{matrix}\vdots\\ \mathcal{D}_2\\ \Gamma' \Rightarrow \psi, \Delta'\end{matrix}}{\Gamma, \Gamma' \Rightarrow \varphi \circ \psi, \Delta, \Delta'} \qquad \text{L}\circ\ \dfrac{\begin{matrix}\vdots\\ \mathcal{D}_3\\ \Gamma'', \varphi, \psi \Rightarrow \Delta''\end{matrix}}{\Gamma'', \varphi \circ \psi \Rightarrow \Delta''}}{\Gamma, \Gamma', \Gamma'' \Rightarrow \Delta, \Delta', \Delta''}
$$

is replaced by:

$$
\text{Cut}\ \dfrac{\text{Cut}\ \dfrac{\begin{matrix}\vdots\\ \mathcal{D}_1\\ \Gamma \Rightarrow \varphi, \Delta\end{matrix} \qquad \begin{matrix}\vdots\\ \mathcal{D}_3\\ \Gamma'', \varphi, \psi \Rightarrow \Delta''\end{matrix}}{\Gamma, \Gamma'', \psi \Rightarrow \Delta, \Delta''} \qquad \dfrac{\begin{matrix}\vdots\\ \mathcal{D}_2\end{matrix}}{\Gamma' \Rightarrow \psi, \Delta'}}{\Gamma, \Gamma', \Gamma'' \Rightarrow \Delta, \Delta', \Delta''}
$$

Decreasing the level of the Cuts. For the exponentials there are three cases of pairs of rules that make both formulas principals for $!$.

The L !,R ! case:

$$\text{Cut-!}\ \dfrac{\text{L!}\ \dfrac{\begin{array}{c}\vdots\\ \mathcal{D}_1\\ \Gamma, (!\,\varphi)^m, \varphi \Rightarrow \Delta\end{array}}{\Gamma, (!\,\varphi)^{m+1} \Rightarrow \Delta} \qquad \text{R!}\ \dfrac{\begin{array}{c}\vdots\\ \mathcal{D}_2\\ !\,\Gamma' \Rightarrow \varphi, ?\,\Delta'\end{array}}{!\,\Gamma' \Rightarrow !\,\varphi, ?\,\Delta'}}{\Gamma, !\,\Gamma' \Rightarrow \Delta, ?\,\Delta'}$$

If L ! is introducing the first $!\,\varphi$ of the sequent then what we actually have looks like this:

$$\text{Cut-!}\ \dfrac{\text{L!}\ \dfrac{\begin{array}{c}\vdots\\ \mathcal{D}_1\\ \Gamma, \varphi \Rightarrow \Delta\end{array}}{\Gamma, !\,\varphi \Rightarrow \Delta} \qquad \text{R!}\ \dfrac{\begin{array}{c}\vdots\\ \mathcal{D}_2\\ !\,\Gamma' \Rightarrow \varphi, ?\,\Delta'\end{array}}{!\,\Gamma' \Rightarrow !\,\varphi, ?\,\Delta'}}{\Gamma, !\,\Gamma' \Rightarrow \Delta, ?\,\Delta'}$$

Which is replaced by:

$$\text{Cut}\ \dfrac{\begin{array}{c}\vdots\\ \mathcal{D}_1\\ \Gamma, \varphi \Rightarrow \Delta\end{array} \qquad \begin{array}{c}\vdots\\ \mathcal{D}_2\\ !\,\Gamma' \Rightarrow \varphi, ?\,\Delta'\end{array}}{\Gamma, !\,\Gamma' \Rightarrow \Delta, ?\,\Delta'}$$

Otherwise it has to be rearranged like this:

$$\dfrac{\text{Several Contractions}\ \dfrac{\text{Cut}\ \dfrac{\text{Cut-!}\ \dfrac{\begin{array}{c}\vdots\\ \mathcal{D}_1\\ \Gamma, (!\,\varphi)^m, \varphi \Rightarrow \Delta\end{array} \qquad \text{R!}\ \dfrac{\begin{array}{c}\vdots\\ \mathcal{D}_2\\ !\,\Gamma' \Rightarrow \varphi, ?\,\Delta'\end{array}}{!\,\Gamma' \Rightarrow !\,\varphi, ?\,\Delta'}}{\Gamma, !\,\Gamma', \varphi \Rightarrow \Delta, ?\,\Delta'} \qquad \begin{array}{c}\vdots\\ \mathcal{D}_2\\ !\,\Gamma' \Rightarrow \varphi, ?\,\Delta'\end{array}}{\Gamma, !\,\Gamma', !\,\Gamma' \Rightarrow \Delta, ?\,\Delta', ?\,\Delta'}}{\vdots}}{\Gamma, !\,\Gamma' \Rightarrow \Delta, ?\,\Delta'}$$

Note that even if we have introduced a new Cut, the level of the Cut on $!\,\varphi$ has decreased.

The W !, R ! **case**: this is almost identical and left to the reader.

The C !, R ! case: If we have:

$$
\dfrac{\dfrac{\dfrac{\vdots}{\mathcal{D}_1}}{\Gamma, (!\,\varphi)^n \Rightarrow \Delta}\ \text{C!}\quad \text{and}\quad \dfrac{}{\Gamma, (!\,\varphi)^{n-1} \Rightarrow \Delta}\qquad \dfrac{\dfrac{\vdots}{\mathcal{D}_2}}{!\,\Gamma' \Rightarrow \varphi, ?\,\Delta'}}{}
$$

$$
\dfrac{\dfrac{\begin{matrix}\vdots\\ \mathcal{D}_1\\ \Gamma, (!\,\varphi)^n \Rightarrow \Delta\end{matrix}}{\Gamma, (!\,\varphi)^{n-1} \Rightarrow \Delta}\,\text{C!} \qquad \dfrac{\begin{matrix}\vdots\\ \mathcal{D}_2\\ !\,\Gamma' \Rightarrow \varphi, ?\,\Delta'\end{matrix}}{!\,\Gamma' \Rightarrow !\,\varphi, ?\,\Delta'}\,\text{R!}}{\Gamma, !\,\Gamma' \Rightarrow \Delta\,?\,\Delta'}\,\text{Cut-!}
$$

One can replace it by:

$$
\dfrac{\begin{matrix}\vdots\\ \mathcal{D}_1\\ \Gamma, (!\,\varphi)^n \Rightarrow \Delta\end{matrix} \qquad \dfrac{\begin{matrix}\vdots\\ \mathcal{D}_2\\ !\,\Gamma' \Rightarrow \varphi, ?\,\Delta'\end{matrix}}{!\,\Gamma' \Rightarrow !\,\varphi, ?\,\Delta'}\,\text{R!}}{\Gamma, !\,\Gamma' \Rightarrow \Delta\,?\,\Delta'}\,\text{Cut-!}
$$

Which again has a lesser level.

(b) If at least one of the formulas is not principal, you have to permute the order of the application of Cut and the rule where the Cut-formula is not principal. If the connective is not an exponential the procedure is the same as in the first base step.

Let's pay attention to the exponentials. When we have a Cut- ! (or a Cut- ?) that comes from a two-premise multiplicative rule it can get complicated. We will show how to do it for the case of $\circ$ and leave the remaining cases to the reader. Take the following derivation:

$$
\dfrac{\dfrac{\begin{matrix}\vdots\\ \mathcal{D}_1\\ \Gamma, (!\,\varphi)^m \Rightarrow \Delta, \psi\end{matrix} \qquad \begin{matrix}\vdots\\ \mathcal{D}_2\\ \Gamma', (!\,\varphi)^n \Rightarrow \Delta', \chi\end{matrix}}{\Gamma, \Gamma', (!\,\varphi)^{m+n} \Rightarrow \Delta, \Delta', \psi \circ \chi}\,\text{R}\circ \qquad \begin{matrix}\vdots\\ \mathcal{D}_3\\ \Gamma'' \Rightarrow !\,\varphi, \Delta''\end{matrix}}{\Gamma, \Gamma', \Gamma'' \Rightarrow \Delta, \Delta', \Delta''}\,\text{Cut-!}
$$

where $m,n \geqslant 1$.

If $m = 0$ or $n = 0$ we permute Cut upwards. Otherwise, there are two possibilities. (a) If $!\,\varphi$ is not a principal formula on the right premise (the one that comes from $\mathcal{D}_3$), we can simply reduce the level of the the Cut by cutting the formula some steps earlier. (b) If $!\,\varphi$ is the principal formula on the right premise, we know that Γ'' has to be $!\,\Gamma'''$ and Δ'' has to be $?\,\Delta'''$. So what we have to do is the following:

$$
\dfrac{\dfrac{\dfrac{\vdots}{\mathcal{D}_1} \quad \dfrac{\vdots}{\mathcal{D}_3}}{\dfrac{\Gamma, (!\varphi)^m \Rightarrow \Delta, \psi \qquad !\Gamma''' \Rightarrow !\varphi, ?\Delta'''}{\Gamma, !\Gamma''' \Rightarrow \Delta, ?\Delta''', \psi}\ \text{Cut-!}} \qquad \dfrac{\dfrac{\vdots}{\mathcal{D}_2} \quad \dfrac{\vdots}{\mathcal{D}_3}}{\dfrac{\Gamma', (!\varphi)^n \Rightarrow \Delta, \chi \qquad !\Gamma''' \Rightarrow !\varphi, ?\Delta'''}{\Gamma', !\Gamma''' \Rightarrow \Delta', ?\Delta''', \chi}\ \text{Cut-!}}}{\dfrac{\Gamma, \Gamma', !\Gamma''', !\Gamma''' \Rightarrow \Delta, \Delta', ?\Delta''', ?\Delta''', \psi \circ \chi}{\dfrac{\vdots}{\Gamma, \Gamma', !\Gamma''', \Rightarrow \Delta, \Delta', ?\Delta''', \psi \circ \chi}}\ \text{Several Contractions}}\ \text{R}\circ
$$

Again, note that we reduced the level of the Cuts on the $!\,\varphi$ formula.

1.2.3. Conclusion: Cut on derivations of grade m may be substituted with Cut on derivations of lesser level.

1.3. Conclusion: Cut is eliminable for deductions of any grade.

□

It is possible to prove a stronger result for **CLL** called strong normalizability. That is, first you show that Cut is eliminable via reduction and then you show that every reduction procedure is finite. Again, we refer the reader to Roorda (1989) for a clear exposition of this result.

Exercise 101 Complete the proof of Cut-elimination for the remaining connectives.

5.2.4.3 Admisibility and Derivability

As should be expected, neither Weakening nor Contraction are admissible in **CLL**. To witness, proving the following is immediate:

$$
\dfrac{\dfrac{\dfrac{\varphi \Rightarrow \varphi}{\Rightarrow \varphi, \neg\varphi}\ \text{R}\neg}{\Rightarrow \varphi \vee \neg\varphi, \neg\varphi}\ \text{R}\vee}{\Rightarrow \varphi \vee \neg\varphi, \varphi \vee \neg\varphi}\ \text{R}\vee
$$

Yet there is no way to prove $\Rightarrow \varphi \vee \neg\varphi$. It is also immediate to note that $p \Rightarrow p$ is provable, yet $p, q \Rightarrow p$ is not. Nevertheless, one can make these rules admissible by adding extra rules for the binary connectives at the cost of losing the eliminability of Cut. We will show that giving additive rules to $\rightarrow$, $\circ$ or $+$ makes Weakening and Contraction derivable, yet Cut is no longer eliminable.

We will show that if one wants to have eliminability of Cut we need to have the two separate connectives in the language, meaning the additive and the multiplicative rules for different symbols. We follow Schellinx's (1991) investigations on negation and linear implications to illustrate this point. We will present this result with implications because the original result was presented in this terms, but also because we want to give the reader some familiarity with linear implication, which is a connective we barely mentioned in the previous sections. Yet, because implications are interderivable with conjunctions and disjunctions nothing really hinges on this particular connective.

Take the following set of rules to characterize Classical Logic:

Structural Rules

$$\text{Id}\ \frac{}{\varphi \Rightarrow \varphi} \qquad \text{Cut}\ \frac{\Gamma \Rightarrow \varphi, \Delta \qquad \Pi, \varphi \Rightarrow \Sigma}{\Gamma, \Pi \Rightarrow \Delta, \Sigma}$$

$$\text{LW}\ \frac{\Gamma \Rightarrow \Delta}{\Gamma, \varphi \Rightarrow \Delta} \qquad \text{RW}\ \frac{\Gamma \Rightarrow \Delta}{\Gamma \Rightarrow \varphi, \Delta}$$

$$\text{LC}\ \frac{\Gamma, \varphi, \varphi \Rightarrow \Delta}{\Gamma, \varphi \Rightarrow \Delta} \qquad \text{RC}\ \frac{\Gamma \Rightarrow \varphi, \varphi, \Delta}{\Gamma \Rightarrow \varphi, \Delta}$$

Operational Rules

$$L\bot\ \frac{}{\bot \Rightarrow} \qquad \text{R}\bot\ \frac{\Gamma \Rightarrow \Delta}{\Gamma \Rightarrow \bot, \Delta}$$

$$\text{L}\to\ \frac{\Gamma \Rightarrow \varphi, \Delta \qquad \Pi, \psi \Rightarrow \Sigma}{\Gamma, \Pi, \varphi \to \psi \Rightarrow \Delta, \Sigma} \qquad \text{R}\to\ \frac{\Gamma, \varphi \Rightarrow \psi, \Delta}{\Gamma \Rightarrow \varphi \to \psi, \Delta}$$

Let's call this calculus $\mathbf{CL}^M$. Note that in $\mathbf{CL}^M$ the rules for the implication are multiplicative and the additive rules are derivable in the presence of Weakening and Contraction. These rules are enough to characterize Classical Logic, in the sense that all formulas involving classical conjunctions and disjunctions are definable using material conditional and classical negation (defined in turn as $\varphi \to \bot$). At the same time, it is straightforward to prove that Cut is eliminable in this calculus. In fact, we have already proved it in Chap. 2 with a primitive negation. Subtracting the rules for negation that are primitive in **LK** but adding the the rules for $\bot$ has the effect of preserving Cut-elimination. Note that R$\bot$ is also eliminable, as every application of it can be replaced by an application of WR. Now, compare $\mathbf{CL}^M$ with what we will call $\mathbf{CL}^{\to}$, that is Identity, Cut, rules for $\bot$ and both the multiplicative and the additive rules for the conditional, $\to$:

Structural Rules

$$\text{Id}\ \frac{}{\varphi \Rightarrow \varphi} \qquad \text{Cut}\ \frac{\Gamma \Rightarrow \varphi, \Delta \qquad \Pi, \varphi \Rightarrow \Sigma}{\Gamma, \Pi \Rightarrow \Delta, \Sigma}$$

Operational Rules

$$L\bot\ \frac{}{\bot \Rightarrow} \qquad \text{R}\bot\ \frac{\Gamma \Rightarrow \Delta}{\Gamma \Rightarrow \bot, \Delta}$$

$$\text{Lm}\!\to \dfrac{\Gamma \Rightarrow \varphi, \Delta \qquad \Pi, \psi \Rightarrow \Sigma}{\Gamma, \Pi, \varphi \to \psi \Rightarrow \Delta, \Sigma} \qquad \text{Rm}\!\to \dfrac{\Gamma, \varphi \Rightarrow \psi, \Delta}{\Gamma \Rightarrow \varphi \to \psi, \Delta}$$

$$\text{La}\!\to \dfrac{\Gamma \Rightarrow \varphi, \Delta \qquad \Gamma, \psi \Rightarrow \Delta}{\Gamma, \varphi \to \psi \Rightarrow \Delta} \qquad \text{Ra}\!\to_1 \dfrac{\Gamma \Rightarrow \psi, \Delta}{\Gamma \Rightarrow \varphi \to \psi, \Delta}$$

$$\text{Ra}\!\to_2 \dfrac{\Gamma, \varphi \Rightarrow \Delta}{\Gamma \Rightarrow \varphi \to \psi, \Delta}$$

We can prove the following result:

Theorem 5.2.33 $\mathbf{CL}^{\to}$ *is inferentially and metainferentially identical to* **LK***, yet Cut is not eliminable.*

Proof To prove that $\mathbf{CL}^{\to}$ has the same provable sequents and derivable rules than **LK** we have to show that with this set of rules Weakening and Contraction are derivable. We need Cut to derive the four rules:

Weakening Right

$$\text{Cut}\ \dfrac{\text{Ra}\!\to_2 \dfrac{\bot \Rightarrow}{\Rightarrow \bot \to \psi} \qquad \text{Lm}\!\to \dfrac{\text{R}\bot\ \dfrac{\Gamma \Rightarrow \Delta}{\Gamma \Rightarrow \bot, \Delta} \qquad \psi \Rightarrow \psi}{\Gamma, \bot \to \psi \Rightarrow \Delta, \psi}}{\Gamma \Rightarrow \Delta, \psi}$$

Weakening Left

$$\text{Cut}\ \dfrac{\text{Lm}\!\to \dfrac{\varphi \Rightarrow \varphi \qquad \bot \Rightarrow}{\varphi, \varphi \to \bot \Rightarrow} \qquad \text{Ra}\!\to_1 \dfrac{\text{R}\bot\ \dfrac{\Gamma \Rightarrow \Delta}{\Gamma \Rightarrow \Delta, \bot}}{\Gamma \Rightarrow \Delta, \varphi \to \bot}}{\Gamma, \varphi \Rightarrow \Delta}$$

Contraction Right

$$\text{Cut}\ \dfrac{\text{Cut}\ \dfrac{\text{Ra}\!\to_2 \dfrac{\varphi \Rightarrow \varphi}{\Rightarrow \varphi \to \varphi, \varphi} \qquad \text{Lm}\!\to \dfrac{\varphi \Rightarrow \varphi \qquad \varphi \Rightarrow \varphi}{\varphi, \varphi \to \varphi \Rightarrow \varphi}}{\varphi \Rightarrow \varphi, \varphi} \qquad \Gamma, \varphi, \varphi \Rightarrow \Delta}{\Gamma, \varphi \Rightarrow \Delta}$$

Contraction Left Analogous to RC.

In order to prove that Cut is not eliminable, one needs to show that there are certain sequents that are not provable without using the Cut rule in the derivation. The simplest witness is $p, q \Rightarrow p$. □

Another way to read this theorem is by saying that in this calculus Weakening and Contraction are not derivable without Cut. It is of particular interest because it allows us to see that two calculi (such as this and **LK**, for example) can determine the same logic (in the sense that they prove the same inferences and metainferences) and still have different metainferential properties (for instance, Cut being eliminable in **LK** but not in $\mathbf{CM}^{\rightarrow}$).

As we said at the beginning of the section, this is not a property inherent of conditionals. Because they are interderivable with conjuntions and disjunctions one can prove the same result for these connectives.

Fact 5.2.34 Adding to **CLL** the additive rules for the multiplicative connectives and vice-versa has the effect of making the rules of left and right Contraction derivable for arbitrary formulas.

Proof The following derivations, where the connectives $\wedge$ and $\vee$ have both multiplicative and additive rules, show the (slightly stronger) result, that only one of the two multiplicative rules on each additive connective is needed:

- Left Contraction:

$$\text{Cut}\ \dfrac{\text{R}\wedge\ \dfrac{\varphi \Rightarrow \varphi \qquad \varphi \Rightarrow \varphi}{\varphi \Rightarrow \varphi \wedge \varphi} \qquad \text{Lm}\wedge\ \dfrac{\Gamma, \varphi, \varphi \Rightarrow \Delta}{\Gamma, \varphi \wedge \varphi \Rightarrow \Delta}}{\Gamma, \varphi \Rightarrow \Delta}$$

- Right Contraction:

$$\text{Cut}\ \dfrac{\text{Rm}\vee\ \dfrac{\Gamma \Rightarrow \psi, \psi, \Delta}{\Gamma \Rightarrow \psi \vee \psi, \Delta} \qquad \text{R}\vee\ \dfrac{\psi \Rightarrow \psi \qquad \psi \Rightarrow \psi}{\psi \vee \psi \Rightarrow \psi}}{\Gamma \Rightarrow \psi, \Delta}$$

□

Allowing both sets of rules for each constants makes the rules of Weakening derivable.

Fact 5.2.35 Adding to **CLL** the additive rules for the multiplicative constants and vice-versa has the effect of making the rules of left and right Weakening derivable for arbitrary formulas.

Proof Consider the following derivations, where the connectives $\top$ and 0 have both multiplicative and additive rules.

- Left Weakening:

$$\text{Cut}\ \dfrac{\text{R}\top\ \dfrac{}{\varphi \Rightarrow \top} \qquad \text{Lm}\top\ \dfrac{\Gamma \Rightarrow \Delta}{\Gamma, \top \Rightarrow \Delta}}{\Gamma, \varphi \Rightarrow \Delta}$$

- Right Weakening:

$$\text{Cut}\ \dfrac{\text{L}0\ \dfrac{}{0 \Rightarrow \varphi} \qquad \text{Rm}0\ \dfrac{\Gamma \Rightarrow \Delta}{\Gamma \Rightarrow 0, \Delta}}{\Gamma \Rightarrow \varphi, \Delta}$$

□

Now, if we want to have a calculus that enjoys the Cut-elimination property, we have to distinguish between multiplicative and additive rules and thus have two implications in the language.

Structural Rules

$$\text{Id}\ \frac{}{A \Rightarrow A} \qquad \text{Cut}\ \frac{\Gamma \Rightarrow \varphi, \Delta \qquad \Pi, \varphi \Rightarrow \Sigma}{\Gamma, \Pi \Rightarrow \Delta, \Sigma}$$

Operational Rules

$$\text{L}\bot\ \frac{}{\bot \Rightarrow} \qquad \text{L0}\ \frac{}{\Gamma, 0 \Rightarrow \Delta}$$

$$\text{R}\bot\ \frac{\Gamma \Rightarrow \Delta}{\Gamma \Rightarrow \bot, \Delta}$$

$$\text{L}\supset\ \frac{\Gamma \Rightarrow \varphi, \Delta \qquad \Gamma, \psi \Rightarrow \Delta}{\Gamma, \varphi \supset \psi \Rightarrow \Delta}$$

$$\text{L}\!\to\ \frac{\Gamma \Rightarrow \varphi, \Delta \qquad \Pi, \psi \Rightarrow \Sigma}{\Gamma, \Pi, \varphi \to \psi \Rightarrow \Delta, \Sigma}$$

$$\text{R}\supset\ \frac{\Gamma \Rightarrow \psi, \Delta}{\Gamma \Rightarrow \varphi \supset \psi, \Delta}$$

$$\text{R}\!\to\ \frac{\Gamma, \varphi \Rightarrow \psi, \Delta}{\Gamma \Rightarrow \varphi \to \psi, \Delta}$$

$$\text{R}\supset\ \frac{\Gamma, \varphi \Rightarrow \Delta}{\Gamma \Rightarrow \varphi \supset \psi, \Delta}$$

The proof of Cut-elimination was already provided in Sect. 5.2.4.2; there is no new vocabulary to add to the proof we already gave. And to show that this amounts to the whole fragment of **CLL** without exponentials one has to show that the rest of the connectives can be defined with $\to$, $\supset$, 0 and $\bot$. These definitions are what one would expect:

- $\varphi + \psi \Leftrightarrow (\varphi \to \bot) \to \psi$
- $\varphi \vee \psi \Leftrightarrow (\varphi \supset 0) \supset \psi$
- $\varphi \circ \psi \Leftrightarrow ((\varphi \to (\psi \to \bot)) \to \bot$
- $\varphi \wedge \psi \Leftrightarrow ((\varphi \supset (\psi \supset 0)) \supset 0$
- $1 \Leftrightarrow \bot \to \bot$
- $\top \Leftrightarrow 0 \supset 0$

What is more interesting about this calculus is that these two implications can co-exist without collapsing, yet the negations will collapse into one. Formally, this means the following:

Fact 5.2.36
$\varphi \to \bot \Leftrightarrow \varphi \supset 0$

Note, nevertheless, that as expected the collapse doesn't happen between both falsity constants, but with the defined negation.

Exercises

Exercise 102 Prove that, if we take $\mathbf{CM}^{\rightarrow}$ with the axiom for 0 instead of $\bot$, we obtain a system where Weakening and Contraction are also derivable while Cut is not eliminable.

Exercise 103 Prove these equivalences mentioned above:

- $\varphi + \psi \Leftrightarrow (\varphi \rightarrow \bot) \rightarrow \psi$
- $\varphi \vee \psi \Leftrightarrow (\varphi \supset 0) \supset \psi$
- $\varphi \circ \psi \Leftrightarrow ((\varphi \rightarrow (\psi \rightarrow \bot)) \rightarrow \bot$
- $\varphi \wedge \psi \Leftrightarrow ((\varphi \supset (\psi \supset 0)) \supset 0$
- $1 \Leftrightarrow \bot \rightarrow \bot$
- $\top \Leftrightarrow 0 \supset 0$

Before the closing of this section, just a few comments on invertibility and decidability. Invertibility is a property of R $\rightarrow$, L$\circ$, R$+$, R$\neg$, L$\neg$, R$\bot$, L1, L $\supset$, R$\wedge$ and L$\vee$, and one can prove it by induction on the length of the proof such as we did in Sect. 2.2.3 for R$\wedge$. We leave the proof of the other rules as an exercise to the reader (Exercise 104).

Decidability for the whole logic is an open debate. Lincoln et al. (1992) and later on Kanovuch (2016) showed its undecidability. Nevertheless, Bimbó and Dunn (2022) published a proof of decidability as well as a claim that there were mistakes in these proofs.

Exercise 104 Prove the invertibility of R $\rightarrow$, L$\circ$, R$+$, R$\neg$, L$\neg$, R$\bot$, L1, L $\supset$, R$\wedge$ and L$\vee$.

5.2.4.4 The Multiplicative Connectives and the Structure

Before the closing of this section, we would like to make a brief mention to something that may come out in modern discussions on substructural logics. A feature that is sometimes taken as philosophically interesting about multiplicative connectives is that they relate to some distinctive structural features of sequents. Let's take conjunction as an example. A sequent of the form:

$$\varphi_1, \ldots, \varphi_n \Rightarrow \Delta$$

Is read canonically as: "From φ_1 and ... and φ_n, it follows that Δ". So the comma that separates the formulas on the antecedent of a sequent is read conjunctively: 'and'. Moreover, we have that in **CLL** the following rule is derivable:[8]

$$\frac{\Gamma, \varphi, \psi \Rightarrow \Delta}{\Gamma, \varphi \circ \psi \Rightarrow \Delta}$$

[8] The double line abbreviates two derivations: one from top to bottom, and the other from bottom to top. The proof of this double derivation is more or less straightforward. In its top-down reading, it is just an application of L$\circ$; for the converse, use R$\circ$ and Cut.

Where the only difference between the upper and lower sequent is that in the first there are two formulas separated by a comma (on the antecedent), while in the second the same two formulas are compounds of a multiplicative conjunction. This gives justification to the philosophical claim that the left-hand comma *is*, in a certain sense, a multiplicative conjunction.

In a very similar vein one can find justifications for the following claims:

- The comma on the right *is* a multiplicative disjunction.
- The empty multiset on the left *is* 1.
- The empty multiset on the right *is* $\perp$.
- The availability of the rules of Weakening and Contraction on the left *is* !.
- The availability of rules of Weakening and Contraction on the right *is* ?.
- The turnstile *is* the multiplicative conditional.
- The symmetry between the left and the right side of the sequent *is* negation.

If interested, the reader may find instructive to try to justify these claims as we did with conjunction. We refer to Schroeder-Heister (2024), section 2.4, for more on this topic.

5.3 Some Other Non-contractive and Non-monotonic Logics

In his explication of Linear Logic, Girard saw the rules of Weakening and Contraction as two sides of the same coin. The exponentials in that logic expressed the availability of *both* structural rules at each side of the turnstile (! at the left, ? at the right). As Girard would say, "Weakening without Contraction is dishonest", whereas "Contraction without Weakening is marvelous" (Girard, 1995). Yet in the philosophical logic tradition, Weakening and Contraction were usually treated separately. In this section we will have a look at some logics that do precisely this.

5.3.1 Non-Contractive Logics

5.3.1.1 Affine Logic and Łukasiewicz Logic

In this section, we will look at some properties of a fragment of $\mathbb{CLL}$. In particular, we will start by presenting Affine Logic and its relation to the Łukasiewicz family of logics.

Affine Logic is the multiplicative and additive fragment of **CLL** plus Weakening. As Prijatelj (1996) shows, one can extend Affine Logic with bounded Contraction and some axioms to get the well-known family of Łukasiewicz logics. To be more precise, one can get a calculus $\mathbf{Ł}_n$ such that if a sequent $\Gamma \Rightarrow \Delta$ is provable in it, then every valuation v in the class that characterices $\mathbf{Ł}_n$ connectives, gives 1 to the sentence $\neg\gamma_1 + \cdots + \neg\gamma_n + \delta_1 \cdots + \delta_m$, where each $\gamma_i \in \Gamma$ and each $\delta_i \in \Delta$.

Łukasiewicz logics are a family of many-valued logics that go from 3-valued to infinitely many-valued logics. It was originally motivated to treat contingent future statements. In particular, 3-Łukasiewicz differs from $\mathbb{K}3$ mainly because it has what we can call a 'non-contractive' conditional. This feature extends to the whole family of n-Łukasiewicz logics. That is, the following inference is invalid in n-Łukasiewicz: $\Rightarrow (\varphi \to (\varphi \to_{n-1} \psi)) \to (\varphi \to_{n-2} \psi)$. For example, in 3-Łukasiewicz $\Rightarrow (\varphi \to (\varphi \to \psi)) \to (\varphi \to \psi)$ fails; in 4-Łukasiewicz $\Rightarrow (\varphi \to (\varphi \to (\varphi \to \psi))) \to (\varphi \to (\varphi \to \psi))$ fails, and so on. We will restrict the presentation to the family of finite Łukasiewicz logics for simplicity, but in Prijatelj (1996) the infinitely many-valued Łukasiewicz logic is covered. In what follows we will call $\mathbb{L}_n$ each n-Łukasiewicz logic where $n \in \mathbb{N} - \{1, 2\}$.

The trick to give a sequent calculus for some $\mathbb{L}_n$ is to restrict the Contraction rules accordingly. The mnemonics are a bit tricky, but manageable: if you let the calculus contract two formulas into one, then you get 3-valued Łukasiewicz, called $\mathbb{L}_3$. If you let the calculus contract three formulas into two, you get $\mathbb{L}_4$. To make it clear, take Affine Logic and add the following class of schematic rules, that we can call $\mathbf{C^n}$:

$$\mathrm{L}C^n\ \frac{\Gamma, \varphi^n \Rightarrow \Delta}{\Gamma, \varphi^{n-1} \Rightarrow \Delta} \qquad\qquad \mathrm{R}C^n\ \frac{\Gamma \Rightarrow \Delta, \varphi^n}{\Gamma \Rightarrow \Delta, \varphi^{n-1}}$$

Where φ^n is read as n copies of φ. The calculus $\mathbf{Ł}_{n+1}$ will be determined by the Affine calculus plus $\mathbf{C^n}$, resulting in $\mathbb{L}_{n+1}$. All these calculi are sound with respect to the semantics that we will present in short. Yet to get completeness we need to add some axioms. Here we show the axiomatization needed to get $\mathbb{L}_3$ and refer to the reader to Prijatelj (1996) to understand the general strategy to get the axioms for each $\mathbb{L}_n$.

(a) $2(\psi \wedge \neg\psi) \wedge (\neg\delta)^2 \Rightarrow 2((\psi \vee \delta) \wedge \neg(\psi \vee \delta))$
(b) $2(\psi \wedge \neg\psi) \wedge 2(\delta \wedge \neg\delta) \Rightarrow 2((\psi \vee \delta) \wedge \neg(\psi \vee \delta))$
(c) $2(\psi \wedge \neg\psi) \circ (\neg\delta)^2 \Rightarrow 2((\psi + \delta) \wedge \neg(\psi + \delta))$
(d) $2(\psi \wedge \neg\psi) \circ 2(\delta \wedge \neg\delta) \Rightarrow \psi + \delta$
(e) $\Rightarrow p^2 \vee (2(p \wedge \neg p))^2 \vee (\neg p)^2$

Where ps are atomic formulas and $\varphi^2 = \varphi \circ \varphi$ and $2(\varphi) = \varphi + \varphi$.

The point of axioms (a) to (d) is to encode the correct definition of the operations corresponding to conjunctions and disjunctions. Notice that in a way (e) works as an analog of the Law of Excluded Middle. That is, it states that any formula has value 1 or 0 or $\frac{1}{2}$.

To see how these calculi determine each $\mathbb{L}_n$ we will present the semantics that make this evident. Yet, we will skip the proof of soundness and completeness, which, again, can be found in Prijatelj (1996).

To get the corresponding class of valuations V we will start by defining a set of truth-values. For each $n \geqslant 2$ where $n \in \mathbb{N}$: $S_n = \{\frac{x}{n} | 0 \leqslant \frac{x}{n} \leqslant 1 \text{ and } x \in \mathbb{N}\}$. That is, if $n = 2$, then $S_n = \{0, \frac{1}{2}, 1\}$. Therefore, n will determine an $n + 1$ valued Łukasiewicz logic. v will be a function that assigns for every propositional letter p a value $x \in S_n$. We will extend v recursively in the following way:

- $v(\neg\varphi) = 1 - v(\varphi)$
- $v(\varphi + \psi) = \min\{v(\varphi) + v(\psi), 1\}$
- $v(\varphi \vee \psi) = \max\{v(\varphi), v(\psi)\}$
- $v(\varphi \circ \psi) = \max\{v(\varphi) + v(\psi) - 1, 0\}$
- $v(\varphi \wedge \psi) = \min\{v(\varphi), v(\psi)\}$
- $v(\varphi \supset \psi) = \max\{1 - v(\varphi), v(\psi)\}$
- $v(\varphi \rightarrow \psi) = \min\{1 - v(\varphi) + v(\psi), 1\}$

Then we can define the notions of satisfaction and validity:

Definition 5.3.1 A sequent $\Gamma \Rightarrow \Delta$ is satisfied by a valuation v if and only if $v(\neg\gamma_1 + \cdots + \neg\gamma_n + \delta_1 \cdots + \delta_m)$=1, where each $\gamma_i \in \Gamma$ and each $\delta_i \in \Delta$. A sequent is valid if and only if it is satisfied by every valuation v of the corresponding class V.

With all the definitions in place, we can see in which sense this is Ł$_n$ for some n. The answer is that **Ł**$_n$ captures Ł$_n$ in a similar sense that we said in Sect. 3.4 that $\mathbb{ST}$ is $\mathbb{LP}$. We can prove that every probable sequent of **Ł**$_n$ corresponds to a sentence that is true in every valuation v in the semantics for Ł$_n$.

Łukasiewicz logics are farely famous in the literature of non-classical logic for its many uses. Nevertheless, as we will see in the next chapter, they cannot be used to deal with semantic paradoxes, because, as axiom (e) allows us to see, these logics are able to express a strong negation in the language and we cannot have, for example, a transparent truth predicate. Finally, it is also worth mentioning that here Cut is not eliminable, because of the extra axioms added (Prijatelj, 1996).

In the next section, we will present a calculus which can be used to characterize $\mathbb{CL}$, but also to characterize a non-contractive logic. This particular logic will have an interesting and manageable semantics like the logics we just presented and will be able to address semantic paradoxes.

5.3.1.2 Non-contractive Classical Logic

In this section, we will see what we get when we combine additive and multiplicative rules and have Weakening built in the identity axiom. Take the following set of rules:

Structural Rules

$$\text{Id}\ \frac{}{\Gamma, p \Rightarrow p, \Delta}$$

Operational Rules

$$\text{L}\neg\ \frac{\Gamma \Rightarrow \varphi, \Delta}{\Gamma, \neg\varphi \Rightarrow \Delta} \qquad \text{R}\neg\ \frac{\Gamma, \varphi \Rightarrow \Delta}{\Gamma \Rightarrow \neg\varphi, \Delta}$$

$$\text{L}\wedge\ \frac{\Gamma, \varphi, \psi \Rightarrow \Delta}{\Gamma, \varphi \wedge \psi \Rightarrow \Delta} \qquad \text{R}\wedge\ \frac{\Gamma \Rightarrow \varphi, \Delta \qquad \Gamma \Rightarrow \psi, \Delta}{\Gamma \Rightarrow \varphi \wedge \psi, \Delta}$$

$$\mathrm{L}\vee\ \frac{\Gamma, \varphi \Rightarrow \Delta \qquad \Gamma, \psi \Rightarrow \Delta}{\Gamma, \varphi \vee \psi \Rightarrow \Delta} \qquad\qquad \mathrm{R}\vee\ \frac{\Gamma \Rightarrow \varphi, \psi, \Delta}{\Gamma \Rightarrow \varphi \vee \psi, \Delta}$$

This calculus is called **G3c** but with a primitive negation instead of a falsum constant and it was introduced originally by Wieckowski (2022) and named by Troelstra and Schwichtenberg (1996). **G3c** is inferentially Classical Logic. Metainferentially, it can also deliver Classical Logic if one considers its admissible rules, since one can prove that Weakening, Contraction, Cut and the invertible rules are admissible. We will omit these proofs for reasons of space.[9] However, Contraction and Cut are not derivable in it. Thus, the metainferential Logic it characterizes is Substructural.

As one can see, the connectives of this calculus are defined by a two-premise additive rule and a single premise multiplicative rule. And in a similar sense as it happens for $\mathbb{ST}$ with its calculus $\mathbf{LK}^{-c}_{+I}$, it does not exclude some non-classical models. In particular, as Rosenblatt (2019) shows it is possible to provide what we will call V_{NCp} models, that is models such that the consequence relation is not contractive and not transitive.[10] V_{NCp} models are defined by means two different three valued valuational schemes with the set $\{1, \frac{1}{2}, 0\}$ as truth-values. We will consider both **SK** operations and **Ł**$_3$ operations for conjunction and disjunction, as defined semantically in Sect. 5.3.1.1. To avoid graphicall confusion, during this section we will refer to the functions that gives the **SK** truth conditions for conjunction and disyunction as $f^{\wedge}$ and $f^{\vee}$ respectively, and to the functions of **Ł**$_3$ as f° and f^{+}

Definition 5.3.2 A V_{NCp} model is an ordered pair of valuations $\langle v^-, v^+\rangle$, where both v^- and $v^+ : \mathcal{L} \longmapsto \{0, \frac{1}{2}, 1\}$ are such that:

- For every atomic formula p, $v^+(p) = v^-(p) \in \{0, \frac{1}{2}, 1\}$.
- $v^-(\neg\varphi) = 1 - v^+(\varphi)$ and $v^+(\neg\varphi) = 1 - v^-(\varphi)$.
- $v^-(\varphi \wedge \psi) = f^{\circ}(v^-(\varphi), v^-(\psi))$ and $v^+(\varphi \wedge \psi) = f^{\wedge}(v^+(\varphi), v^+(\psi))$.
- $v^-(\varphi \vee \psi) = f^{\vee}(v^-(\varphi), v^-(\psi))$ and $v^+(\varphi \vee \psi) = f^{+}(v^+(\varphi), v^+(\psi))$.

Then, the definition of validity runs as follows:

Definition 5.3.3 An inference $\Gamma \Rightarrow \Delta$ is valid in $\mathbf{NC}^p$ if and only if for all V_{NCp} models

(a) If $v^-(f^{\circ}[v^-(\gamma)]) = 1$ for all $\gamma \in \Gamma$, then $v^+(f^{+}[v^+(\delta)]) = 1$ for all $\delta \in \Delta$ and
(b) If $v^+(f^{+}[v^+(\delta)]) = 0$ for all $\delta \in \Delta$, then $v^-(f^{\circ}[v^-(\gamma)]) = 0$ for all $\gamma \in \Gamma$.

That is, truth preservation from the (multiplicative) conjunction of the premises to the (multiplicative) disjunction of the conclusions and falsity preservation of the (multiplicative) disjunction of the conclusions to the (multiplicative) conjunction of the premises. A sentence φ will be a tautology if and only if $v^+(\varphi) = 1$

[9] See Indrzejczak (2021) for a proof of each of these theorems.

[10] The applications of this logic will be discussed in Chap. 6.

in every V_{NCp}- model and a contradiction if and only if $v^{-}(\varphi) = 0$ in every V_{NCp}- model. The value a formula takes in the negative valuation is irrelevant to determinate whether a sentence is a tautology or not, and the same goes for the positive valuations and contradictions. This logic is both non-contractive and non-transitive and it allows us to see that there are also semantics that work via well-know matrices (at least in some sense) that can capture non-contractive logics.

5.3.2 A Non-monotonic Logic

The most usual motivations for dropping Weakening come from two projects associated with the formalization of natural language:

- The *relevance* logicians, to whom valid inference should preserve a certain connection of content between premises and conclusions, and therefore it is incorrect to conclude something about an atomic proposition p from a set of premises in which the proposition p is missing (see Mares, 2024);
- The *inductive* logicians, to whom there are certain forms of non-deductive reasoning that can be brought to a formal analysis, and so they introduce systems where, for instance, an inference previously deemed valid may turn out to be invalid in the presence of new information. These are the most commonly called in the literature *non-monotonic logics* (see Strasser & Antonelli, 2024).

Despite their intrinsic interest (and because of the vast literature surrounding them) we will not cover these logics here. The reason is that, although they can be seen as substructural logics, they are usually not motivated as such. Instead we will present **CLNW** (Classical Logic, (with) No Weakening), a calculus introduced in Da Ré (2021) as an immediate substructural variant of **LK** that holds Contraction but drops Weakening.

CLNW is an extension of **CLL** that does the opposite of the ones we saw in the previous section: instead of adding only Weakening, we will add Contraction but leave Weakening out. Even though the wide family of Relevance logics are usually presented as logics without Weakening, this logic is not part of that family.

Da Ré in (2021) introduced this logic to show that there can be a non-monotonic, substructural solution to the semantic paradoxes (see Sect. 6.2.3.3). It may be obtained from the rules of Linear Logic by taking:

- All the additive rules for conjunction, disjunction and conditional, for which we shall use the symbols $\wedge, \vee, \supset$.
- An (eliminable; see Da Ré, 2021, remark 3.2) additive form of Cut:

$$\text{Cut}^{A}\ \frac{\Gamma \Rightarrow \varphi, \Delta \qquad \Gamma, \varphi \Rightarrow \Delta}{\Gamma \Rightarrow \Delta}$$

- The rules for negation.
- The rules of Contraction.

This calculus has very unique properties. For a start, we have that:

Theorem 5.3.4 *All provable sequents in* **CLNW** *have at most two formulas.*

Proof See Da Ré (2021), Proposition 3.8. □

Moreover, as we are not introducing new rules to **LK**, it is immediate to notice that this calculus is sound with respect to classical valuations. In fact, we may prove the following fact:

Theorem 5.3.5 **CLNW** *is sound with respect to* $\mathbb{CL}$.

Proof Immediate given Theorem 2.2.7. □

Corollary 5.3.6 *For every sequent* $\Gamma \Rightarrow \Delta$ *provable in* **CLNW**, *the following properties hold:*

- *If* $\Gamma \neq \emptyset$ *then* $\Gamma \vDash_{\mathbb{K}3} \Delta$
- *If* $\Delta \neq \emptyset$ *then* $\Gamma \vDash_{\mathbb{LP}} \Delta$

Despite the above, and even though its name may suggest otherwise, this logic does not coincide with $\mathbb{CL}$ in the at-most-two-formulas fragment. For instance:

$$\Rightarrow (p \vee (q \wedge r)) \supset ((p \vee q) \wedge (p \vee r))$$

is provable in **LK**, but not in **CLNW**.

This logic is paraconsistent, because

$$\varphi, \neg\varphi \Rightarrow \psi$$

cannot be derived (for it has three formulas!). In fact, the inference:

$$\varphi \wedge \neg\varphi \Rightarrow \psi$$

is also not provable, because even when it has two formulas, its cut-free proof in **LK** uses Weakening:

$$\dfrac{\dfrac{\dfrac{\dfrac{\dfrac{\varphi \Rightarrow \varphi}{\varphi, \neg\varphi \Rightarrow}\text{L}\neg}{\varphi \wedge \neg\varphi, \neg\varphi \Rightarrow}\text{L}\wedge}{\varphi \wedge \neg\varphi, \varphi \wedge \neg\varphi \Rightarrow}\text{L}\wedge}{\varphi \wedge \neg\varphi \Rightarrow}\text{LC}}{\varphi \wedge \neg\varphi \Rightarrow \psi}\text{RW}$$

It is also metainferentially paraconsistent (according to the definition of this concept delivered by Barrio et al., 2015), insofar as the principle of Meta Explosion is not derivable:

$$\dfrac{\Rightarrow \varphi \qquad \varphi \Rightarrow}{\Rightarrow \psi}$$

In fact, even if the empty sequent is provable in this logic (or a theory based on it), it is never the case that you can prove $\psi \Rightarrow$ or $\Rightarrow \psi$ from it, since the logic lacks Weakening. And this is a general feature of the Linear empty set. As Girard clarifies:

> I called 'false' the constant 0 for mnemonic reasons. But this constant is not contradictory in the absence of weakening. In the same way, the empty sequent $\Rightarrow$ is not quite absurd. Girard (2011, p. 190) (Our notation.)

Concerning his system, Da Ré states:

> One important feature of the theory is that the empty sequent is not the same as plain triviality: the metainference from $\Rightarrow$ to $\Rightarrow \varphi$ is not derivable, since this derivation involves Weakening. The intuition that could sustain this claim is that if we do not have premises, then we are not assuming anything, so nothing will follow from not assuming anything. In other words, the intuitive reading of the logical consequence relation involves something of the following: from nothing, nothing follows (this is what Paoli calls the "Hobbesian" reading of the logical consequence). Da Ré (2021, p. 383)

Although it is important to bear in mind that, as he himself acknowledges, this logic still lacks a precise semantic reading. See note 43 in Da Ré (2021).

5.4 Conclusions

In this chapter we presented Classical Linear Logic and explained why one can see it as a version of Classical Logic where Weakening and Contraction are handled in a controlled way. We presented a semantics for $\mathbb{CLL}$, namely phase space semantics and then we moved to its metainferential properties. Then we presented several scattered logics where either Weakening or Contraction fails. We focused on how it is possible to extend Affine Logic to get the family of finitely many-valued Łukasiewicz logics. And then we introduced $\mathbf{NC}^p$, a non-contractive version of $\mathbb{CL}$. Finally, we presented a non-diluting logic, **CLNW**, the multiplicative fragment of **CLL** plus Contraction.

Bibliography

Avron, A. (1988). The semantics and proof theory of linear logic. *Theoretical Computer Science, 57*(2), 161–184.

Barrio, E. A., Rosenblatt, L., & Tajer, D. (2015). The logics of strict-tolerant logic. *Journal of Philosophical Logic, 44*(5), 551–571.

Bimbó, K., & Dunn, J. M. (2022). Modalities in lattice-R. *Relevance logics and other tools for reasoning. Essays in honor of J. Michael Dunn* (pp. 89–127). London: College Publications.

Girard, J.-Y. (1987). Linear logic. *Theoretical Computer Science, 50*(1), 1–101.

Girard, J.-Y. (1995). Linear logic: Its syntax and semantics. In J.-Y. Girard, Y. Lafont, & L. Regnier (Eds.), *Advances in linear logic*. London mathematical society lecture note series (pp. 1–42). Cambridge University Press.

Girard, J.-Y. (2011). *The blind spot: Lectures on logic*. European Mathematical Society.
Indrzejczak, A. (2021). *Sequents and trees*. Birkhäuser.
Kanovuch, M. (2016). The undecidability theorem for the horn-like fragment of linear logic (Revisited). *Mathematical Structures in Computer Science, 26*(5), 719–744.
Lincoln, P., Mitchell, J., Scedrov, A., & Shankar, N. (1992). Decision problems for propositional linear logic. *Annals of Pure and Applied Logic, 56*(1–3), 239–311.
Mares, E. (2024). Relevance logic. In E. N. Zalta, & U. Nodelman (Eds.), *The Stanford encyclopedia of philosophy* (Summer 2024 ed.). Metaphysics Research Lab, Stanford University.
Okada, M. (1998). An introduction to linear logic: Expressiveness and phase semantics. In *Theories of types and proofs* (Vol. 2, pp. 255–296). Mathematical Society of Japan.
Paoli, F. (2002). *Substructural logics: A primer*. Springer.
Paoli, F. (2013). *Substructural Logics: a Primer* (Vol. 13). Springer Science & Business Media.
Prijatelj, A. (1996). Bounded contraction and gentzen-style formulation of łukasiewicz logics. *Studia Logica, 57*(2–3), 437–456.
Ré, B. D. (2021). Structural weakening and paradoxes. *Notre Dame Journal of Formal Logic, 62*(2), 369–398.
Restall, G. (2000). *An introduction to substructural logics*. Routledge.
Roorda, D. (1989). *Investigations into classical linear logic*. Institute for Language, Logic and Information.
Rosenblatt, L. (2019). Noncontractive classical logic. *Notre Dame Journal of Formal Logic, 60*(4), 559–585.
Schellinx's, H. (1991). Some syntactical observations on linear logic. *Journal of Logic and Computation, 1*(4), 537–559.
Schroeder-Heister, P. (2024). Proof-theoretic semantics. In E. N. Zalta, &U. Nodelman (Eds.), *The Stanford encyclopedia of philosophy* (Summer 2024 ed.). Metaphysics Research Lab, Stanford University.
Strasser, C., & Antonelli, G. A. (2024). Non-monotonic Logic. In E. N. Zalta & U. Nodelman (Eds.), *The Stanford encyclopedia of philosophy* (Summer 2024 ed.). Metaphysics Research Lab, Stanford University.
Troelstra, A. S. (1992). *Lectures on linear logic*. Center for the Study of Language and Information Publications.
Troelstra, A. S., & Schwichtenberg, H. (1996). *Basic proof theory*. Cambridge University Press.
Wieckowski, B. (2022). Oiva ketonen, investigations into the predicate calculus, vol. 3 of Logic Phds, Sara Negri, and Jan von Plato, (Eds.), College Publications, 2022, pp. 130+vii; isbn 978-1-84890-407-1. *Studia Logica*, 1–4.

Chapter 6
Philosophical Applications

Edson Bezerra and Jonathan Erenfryd

Abstract Substructural logics are interesting tools to analyse philosophical problems that concern logic. In this chapter, we will take a philosophical approach to the logics presented earlier in this book. We will show how substructural approaches provide uniform solutions to the paradoxes involving the truth predicate—Liar and Curry paradoxes—and the paradox involving the validity predicate—the validity paradox. Concretely, we will consider non-transitive, non-reflexive, non-contractive and non-monotonic solutions to these antinomies. Giving a uniform solution to the semantic paradoxes is arguably a virtue of substructural logics when it comes to their applications in the face of other non-classical approaches, such as $\mathbb{LP}$ and $\mathbb{K}3$. However, this is not enough. For this reason, we explore philosophical interpretations that make sense of the failures of structural properties of these logics. In particular, we argue that these substructural logics can receive dialogical interpretations. Besides blocking the paradoxes of truth and validity and having interesting philosophical interpretations, it is required of these substructural theories that their validity predicate be sound with respect to their base logic, internalizing in their object-language their valid metainferences. Then, we address these questions for some of the substructural solutions. We will be particularly concerned with the most discussed validity theories, such as non-contractive and non-transitive.

6.1 Applying Substructural Logics

In this chapter, we will take a philosophical approach to the substructural logics presented earlier in this book. Examining how these logics address some philosophical issues is essential to fully understand their motivations and appreciate their fruitfulness.

The chapter is organized as follows. In Sect. 6.2, we discuss the relevance of expressing semantic concepts in a theory **T**. In particular, we will explore the

E. Bezerra (✉) · J. Erenfryd
IIF-SADAF-CONICET, Bulnes, Aires, Argentina

P. Teijeiro, E. A. Barrio (eds.), *Metainferences in Substructural Logics*, Synthese Library 528, https://doi.org/10.1007/978-3-032-05449-4_6

truth and the validity predicates. We will see that adding these predicates to the language of such a theory is problematic. One possible take on this problem is that the Liar paradox shows that **T** should give up some rules governing the negation connective; Curry's paradox shows that **T** should give up principles governing the conditional; and the validity paradox shows that **T** should give up structural rules, because the derivation of the validity paradox makes no use of the logical connectives. We explore how substructural approaches can handle both the truth and the validity predicates, avoiding the triviality results and providing a uniform solution to paradox, presumably an advantage for substructural solutions in the face of other non-classical approaches, such as $\mathbb{LP}$ and $\mathbb{K}3$. We consider non-transitive (Sect. 6.2.3.1), non-reflexive (Sect. 6.2.3.2), non-contractive and non-monotonic logics (Sect. 6.2.3.3).

Giving a uniform solution to the semantic paradoxes is arguably a virtue of substructural logics when it comes to their applications. This, however, is not enough. One may reasonably require additional motivations for adopting these logics. Concretely, one may ask how can we understand the consequence relation of these logics from an intuitive perspective. In other words, one could ask what kind of logical consequence do these logics formalize. We will address these questions for each of the substructural solutions, presenting philosophical interpretations for these logics. Additionally, in Sect. 6.2.3.4, we argue that most substructural logics can receive dialogical interpretations.

Besides blocking the semantic paradoxes and having interesting philosophical interpretations, theories extended with a validity predicate can still face an additional challenge: it is required of these theories that their validity predicate be sound with respect to their base logic, including now their valid metainferences. Thus, these theories should internalize in their object-language not just their valid inferences, but their valid metainferences too. In Sect. 6.3, we discuss the problems related to internalization of validities.

Before we undertake our task, a word regarding our approach is in order. When dealing with paradoxes and their solutions, we will rely on a sequent-style presentation of them. There is a bunch of reasons for going this way. Of course, sequent calculi were used in previous chapters, so this makes our treatment uniform concerning the rest of the book. This will let the reader appreciate how the technicalities seen in previous chapters get a concrete application in philosophical issues. Furthermore, it is common in the literature on substructural logics to draw on this kind of tool, so the reader will become familiar with common usage in the field. Moreover, sequent-style presentations offer some more advantages. For one thing, when it comes to providing models for our theories, different substructural logics make use of different formal tools, where the machinery might even get convoluted. Because of this, it is not always easy to compare and evaluate different approaches from the model-theoretic point of view. Appealing to sequent calculus allows for a more uniform treatment while making explicit, in the form of a metarule, which structural features hold or fail to hold in a given logic.

6.2 Semantic Paradoxes and Substructural Solutions

6.2.1 *Expressing Truth and Validity*

In this section, we will show how some substructural logics introduced in the previous chapters can be used as a basis for theories of truth and validity. The theories we will investigate extend the propositional language $\mathcal{L}$ with the sentence predicates of truth and validity, together with devices that provide sentence names and a self-reference procedure. Before spelling out the details of our construction, we discuss how these concepts were traditionally investigated in the literature.

A sufficiently expressive framework must have a language capable of talking about its syntax and semantics.[1] Such a formal theory would enable the expression of philosophical and scientific concepts unambiguously (Tarski, 1956; Fitch, 1964). The semantic concepts of 'truth' and 'validity' have philosophical significance because they are fundamental concepts of reasoning and theorizing in general.

Let **T** be a semantic theory that intends to capture semantic concepts such as those of 'truth' and 'validity'. This approach is called 'naïve approach' to semantics (to truth and validity, respectively). As is widely known, Tarski defended the view that any attempt to define the truth sentential predicate Tr for some language must fulfill the requirement of implying the instances of the *T-schema*:

$$Tr(\ulcorner \varphi \urcorner) \leftrightarrow \varphi$$

Where $\ulcorner \varphi \urcorner$ stands for the name of φ. $Tr(\ulcorner \varphi \urcorner)$ means that φ *is true*. It is in this way that it can be expected from semantic theories like **T** that they get to talk about their own syntax and semantics. In turn, these theories are usually couched in first-order arithmetical languages, because these languages provide codes (e.g., Gödel names) for all the sentences of the corresponding language, and they are able to prove the well-known Diagonalization Lemma, that can be stated as follows:

Diagonalization Lemma. Let **Ar** be a first-order arithmetical theory, and $\vdash$ its consequence relation. For every formula $\varphi(x)$ where x is the only free variable occurring in it, there is a sentence γ such that $\mathbf{Ar} \vdash \gamma \leftrightarrow \varphi(\ulcorner \gamma \urcorner)$.

So, if **T** extends **Ar**, then the Diagonalization Lemma formalizes self-reference in arithmetical theories (Smoryński, 1985).[2] Note, however, that arithmetical languages are not a necessary condition for the formulation of theories like **T**. For instance, Kremer (1988) adopts an uninterpreted first-order language $\mathcal{L}_1^+$ equipped

[1] These languages are traditionally called 'semantically closed languages'.

[2] According to Picollo (2020), the notion of self-reference did not receive proper treatment in the literature. In her work, she presents a rigorous characterization of this notion in arithmetical truth theories. Since we will deal with theories that extend the propositional language with the truth and validity predicates, the naming device that we will use is much simpler than the standard arithmetical coding.

with a quotation device $\ulcorner \cdot \urcorner$ that enables forming singular terms from sentences of $\mathcal{L}_1^+$.[3] Then, $\ulcorner \varphi \urcorner$ is the name of φ, for every first-order sentence φ. From a proof-theoretical perspective, this procedure allows to formulate rules for the truth predicate that ground the possibility to infer $Tr(\ulcorner \varphi \urcorner)$ from φ.

Another possible strategy using first-order languages is taken by Ripley (2012). In her approach, some individual constants are 'distinguished names'. Such a naming device is made possible by means of a 1-1 function τ from names to formulas of $\mathcal{L}_1^+$ such that, for every distinguished name n, n denotes $\tau(n)$ in every model. This means that the set of formulas of $\mathcal{L}_1^+$ is a subset of the domain of every model, and only infinite models are admitted. The difference between Ripley's strategy and Kremer's is that in the former the naming device is not purely syntactic. As Ripley observes, the connection between a sentence φ and its corresponding distinguished name depends on the function τ.

Given that Kremer's and Ripley's approaches are formulated in a non-arithmetical language, the Diagonalization Lemma is not provable in their semantic theories, and so they do not guarantee the existence of self-referential sentences in the language $\mathcal{L}_1^+$. Although they do not prove this lemma, they allow the formulation of these sentences via metatheoretical stipulation, showing a way to dispense with arithmetical languages in dealing with semantic concepts. In this chapter, we will adopt $\ulcorner \cdot \urcorner$ as a naming device. However, our approach is more restricted than Kremer's.

Since the previous chapters were carried out in a propositional language, we will here consider a strategy for dealing with the self-referential paradoxes in such a language. Let $\mathcal{L}^+$ be the language obtained by extending a propositional language with $\ulcorner \cdot \urcorner$ and the predicates for truth and validity, Tr and Val. Now, we adapt Barrio et al.'s strategy (2017) to obtain self-referential statements in $\mathcal{L}^+$. Let $Var^* \subseteq Var$ be an infinite proper subset of propositional variables. Every member of Var^* is metalinguistically marked with an $*$. Now, for each p^*, we have the following rules that govern self-reference:

$$\frac{\Gamma, p^* \Rightarrow \Delta}{\Gamma, A_{p^*} \Rightarrow \Delta} \qquad \frac{\Gamma \Rightarrow p^*, \Delta}{\Gamma \Rightarrow A_{p^*}, \Delta}$$

Where A_{p^*} is a sentence containing at least one occurrence of $Tr(\ulcorner \varphi \urcorner)$ or $Val(\ulcorner \varphi \urcorner, \ulcorner \psi \urcorner)$, where in the former p^* is a subformula of φ, and in the latter p^* is a subformula of φ or ψ. Last, for each sentence A_{p^*}, there is only one p^* that is intersubstitutable with it. This procedure, combined with Tr and Val, generates some paradoxical sentences relevant for our analysis.

[3] The subscript in $\mathcal{L}_1^+$ conveys that we are dealing with a first-order language. Moreover, it is worth noting that the sentence names in $\mathcal{L}_1^+$ are not Gödel names because Kremer is not dealing with an arithmetical language. The same applies to the other non-arithmetical approaches we will deal with in this chapter.

Starting with a sequent calculus—that naïvely one might expect to be a non-trivial extension of **LK**–, one introduces a truth and a validity predicate by adding some rules that govern them, and the rules governing self-reference. The truth predicate obeys the following rules:

$$LTr\frac{\Gamma, \varphi \Rightarrow \Delta}{\Gamma, Tr(\ulcorner\varphi\urcorner) \Rightarrow \Delta} \qquad RTr\frac{\Gamma \Rightarrow \varphi, \Delta}{\Gamma \Rightarrow Tr(\ulcorner\varphi\urcorner), \Delta}$$

A validity statement of the form $Val(\ulcorner\varphi\urcorner, \ulcorner\psi\urcorner)$ says that *the argument from* φ *to* ψ *is valid.* This predicate is governed by two principles formalized as follows:

$$\text{VP}\frac{\varphi \Rightarrow \psi}{\Rightarrow Val(\ulcorner\varphi\urcorner, \ulcorner\psi\urcorner)} \qquad \text{VD}\frac{}{\varphi, Val(\ulcorner\varphi\urcorner, \ulcorner\psi\urcorner) \Rightarrow \psi}$$

The first rule, called 'Validity Proof' (VP), says that every provable inference is naïvely valid. The second, called 'Validity Detachment' (VD), states that validity is separable, in the sense that given a valid inference and its premises, its conclusion follows. The rule VD is an horizontalized version of the following:

$$\text{VD}'\frac{\Rightarrow \varphi \qquad \Rightarrow Val(\ulcorner\varphi\urcorner, \ulcorner\psi\urcorner)}{\Rightarrow \psi}$$

As one can see, VD' expresses a form of Meta Modus Ponens for the validity predicate, and it can be proved from VD and Cut (Exercise 106). Thus, the choice between VD and VD' is not innocuous. For example, the semantic theories based on non-transitive logics have to adopt VD instead of VD'.

The presence of both predicates in a semantic theory enables the formulation of the principle of 'validity as truth preservation' (VTP) that can be stated as follows:

(VTP) An argument from Γ to φ is valid if and only if the truth of every γ in Γ implies the truth of φ.

This principle is well accepted in the literature and is presented in many elementary logic textbooks.[4] A naïve semantic theory is supposed to handle both predicates in its object-language. So, one would expect to obtain such a theory in the context of Classical Logic. Indeed, Classical Logic has been regarded as a form of common sense. It is used in many of our scientific theories, such as mathematics, and is even considered by many logicians to be the one true logic (see, for instance, Williamson, 2002). So, it would be perfect if such a logic were consistent with semantic predicates. Yet, as we are about to see, semantic paradoxes of truth and validity show that it is impossible to obtain a semantically closed theory while retaining Classical Logic. This inconsistency result pushed various logicians

[4] Although this principle may be well-known and accepted, it was challenged in the literature. We refer the reader to Field (1991, 2008) for elaborated objections against the idea that validity is truth-preserving.

to look for alternatives by modifying the behaviour of the logical connectives. However, this is also insufficient. Well-entrenched ideas regarding the very structure of consequence must be given up.

Exercises

Exercise 106 Show that VD' can be proved from VD and Cut.

6.2.2 Semantic Paradoxes

Obtaining a naïve semantic theory is not straightforward. In what follows, we will present the semantic paradoxes involving the notions just considered. In particular, we will introduce the Liar and Curry paradoxes, which arise from adding the predicate Tr, and the validity paradox involving the predicate Val.

In his seminal work, Tarski (1956) shows that the predicate Tr for $\mathbf{T}$ is undefinable in $\mathbf{T}$. More precisely, Tarski's results establish that arithmetical theories based on Classical Logic are incompatible with a truth predicate satisfying the T-schema. By the Diagonalization Lemma, one obtains the sentence λ, that is equivalent to its negation:

$$\lambda \leftrightarrow \neg Tr(\ulcorner \lambda \urcorner)$$

λ is known as the Liar sentence, and proves that the truth predicate is inconsistent with $\mathbf{Ar}$ when this theory is based on Classical Logic. Here, on the other hand, we are working with sequent calculi theories $\mathbf{T}$ in the language $\mathcal{L}^+$, and the self-reference device we described above. For convenience, let $\lambda \in Var^*$ and let the following rules allow the intersubstitutivity of λ with $\neg Tr(\ulcorner \lambda \urcorner)$ (let us call this 'Liar sentence intersubstitutivity': 'LSI'):

$$\text{LLSI}\ \frac{\Gamma, \lambda \Rightarrow \Delta}{\Gamma, \neg Tr(\ulcorner \lambda \urcorner) \Rightarrow \Delta} \qquad \text{RLSI}\ \frac{\Gamma \Rightarrow \lambda, \Delta}{\Gamma \Rightarrow \neg Tr(\ulcorner \lambda \urcorner), \Delta}$$

Intuitively, the sentence λ 'says of itself' that it is not true. This sentence λ generates the so-called Liar paradox. Suppose that λ is true. But it says it is not true. So, it is not the case that it is true. Now, suppose the sentence is not true. Then, what it states is not the case, so the sentence must be true. The corresponding paradox can be presented in the following way:

$$\text{LW}\ \dfrac{\text{RW}\ \dfrac{\text{Cut}\ \dfrac{\text{RC}\ \dfrac{\text{RLSI}\ \dfrac{\text{R}\neg\ \dfrac{\text{L}Tr\ \dfrac{\text{Id}\ \dfrac{}{\lambda \Rightarrow \lambda}}{Tr(\ulcorner \lambda \urcorner) \Rightarrow \lambda}}{\Rightarrow \neg Tr(\ulcorner \lambda \urcorner), \lambda}}{\Rightarrow \lambda, \lambda}}{\Rightarrow \lambda} \qquad \text{LC}\ \dfrac{\text{LLSI}\ \dfrac{\text{L}\neg\ \dfrac{\text{R}Tr\ \dfrac{\text{Id}\ \dfrac{}{\lambda \Rightarrow \lambda}}{\lambda \Rightarrow Tr(\ulcorner \lambda \urcorner)}}{\neg Tr(\ulcorner \lambda \urcorner), \lambda \Rightarrow}}{\lambda, \lambda \Rightarrow}}{\lambda \Rightarrow}}{\Rightarrow}}{\Rightarrow \varphi}}{\psi \Rightarrow \varphi}$$

So, the corresponding theory becomes trivial, since anything will entail anything else. In this way, the paradox shows that the predicate Tr and semantic theories based on Classical Logic are incompatible.

If one wishes to block the paradox and retain classicality, it seems mandatory to abandon the *naïveté* of the truth predicate, for instance, by adopting Tarski's distinction between object-language and metalanguage (Tarski, 1956). The truth predicate of **T** is only definable in its metalanguage. As a result, this induces a transfinite hierarchy of metalanguages. This, however, may be resisted on several grounds (e.g., Field, 1972; Kripke, 1975; Priest, 2006; and Halbach, 2014).[5] If, on the face of it, one wants to keep Tr in the object-language, one could weaken the deductive strength of logic.

A reasonable option then is to weaken negation. The Liar causes trouble when one assumes that it is true, and when one assumes that it is false. But perhaps the Liar is none. So, by denying the truth of λ one is not thereby accepting its falsity. Perhaps there are truth-value gaps. According to paracomplete theorists, the Principle of Excluded middle will not hold without restriction, and this will be reflected in the behavior of negation. Dually, paraconsistent solutions will see the Liar to be both true and false. So, by accepting the truth of λ one is not thereby denying its falsity. However, this will not cause trouble, since the Principle of Explosion is rejected, and so an inconsistent theory is not thereby trivial. More rigorously, in the derivation of the Liar paradox, the paracomplete theorist denies the derivation of $\Rightarrow \lambda, \neg\lambda$, and the paraconsistent theorist denies the derivation of $\lambda, \neg\lambda \Rightarrow$. Then, both paracomplete and paraconsistent theorists block the paradox by weakening negation.

However, Curry paradox shows that restrictions on negation do not suffice (Beall & Murzi, 2013). Let $\kappa \in Var^*$ be such that κ is intersubstitutable with $Tr(\ulcorner\kappa\urcorner) \to \bot$ ('Curry intersubstitutivity': 'CI'):[6]

$$\text{RCI}\ \frac{\Gamma \Rightarrow \kappa, \Delta}{\Gamma \Rightarrow Tr(\ulcorner\kappa\urcorner) \to \bot, \Delta}$$

The Curry sentence intuitively says 'if I am true, then everything is'. Consider the following derivation:[7]

$$\text{MMP}\ \dfrac{\text{MMP}\ \dfrac{\text{CI}\ \dfrac{\text{Id}\ \dfrac{}{\kappa \Rightarrow \kappa}}{\kappa \Rightarrow Tr(\ulcorner\kappa\urcorner) \to \bot} \quad \text{R}Tr\ \dfrac{\text{Id}\ \dfrac{}{\kappa \Rightarrow \kappa}}{\kappa \Rightarrow Tr(\ulcorner\kappa\urcorner)}}{\text{R}{\to}\ \dfrac{\text{L}Tr\ \dfrac{\text{LC}\ \dfrac{\kappa, \kappa \Rightarrow \bot}{\kappa \Rightarrow \bot}}{Tr(\ulcorner\kappa\urcorner) \Rightarrow \bot}}{\Rightarrow Tr(\ulcorner\kappa\urcorner) \to \bot}} \qquad \text{R}Tr\ \dfrac{\text{CI}\ \dfrac{\begin{matrix}\vdots\\ \Rightarrow Tr(\ulcorner\kappa\urcorner) \to \bot\end{matrix}}{\Rightarrow \kappa}}{\Rightarrow Tr(\ulcorner\kappa\urcorner)}}{\Rightarrow \bot}$$

---

[5] We invite the reader to consult (Barrio, 2014a) for a good presentation of theories of truth.

[6] We present RCI only, since it is the only rule of intersubstitutivity used in the derivation of the Curry paradox.

[7] 'MMP' stands for Meta Modus Ponens.

As Beall and Murzi (2013) argue, the Curry paradox forces the paraconsistent and the paracomplete logicians to weaken the conditional. To block Curry paradox, one also has to give up MMP or R $\rightarrow$. Both $\mathbb{LP}$ and $\mathbb{K}3$ approaches, respectively, block Liar and Curry paradoxes. The former gives up MMP and the later gives up R $\rightarrow$. The costs of both solutions are considerably weak conditionals (Barrio, 2014b; Ripley, 2013). The solutions that proceed in such a way, by reducing the inferential power of the different logical connectives, are called 'operational solutions' (Beall & Murzi, 2013).

The weakening of the logical constants may be seen as problematic. As Ripley (2013) acknowledges, this exposes the structuralist to the charge presented by Quine (1986), according to which changing logic is changing the subject. Paradoxes were presented for connectives obeying certain rules. A connective that does not obey these rules is not the same connective, despite the fact that the same symbol is used.

Furthermore, by changing operational rules, paradoxes are dealt with in a piecemeal way. Given the similarity in the shape of the semantic paradoxes, it may be suspected that there is a common root that allows all of them to arise in the first place. Arguably, piecemeal approaches, like the operational ones, miss this general point. Thus, paracomplete and paraconsistent solutions would not offer uniform solutions to the paradoxes because different paradoxes make use of different connectives. Even more so, some paradoxes do not involve any connective, such as the validity paradox. In the latter, the corresponding derivation makes use of no sentential connectives, and so the operational solutions will fail to block it.

Let $\pi \in Var^*$ be such that π is intersubstitutable with $Val(\ulcorner \pi \urcorner, \ulcorner \bot \urcorner)$ ('vCurry intersubstitutivity': 'vCI'), as the following rules state:

$$\text{LvCI}\ \frac{\Gamma, \pi \Rightarrow \Delta}{\Gamma, Val(\ulcorner \pi \urcorner, \ulcorner \bot \urcorner) \Rightarrow \Delta} \qquad \text{RvCI}\ \frac{\Gamma \Rightarrow \pi, \Delta}{\Gamma \Rightarrow Val(\ulcorner \pi \urcorner, \ulcorner \bot \urcorner), \Delta}$$

The sentence π is usually called 'validity Curry' (vCurry, for short). Intuitively, it says 'Anything follows from this sentence'. Now, consider the following derivation:

$$\text{Cut}\ \dfrac{\text{VP}\ \dfrac{\text{LC}\ \dfrac{\text{Cut}\ \dfrac{\text{RvCI}\ \dfrac{\text{Id}\ \dfrac{}{\pi \Rightarrow \pi}}{\pi \Rightarrow Val(\ulcorner \pi \urcorner, \ulcorner \bot \urcorner)} \qquad \text{VD}\ \dfrac{}{\pi, Val(\ulcorner \pi \urcorner, \ulcorner \bot \urcorner) \Rightarrow \bot}}{\pi, \pi \Rightarrow \bot}}{\pi \Rightarrow \bot}}{\Rightarrow Val(\ulcorner \pi \urcorner, \ulcorner \bot \urcorner)} \qquad \text{LvCI}\ \dfrac{\begin{array}{c}\vdots\\ \pi \Rightarrow \bot\end{array}}{Val(\ulcorner \pi \urcorner, \ulcorner \bot \urcorner) \Rightarrow \bot}}{\Rightarrow \bot}$$

As the derivation above shows, no operational rule is needed. There are no other rules beyond those governing the validity predicate and the structural rules that play a role in it. Consequently, there are two ways out of this problem available for structural logics. One is to deny the legitimacy of the paradox. The other one is to change the means by which self-reference is achieved.

There has been some skepticism around validity Curry. Remarkably, according to Ketland (2012) there is no paradox involving the validity predicate. Ketland argues that the validity of φ is defined as truth in all reinterpretations of the non-logical constants of φ. So, the validity predicate can only be applied to logically provable formulas by the Completeness theorem of first-order logic. Since the validity paradox makes essential use of the Diagonalization Lemma, the derivation of this paradox cannot be carried out, given that it relies on arithmetical resources.

It might be argued, however, that Ketland works with a notion that is too restrictive, so there is still a problem with some way of understanding the validity predicate. Indeed, we find arguments in the literature against the idea that logical validity captures the intuitive notion of validity. For example, Etchemendy's (1999) point of view on the matter is as follows. The proof-theoretic definition fails because many deductive systems are adequate for the same logic, and it is arbitrary to point out one as the single system that captures intuitive validity. Concerning the model-theoretic definition of validity, Etchemendy argues that it can be understood in two possible ways: representationally or interpretationally. In the first case, the validity of a sentence φ is understood as truth in all possible configurations of the world. In the second case, the validity of a sentence φ is understood as truth in all variations of the non-logical vocabulary. Etchemendy argues that logical validity fails to satisfy these two perspectives. In the representational case, logical validity does not capture the formality of logic. In the interpretational case, it fails to justify the distinction between logical and non-logical vocabulary. Given these reasons, he argues that model-theoretic validity does not capture intuitive validity.

Field (2008) also argues that the formal definitions of logical validity fail to capture intuitive validity, with arguments similar to Etchemendy's. Regarding model-theoretic validity, he argues that this definition takes models with sets as their domains. However, the actual world is not a set, so model-theoretic truth does not capture actual truth. For this reason, intuitive validity is not reducible to their formalizations. In general, as Murzi (2014) and Andrade-Lotero and Novaes (2012) argue, intuitive validity is a broader notion than that of logical validity because it validates some inferences that are not valid from the logical point of view (e.g., set-theoretic inferences, analytic inferences, etc).[8]

Regarding the second strategy left open for the operational accounts on the face of validity paradox, Barrio et al. (2017) and Pailos (2020b) show that it is possible to provide operational solutions to the semantic paradoxes if one weakens the self-referential devices of the theory.[9] In all the derivations above, the rules of intersubstitutivity were applied to the paradoxical sentences. For example, LvCI and RvCI allow the replacement of π by $Val(\ulcorner\pi\urcorner, \ulcorner\bot\urcorner)$ and vice versa. Such an intersubstutivity is a form of self-reference.

As they show, paraconsistent semantic theories are trivial in the presence of a robust self-referential procedure. Thus, it is necessary to adopt a weak self-referential principle. Let Var^* be the set of propositional variables defined above. Consider now the following biconditional:

[8] So, the predicate can be understood as expressing a broader concept than that of formal validity. For instance, it could be that of informal or absolute provability (see Myhill, 1960).

[9] Their results explore Goodship's arguments, who defends that non-trivial theories of truth require conditionals that invalidate Modus Ponens (Goodship, 1996).

$$p^* \leftrightarrow A_{p^*}$$

For each sentence A_{p^*}, there is only one biconditional with a different propositional letter in the left side. The set of all this sentences contains the paradoxical statements, such as the Liar sentence.

With this machinery, they show that paraconsistent detachment-free semantic theories can be non-trivially extended with a weak self-referential principle. To see how this weaker version of the validity paradox is blocked, consider the sentence π such that $\pi \leftrightarrow Val(\ulcorner\pi\urcorner, \ulcorner\bot\urcorner)$, where $\pi \in Var^*$ and $Val(\ulcorner\pi\urcorner, \ulcorner\bot\urcorner) = A_{p^*}$. Now, consider the following derivation:

$$\text{VP}\,\dfrac{\text{LC}\,\dfrac{\text{VD'}\,\dfrac{\text{MMP}\,\dfrac{\text{Id}\,\dfrac{}{\pi \Rightarrow \pi} \qquad \dfrac{}{\Rightarrow \pi \leftrightarrow Val(\ulcorner\pi\urcorner, \ulcorner\bot\urcorner)}}{\pi \Rightarrow Val(\ulcorner\pi\urcorner, \ulcorner\bot\urcorner)} \qquad \text{Id}\,\dfrac{}{\pi \Rightarrow \pi}}{\pi, \pi \Rightarrow \bot}}{\pi \Rightarrow \bot}}{\Rightarrow Val(\ulcorner\pi\urcorner, \ulcorner\bot\urcorner)}$$

Call this derivation $\mathcal{D}$. Then:

$$\text{VD'}\,\dfrac{\text{MMP}\,\dfrac{\dfrac{}{\Rightarrow \pi \leftrightarrow Val(\ulcorner\pi\urcorner, \ulcorner\bot\urcorner)} \qquad \dfrac{\mathcal{D}}{\Rightarrow Val(\ulcorner\pi\urcorner, \ulcorner\bot\urcorner)}}{\Rightarrow \pi} \qquad \dfrac{\mathcal{D}}{\Rightarrow Val(\ulcorner\pi\urcorner, \ulcorner\bot\urcorner)}}{\Rightarrow \bot}$$

As we can see, the above derivation makes use of the detachment of the conditional. If we have a detachment-free conditional, it cannot be carried out. However, if we used the sequent rules that allow self-reference instead of the biconditional $p^* \leftrightarrow A_{p^*}$, these operational approaches would be trivial. Rosenblatt (2021a) objects to the approaches offered by Barrio et al. (2017) and Pailos (2020b), that non-classical semantic theories cannot depend on the choice of a particular self-reference procedure, on the pain of being philosophically unsatisfactory.

6.2.3 Substructural Solutions and Their Philosophy

In the face of this, it seems as if any truly satisfying approach to paradox should provide a uniform solution by targeting where paradoxes presumably arise: in the very structure of inference. Therefore, we will consider the different substructural approaches to paradox. As we are about to see, each formal theory blocks the corresponding derivations in a specific way. Moreover, it is interesting to consider the reasons why a particular structural principle is to be restricted. Now, this latter aspect takes us beyond the pure formalism, into the realm of their philosophical motivations and interpretations, to which we turn now.

According to Priest (2005), 'pure logics' are formal languages endowed with a consequence relation. Those purely formal languages can be 'applied' to several fields, from computer science to formal linguistics. In our present case, these substructural logics are applied to philosophical problems. Barrio and Da Ré (2018); Barrio et al. (2024) argue that besides Priest's distinction between pure and applied logics, the notion of 'philosophical interpretation' has an important explanatory role. According to Barrio et al. (2024), philosophical interpretations can be understood as giving a conceptual reading of the formalism of a particular logic.[10]

A philosophical interpretation for a logic can be extended to the semantic theories it gives rise to. As Terzian (2015) argues, the formalism should be accompanied by a story that explains the informal intuitions that are being formalized. That is, these formal theories should shed light upon the intuitions about the concepts that we are formalizing. Then, it is to be expected that the semantic theories of truth and validity reflect at least some aspects of our intuitions regarding truth and validity. Without such a philosophical complement, one could question the value of these theories by raising the charge of *ad hocness*. In the present case, this means that logics are preferable when they come accompanied with a philosophical justification for the failure of structural rules. So, in what follows, we will not just present the substructural approaches in terms of the formalisms, but also present philosophical interpretations for them.

6.2.3.1 Cut-free Approach

The Cut-free solutions to the semantic paradoxes we will explore here were presented mainly in Ripley (2012, 2013) and Cobreros et al. (2012, 2015). The reason to focus on these proposals is that they take $\mathbf{LK}^{-c}$, presented in Chap. 3, as their base logical system.[11] The theory of truth and validity based on $\mathbf{LK}^{-c}$ is called **STTV**. The truth predicate is governed by the rules $\mathrm{L}Tr$ and $\mathrm{R}Tr$, and the validity predicate is governed by the rules VP and VD. The choice for the rule VD instead of VD' stems from the Cut-freeness of $\mathbf{LK}^{-c}$, given that VD' presupposes Cut.[12]

STTV blocks the paradoxes involving both predicates given that $\mathbf{LK}^{-c}$ does not have the metarule of Cut. In the case of the Curry paradox, although there is

[10] The conceptual reading encompasses informal interpretations that are not necessarily philosophical. In this sense, the notion of philosophical interpretation is close to what (Benito-Monsalvo, 2022) calls 'external interpretation', having "to do with the philosophical justification of a logic; i.e., with an explanation of why our meaning specification of the logical vocabulary is such that it makes some inferences valid and not others" [pg. 537].

[11] $\mathbb{ST}$ was originally proposed to deal with the Sorites paradox, and was intended to give a formal treatment of vague predicates. Since we are dealing with propositional languages, we will not delve into this. The reader may consult the cited works. Also, we invite the reader to consult (van Rooij, 2011) for the application of $\mathbb{ST}$.

[12] In (Ripley, 2012), Ripley presents a three-sided sequent calculus for her $\mathbb{ST}$-based truth theory, which validates a certain form of Cut. Since we are working with a two-sided sequent calculus, Cut will not hold.

no explicit use of Cut, the cut-freeness of **STTV** also blocks this paradox. In the derivation of the Curry paradox, Meta Modus Ponens is used twice. As the following derivation shows, MMP presupposes Cut:

$$\dfrac{\dfrac{\dfrac{}{\Rightarrow \varphi}\text{Axiom} \qquad \dfrac{}{\varphi, \varphi \to \psi \Rightarrow \psi}\text{MP}}{\varphi \to \psi \Rightarrow \psi}\text{Cut} \qquad \dfrac{}{\Rightarrow \varphi \to \psi}\text{Axiom}}{\Rightarrow \psi}\text{Cut}$$

So, by dispensing Cut, MMP fails in $\mathbf{LK}^{-c}$. For this reason, the Curry paradox is also blocked in **STTV**. Besides avoiding triviality, **STTV** has also enough power to prove some desirable principles. For instance, **STTV** proves VTP as follows:

$$\dfrac{\dfrac{\dfrac{\dfrac{\dfrac{}{\varphi, Val(\ulcorner\varphi\urcorner, \ulcorner\psi\urcorner) \Rightarrow \psi}\text{VD}}{Tr(\ulcorner\varphi\urcorner), Val(\ulcorner\varphi\urcorner, \ulcorner\psi\urcorner) \Rightarrow \psi}LTr}{Tr(\ulcorner\varphi\urcorner), Val(\ulcorner\varphi\urcorner, \ulcorner\psi\urcorner) \Rightarrow Tr(\ulcorner\psi\urcorner)}RTr}{Val(\ulcorner\varphi\urcorner, \ulcorner\psi\urcorner) \Rightarrow Tr(\ulcorner\varphi\urcorner) \to Tr(\ulcorner\psi\urcorner)}\text{R}\to}{\Rightarrow Val(\ulcorner\varphi\urcorner, \ulcorner\psi\urcorner) \to (Tr(\ulcorner\varphi\urcorner) \to Tr(\ulcorner\psi\urcorner))}\text{R}\to$$

As one of its main assets, Ripley (2013) argues that $\mathbb{ST}$-based solutions to paradoxes have the advantage of keeping classicality as much as possible, because $\mathbb{ST}$ and Classical Logic share the same set of valid inferences. This is different, for example, to the structural logics $\mathbb{K}3$ and $\mathbb{LP}$, which need recapture strategies to recover the validity of classical inferences (see Priest, 2006; Beall, 2013).

The interest of these theories, however, extends beyond being a mere technical tool that delivers the desired results. Frankowski (2004) proposes a definition of consequence relation intended to capture the notion of 'plausible reasoning', *p*-consequence (Definition 3.5.5), which formalizes reasoning where the transmission of certainty from the premises to the conclusion may diminish. According to this idea, an inference is valid if we accept all premises and do not reject the conclusion. The formal counterpart of this definition of consequence may be non-transitive. From a semantic perspective, Cobreros et al. (2015) argues that $\mathbb{ST}$ also formalizes this notion because the strict to tolerant consequence allows the conclusion to be more permissive than the premises.

The logic $\mathbb{ST}$ can also be read according to the bilateralist interpretation. Following the bilateralism presented by Restall (2005), Ripley (2013) proposes to understand the failure of Cut (understood, in turn, as a particular sort of generalization of transitivity) in terms of this reading of the consequence relation. For some premises to entail some conclusions means that if one asserts the premises and denies the conclusion, one is 'out of bounds'. Think about a discursive practice. You are out of bounds if something in the conversational position you adopted clashes with some coherence-based norms of assertion and denial governing this practice. In this vein, Ripley proposes to ask *which norms are encoded by the rule of Cut.*

The answer is that Cut encodes a certain sort of 'extensibility'. This becomes clear when the rule is read from the bottom up. Concretely, if asserting the Γ's and denying the Δ's prevents one from denying or asserting φ without going out of bounds, then that is because one was already out of bounds before φ even appeared. In other words, doing something that rules out denying φ, and other things that rule out asserting it, is problematic: there must always be a space open for any φ, either for asserting it or denying it.

But extensibility can be rejected if there are things that cannot be coherently asserted or denied because they are problematic in themselves, as the Liar sentence. Cut is, therefore, not a basic norm of assertion or denial. Nothing in assertion or denial implies that such things cannot exist. Before turning to the other approaches, we will consider a criticism that can be raised against $\mathbb{ST}$.

As argued before, $\mathbb{ST}$-based solutions have the advantage of keeping classicality at the inferential level. However, Pailos (2020a) argues that $\mathbb{ST}$ fails in being fully classical because some classically valid metainferences are not valid in $\mathbb{ST}$. To overcome this limitation, Pailos introduces the logic $\mathbb{TS}/\mathbb{ST}$ that shares the same set of valid metainferences with Classical Logic (Sect. 4.3.3). But, as he observes, $\mathbb{TS}/\mathbb{ST}$ fails in validating some higher-level instance of Cut, such as meta-Cut. Pailos (2020a) and Barrio et al. (2020a) show that the hierarchy of metainferential logics $\mathbb{CM}_\omega$, defined as the union of every logic $\mathbb{CM}_n$ in the system up to level n, recovers classically valid metainferences progressively (Sect. 4.3.3). The metainferences that $\mathbb{CM}_n$ invalidates are recovered in $\mathbb{CM}_{n+1}$.[13]

Pailos' and Barrio et al.'s proposals face some criticisms in the literature. The first objection is presented by Scambler (2020) and can be stated in the following way. Given a theory **T**, there is a distinction between the recognition of an object-language validity, and the acceptance of the corresponding inferential pattern as a valid rule in **T**, that would allow one to attain certain consequences given some premises. Although $\mathbb{CM}_\omega$ validates Cut, every $\mathbb{CM}_n$ invalidates higher-level instances of this metarule. For this reason, Scambler argues that $\mathbb{CM}_\omega$ is not closed under its principles. One is then pushed to admit that the rules said to be valid are not being followed in the sense of being used. So one cannot, after all, reason fully with such a logic. This is the case with transitivity principles in Cut-free theories. Along similar lines, Porter (2023) and Priest (2024) argue that the consequence relation must be structural because structural rules are what enable, among other things, cumulative reasoning. That is, the kind of reasoning that permits exploring consequences. Now, precisely this is affected in a non-transitive logic.

Pailos and Da Ré (2023) have recently answered the challenge from the viewpoint of $\mathbb{CM}_\omega$. They argue that these objections would work only if $\mathbb{CM}_\omega$ invalidated every form of Cut. However, this is not the case. There are indeed calculi designed to capture some valid restricted versions of Cut. Cobreros et al. (2022) presented a labeled sequent calculus capturing every metainferential logic based on the Strong Kleene logics. For each formula in a language $\mathcal{L}$, a 'labeled formula'

[13] $\mathbb{CM}_\omega$ is extended with the validity predicate in (Barrio & Bezerra, 2024; Barrio et al., 2025).

is an expression $x : \varphi$, where $x \in \{s, t\}$, and $x : \Gamma$ (resp., $x : \Delta$) are sets of labeled formulas. The labels work as metalinguistic devices purporting to represent the different ways a valuation satisfies a given formula: strictly or tolerantly. Let us now turn to the form of a sequent. In this approach, the general form of a sequent is:

$$s : \Gamma_1, t : \Gamma_2 \Rightarrow s : \Delta_1, t : \Delta_2$$

And a valuation satisfies a sequent just in case if $v \models^s \gamma$ for all $\gamma \in \Gamma_1$ and $v \models^t \gamma$ for all $\gamma \in \Gamma_2$, then $v \models^s \delta$ for some $\delta \in \Delta_1$ or $v \models^t \delta$ for some $\delta \in \Delta_2$. A sequent is valid if and only if it is satisfied in every valuation. The authors thus provide the calculus **G3SK**, where initial sequents and operational rules work taking into account the mentioned labels. This calculus is shown to be sound and complete regarding the interpretation of Strong Kleene sequents.[14]

It turns out that, in the given interpretation of sequents, we have the following form of Cut:

$$\frac{\Gamma \Rightarrow \Delta, x : \varphi \qquad x : \varphi, \Gamma' \Rightarrow \Delta'}{\Gamma, \Gamma' \Rightarrow \Delta, \Delta'}$$

Still, the corresponding $\mathbb{ST}$ form of Cut will not be admissible regarding the interpretation provided for Strong Kleene sequents. So, the following will not hold:

$$\frac{\Gamma \Rightarrow \Delta, t : \varphi \qquad s : \varphi, \Gamma' \Rightarrow \Delta'}{\Gamma, \Gamma' \Rightarrow \Delta, \Delta'}$$

Cobreros et al. generalize the notion of inference to higher levels and introduce the extension $\mathbf{G3SK}^{\omega}$, which admits still higher-level forms of Cut. Through such a generalization, this labeled sequent calculus can capture $\mathbb{CM}_{\omega}$. As Pailos and Da Ré point out, there are different rules valid in it, and different versions of Cut in particular, making the system transitive to some extent. Hence, the objection which says that it is not possible to reason with $\mathbb{CM}_{\omega}$ seems off the mark.

Having considered Cut-free logics, we will now deal with Identity-free logics.

6.2.3.2 Identity-free Approach

The non-reflexive solutions we will treat here are those investigated by French (2016) and Nicolai and Rossi (2023). French presents non-reflexive solutions based on the sequent calculi $\mathbf{LK}^{-i}$ and $\mathbf{LK}^{-i}_{+E}$ (Sect. 4.1.1). We will present French's approach by focusing on his reading of sequents, as well as considering some independent motivations he introduces for dropping Identity. Nicolai and Rossi base their solution on **G3c**, particularly on the non-reflexive fragment of this system. We

[14] The details of **G3SK** are presented in Cobreros et al. (2022).

will briefly compare Nicolai and Rossi's proposal with Murzi and Rossi's approach (Murzi & Rossi, 2021).

According to French (2016), the source of the problems concerning the derivations of the paradoxes lies in their very beginning, their point of departure. Inspired by the notion of q-consequence given by Malinowski (Definition 4.4.1), French proposes to read a sequent $\Gamma \Rightarrow \Delta$ as telling us that if we do not reject all members in the premises, then we should accept some member of the conclusion. In a nutshell, in this reading, Identity imposes that if a sentence is not rejected, it should be accepted. But accordingly, Identity will fail for any sentence that one should neither reject nor accept, which, according to French, is the case with paradoxical sentences. So semantic theories based on non-reflexive systems, such as $\mathbf{LK}^{-i}$ and $\mathbf{LK}^{-i}_{+E}$, are suitable to block these problematic results.

As one can see, the derivations of the paradoxes presented above begin with instances of Identity containing the paradoxical sentences. According to the interpretation just mentioned, however, we have no reason to accept that paradoxical sentences are the kind of statements we must either accept or reject. This holds for the paradoxes involving the truth predicate, as the Liar and Curry, but also for the validity paradox. So, the corresponding derivations for the paradoxes will not be available, since they cannot even begin.

Nicolai and Rossi (2023) also present a non-reflexive theory to deal with paradoxes. The theory **LPC** extends the arithmetical language with an 'object-linguistic consequence' predicate C, thus obtaining the language $\mathcal{L}_1^C$. Following the methodology of our chapter, **LPC** will be presented in the extended propositional language:

Definition 6.2.1 The system **LPC** in $\mathcal{L}_C^+$ contains the following initial sequents and rules where Γ and Δ are finite multisets of formulas of $\mathcal{L}_C^+$:

$$\text{Id}^- \frac{}{\varphi \Rightarrow \varphi}$$

The rule Id^- has the following restriction: $\varphi \in Var - Var^*$. The rules for conjunction and disjunction are the same as **G3c** (Nicolai & Rossi, 2023), plus the following rules:

$$\top \frac{}{\Gamma \Rightarrow \top, \Delta} \qquad \bot \frac{}{\Gamma, \bot \Rightarrow \Delta}$$

$$\text{L}Cp \frac{\Gamma \Rightarrow \Delta, \varphi \qquad \psi, \Gamma \Rightarrow \Delta}{\Gamma, C(\ulcorner \varphi \urcorner, \ulcorner \psi \urcorner) \Rightarrow \Delta} \qquad \text{R}Cp \frac{\Gamma, \varphi \Rightarrow \psi, \Delta}{\Gamma \Rightarrow C(\ulcorner \varphi \urcorner, \ulcorner \psi \urcorner), \Delta}$$

Let $\rho \in Var^*$, and let the following rules allow the intersubstitutibity of ρ with $C(\ulcorner \rho \urcorner, \ulcorner \bot \urcorner)$ (CpI):

$$\text{LCpI}\ \frac{\Gamma, \rho \Rightarrow \Delta}{\Gamma, C(\ulcorner\rho\urcorner, \ulcorner\bot\urcorner) \Rightarrow \Delta} \qquad \text{RCpI}\ \frac{\Gamma \Rightarrow \rho, \Delta}{\Gamma \Rightarrow C(\ulcorner\rho\urcorner, \ulcorner\bot\urcorner), \Delta}$$

Although the rules LCp and RCp differ from VD and VP, it is possible to prove the same problematic results for the predicate C if no restriction on the structural rules is made. The proof of this triviality result is left as an exercise (Exercise 107). In our framework, it is easy to see how **LPC** blocks the paradox involving C because Id does not hold for atomic formulas that belong to Var^*.

Concerning the arithmetical language $\mathcal{L}_1^C$, Nicolai and Rossi argue that the predicate C cannot be equated with the provability predicate. In arithmetical theories, one can prove the following statement, known as Löb's Theorem:

Löb's Theorem Let **Ar** be a first-order arithmetical theory and $\vdash$ its consequence relation. If $\mathbf{Ar} \vdash Prov(\ulcorner\varphi\urcorner) \rightarrow \varphi$, then $\mathbf{Ar} \vdash \varphi$.

Informally speaking, statements of the form $Prov(\ulcorner\varphi\urcorner) \rightarrow \varphi$ say that provable arithmetical sentences are all true. Löb's Theorem states that the theory can prove $Prov(\ulcorner\varphi\urcorner) \rightarrow \varphi$ only in the trivial case where φ is already provable. On the other hand, it is easy to show that $\Rightarrow C(\ulcorner\top\urcorner, \ulcorner\varphi\urcorner) \rightarrow \varphi$ is provable in **LPC** for $\varphi \in Var - Var^*$. For this reason, C does not collapse with the Arithmetical provability predicate.

The non-reflexive solutions to the semantic paradoxes were mainly explored from the semantic point of view. The semantics for **LPC** is provided by valuations in V^{SK} with the *ts*-consequence relation, and it is combined with a generalization of Kripke's fixed-point semantics (Kripke, 1975).

Murzi and Rossi (2021) also propose a validity theory whose semantics is provided by V^{SK} and the *ts*-consequence relation. The difference between Murzi and Rossi's proposal on the one hand, and Nicolai and Rossi's on the other, is that the former keeps the rules VP and VD of the validity predicate.[15]

Let us now turn to some philosophical motivations and readings for non-reflexivity. As an historical remark, it is interesting to notice that the non-reflexivity of the consequence relation is foreseen in ancient logic. Following Duncombe (2014), it seems to be involved in Aristotle's definition of deduction in his 'Prior Analytics'. According to this idea, the conclusion of a deduction must differ from the premises, and thus Identity will fail.

Now, as mentioned, French (2016) analyses different cases that are partly independent of the paradoxes as reasons for dropping Id. To begin with, if we take at face value the idea that, in a valid argument, the conclusion is true 'in virtue of' the premises being true (an idea used in explaining logical consequence, although in an informal way), then Identity will fail, because the relation 'in virtue of' is irreflexive.

[15] Meadows (2014) also proposes a fixed-point semantics for the validity predicate that results in a non-reflexive semantic theory. His theory, however, is weaker than Nicolai and Rossi's and Murzi and Rossi's because Meadows' proposal does not validate transitivity and monotonicity.

French further considers Swyneshed approach to 'Insolubilia' (Spade, 1979). In particular, he asks to consider the following inference:

The conclusion of this inference is false.

$\therefore$ The conclusion of this inference is false.

According to Swyneshed, this inference has a true premise and a false conclusion. So, it constitutes a counterexample to Identity.

Consider, next, the case of Heterogeneous Logics defined in Humberstone (1988) as logics where premises and conclusions belong to different (formal) languages. To exemplify heterogeneous inferences, French considers a natural language case where premises and conclusion are assessed relative to different dialects: premises are assessed as true relative to Australian English, and the conclusion relative to British English. An argument $\Gamma \Rightarrow \Delta$ is valid whenever if all sentences in Γ (Australian English) are true, then at least one of the sentences in Δ (British English) is true. Then, the inference from 'John lost a thong at the beach' to 'John lost a thong at the beach' will provide a counterexample to Identity when John lost a sandal, but did not lose any swimwear.

Philosophically, it is interesting to consider how non-reflexivity impacts on the understanding of metainferences. We have considered how non-reflexive logics deal with semantic paradoxes. There seems to be nothing wrong, according to logicians who support non-reflexivity, with the reasoning that engenders paradox. The problem is rather with where derivations begin, in asserting sequents like $\lambda \Rightarrow \lambda$. In this sense, instances of Identity that cause trouble are regarded as logically defective.

Concerning metainferences, this interpretation says that, if we take a stance on each premise-sequent, then, if we do not reject all the premises in the conclusion sequent, we should accept some member of the conclusion. The idea is that a metainference makes explicit the fact that, when judging an inference to be valid, its components are being assessed as the kind of statements for which it is allowed to take an attitude. According to French, this allows to recapture classical reasoning as being enthymematic. Consider the metainference with premises $p \Rightarrow p, q \Rightarrow q$ and conclusion $p, q \Rightarrow p \wedge q$. This tells us that, conditional on the supposition that we have taken an attitude towards p and q, if we do not reject p and q, then we should accept $p \wedge q$.

Having considered non-reflexive logics, we will now deal with Contraction-free and Weakening-free logics.

Exercises

Exercise 107 Let **LPC*** be a semantic theory that differs from **LPC** only with respect to the rule Id, which holds for every atomic formula. Show that **LPC*** is trivial.

6.2.3.3 Contraction-free and Weakening-free Approaches

We finally get to Contraction-free and Weakening-free approaches to paradox. These logics have been presented in different versions in Chap. 5. Even though there is a vast tradition in logic and computer science that works without Contraction, many non-contractive logics are not adequate as solutions to the semantic paradoxes due to their strong expressiveness. For example, Łukasiewicz logic $Ł_3$ (Sect. 5.3.1.1) defines classical negation $\sim$ and biconditional $\leftrightarrow$ as follows:

$$\begin{aligned} \sim\varphi &:= \varphi \rightarrow \neg\varphi \\ \varphi \leftrightarrow \psi &:= (\varphi \rightarrow \psi) \wedge (\psi \rightarrow \varphi) \end{aligned}$$

Therefore, it cannot be a basis for a semantic theory, because the following stronger Liar sentence λ^+ is definable:

$$\lambda^+ \leftrightarrow \sim Tr(\ulcorner \lambda^+ \urcorner)$$

And such a sentence would trivialize the theory.[16]

Nonetheless, non-contractive solutions to paradoxes are not new in the literature and, as we will see, they have found many advocates.

It is fair to say that the situation is radically different with approaches that reject Weakening. As Dutilh Novaes and French (2018) observe, Weakening does not play a prominent role in the derivation of the semantic paradoxes and, for this reason, it has not been investigated in depth in the literature on these topics. Furthermore, several issues we will be covering here, which were discussed in the other approaches, have not been dealt with from a non-monotonic perspective. In short, a lot is still to be done. Given the nature of this chapter, we will not focus on this kind of approach, with one notable exception.

Da Ré (2021) shows that it is possible to avoid the paradoxes mentioned by restricting Weakening. As we saw in Sect. 5.3.2, Da Ré presents the non-monotonic calculus **CLNW**. **CLNW** validates Id, Contraction, and a context-sharing version of Cut. How does a semantic theory based on **CLNW** block the paradoxes involving the truth predicate? As shown in Sect. 5.3.2, **CLNW** invalidates a form of Meta Explosion. Then the derivation of the Liar halts when it gets to the empty sequent. And even though this is bad for most theories, in the absence of Weakening it does not imply triviality. The derivation of the Curry paradox cannot be carried out because **CLNW** invalidates any version of MP. Lastly, since **CLNW** invalidates inferences that have more than two formulas (Sect. 5.3.2), one can also add a naïve validity predicate to the semantic theory based on **CLNW** because the applications

[16] The same would happen if one were to add a classicality operator in $\mathbb{ST}$. As Barrio et al. (2020b) show, the presence of this operator trivializes the semantic theory based on $\mathbb{ST}$. To avoid the triviality result, one must adopt a weak self-reference principle.

of Cut in the validity paradox involve more than two formulas. Therefore, **CLNW** is a candidate for a non-trivial theory of truth and validity.

We now turn to Contraction-free approaches. There is a much wider variety of candidates in this realm. Some notable examples that we will not focus on here, because they are not based on logics presented in Chap. 5, are Restall's and Weber's. Restall (1993, 1994) presents logics that block the metarule of Contraction as solutions to set-theoretical paradoxes and paradoxes involving the truth predicate. Weber (2014) presents a validity theory characterized through a fixed-point semantics, which combines a stratified notion of validity and a naïve one.

Zardini (2014) proposes the non-contractive theory of truth **IKT** ('I' stands for the Identity axioms, 'K' for Weakening and 'T' for the truth predicate). The logical theory that bases **IKT** is the multiplicative fragment of **CLL** (Sect. 5.2.1). Additionally, **IKT** contains the characteristic rules for the truth predicate. Since **IKT** does not have the metarule of Contraction, it blocks the paradoxes of truth. Concerning the validity predicate, he shows that **IKT** is consistent with the following two rules:

$$\text{L}Val\ \frac{\Gamma_0 \Rightarrow \Delta_0, \varphi \qquad \psi, \Gamma_1 \Rightarrow \Delta_1}{\Gamma_0, \Gamma_1, Val(\ulcorner\varphi\urcorner, \ulcorner\psi\urcorner) \Rightarrow \Delta_0, \Delta_1} \qquad \text{R}Val\ \frac{\Gamma, \varphi \Rightarrow \Delta, \psi}{\Gamma \Rightarrow \Delta, Val(\ulcorner\varphi\urcorner, \ulcorner\psi\urcorner)}$$

where Γ and Δ in RVal are Val-logical, i.e., containing only formulas of the form $Val(x, y)$, or negations, conjunctions, disjunctions or implications of these formulas. By giving up the metarule of Contraction, from $\pi, \pi \Rightarrow \bot$, we cannot infer $\pi \Rightarrow \bot$, and so the paradox is blocked.

Rosenblatt (2019) also proposes a non-contractive approach to semantic paradoxes. As we saw in Sect. 5.3.1.2, he considers **G3c** as the basic system of his semantic theory of truth. As he shows, the addition of the rules for the truth predicate makes Contraction no longer admissible. Since Contraction is present in all the derivations of the semantic paradoxes, Rosenblatt's proposal blocks uniformly the semantic paradoxes of truth and validity.

Rosenblatt offers a model-theoretic characterization for his non-contractive logic (Definitions 5.3.2 and 5.3.3) as well. With these models, he shows that "whenever valid inferences are concerned, these models yield Classical Logic" (Rosenblatt, 2019, pg. 5). So, in this sense, his theory does not give up Contraction completely. Contraction fails whenever the paradoxical sentences are around.

One could charge against Rosenblatt's proposal that it has counterexamples to Cut. This means that his proposal is non-contractive and Cut-free. So, one could accuse Rosenblatt's proposal of being too costly. Why should one adopt a non-contractive and Cut-free semantic theory instead of one that only dispenses Contraction? Rosenblatt (2021b) argues that all non-contractive approaches to semantic paradoxes dispense some form of Cut. Moreover, following Ripley (2015)'s remarks, he argues that there are formulations of Cut that the non-contractivist should not accept. Particularly, the formulations involving context-sharing premise-sequents:

$$\frac{\Gamma \Rightarrow \varphi \qquad \Gamma, \varphi \Rightarrow \psi}{\Gamma \Rightarrow \psi}$$

As we can see, the above formulation of Cut incorporates Contraction. Thus, Rosenblatt concludes that the failure of Cut should not be surprising for the non-contractivist.[17]

We now turn to the philosophical interpretations of the failure of Contraction. Zardini's approach is based on the idea that, to find an answer to the failure of Contraction, one must step out from formal theories and deal with the metaphysics of truth. The failure of Contraction of a given sentence is caused by the metaphysical property of instability corresponding to certain states of affairs. Take $\neg Tr(\ulcorner\lambda\urcorner)$ as an example. If the corresponding state of affairs holds, by the rules of introduction of Tr, it will lead to the one expressed by $Tr(\ulcorner\lambda\urcorner)$. Notice that the latter is incompatible with the former, to the point that if the latter obtained, the former would not, even though it was just obtaining $\neg Tr(\ulcorner\lambda\urcorner)$ that led to obtaining $Tr(\ulcorner\lambda\urcorner)$. A dual process takes place for the state of affairs expressed by $Tr(\ulcorner\lambda\urcorner)$. According to Zardini, this is just the instability of the first state of affairs. Stable states of affairs are those that would co-exist with all of their consequences, while unstable ones would not. In turn, it is instability that causes failures of Contraction. Counting with $\neg Tr(\ulcorner\lambda\urcorner)$, $\neg Tr(\ulcorner\lambda\urcorner)$, Contraction would imply $\neg Tr(\ulcorner\lambda\urcorner)$. However, this would lead to the state of affairs expressed by $Tr(\ulcorner\lambda\urcorner)$ and would co-obtain with it. Given the instability of the state of affairs expressed by $\neg Tr(\ulcorner\lambda\urcorner)$, this cannot be the case. So, the failure of Contraction in a logical system is a symptom of an unstable metaphysics.

As Zardini acknowledges, the picture sketched is certainly needed of further elaboration and defense. We will not deal with this issue here, but we notice a couple of things. It would be nice to have a more developed picture of what the corresponding state of affairs of a paradoxical sentence such as the Liar would be like, and whether there are other unstable states of affairs. Furthermore, granting the idea of instability, it is still not entirely clear why would such instability cause the failure of Contraction.

As Rosenblatt (2021b) suggested, when one assesses theories with an underlying non-contractive logic, it is important to look at the instances of Contraction that each theory rejects. And it seems that, in the face of a theory rejecting Contraction in full generality, one would be better by rejecting just problematic cases and keeping completely harmless ones. It is not easy to see why, while one token of a non paradoxical sentence has certain consequences, two of them will have different ones (Scharp, 2013).

To get clarification, we turn to another philosophical interpretation of the failure of Contraction: the resource interpretation. Advocated by Mares and Paoli (2014),

[17] It is worth observing that giving up transitivity is not a problem for a non-contractive upholder such as Zardini. As he observes, although he favors a non-contractive logic for semantic paradoxes, failures of Cut appear when dealing with vagueness (Zardini, 2014).

the main idea is that, when someone says that some conclusion follows from the premises, different things are being said. One of them is that, given certain rules of logic, one can extract the information in the conclusion from the information provided by the premises. Understood in this way, premises are concrete resources, and these can be used a specific number of times. This gives independent reasons to reject Contraction, but we would like to have an appropriate explanation of why one cannot contract specifically on paradoxical sentences.

Mares and Paoli add that paradoxes can be understood as fallacies of equivocation. This means there is a distinction to make between multiplicative and additive junctions which, in turn, will turn into certain inferential patterns failing in the absence of Contraction. But, as Rosenblatt argues, there seems to be no reason to believe that the ambiguities are caused by the paradoxical sentences. In turn, he presents an alternative approach.

Rosenblatt (2021b)'s proposal gives a well-defined answer to why Contraction fails, specifically why it does so in the case of paradoxes. Focusing on the Liar, his point is that by allowing structural Contraction to fail, we can make sense of the idea that it is incoherent to accept that the Liar sentence is true and untrue, but that there's nothing incoherent with accepting each of the conjuncts. Note that none of $\neg Tr(\ulcorner\lambda\urcorner)$ and $Tr(\ulcorner\lambda\urcorner)$ are contradictory in themselves. It is not the same to assert one of them, even when it could lead us to disaster, as to assert the contradiction $Tr(\ulcorner\lambda\urcorner) \wedge \neg Tr(\ulcorner\lambda\urcorner)$. The distinction between claiming that the Liar sentence is true (or that the Liar sentence is untrue) and the stronger claim that the Liar sentence is both true and untrue can be maintained if structural Contraction is avoided.

As we have seen up to this point, for each of the structural properties there may be different idiosyncratic readings to make sense of their failures. However, as structural properties, they are still the same kind of thing. So, it would be interesting to consider whether their failures can be explained in a unified way. In the next section, we turn to a philosophical interpretation that seeks to match that goal.

Exercises

Exercise 108 If Weakening is not a valid metarule, according to non-monotonic solutions, the empty sequent is not problematic. Consider whether there is a conceptual reading for the provability of the empty sequent.

6.2.3.4 A (Not So) Common Framework: Dialogical Interpretation

The availability of justifications for the structural rules gives us independent reasons for adopting substructural logics, besides blocking semantic paradoxes. In this section, we analyse the failure of these rules from a unifying dialogical perspective. Although this unifying perspective falls short of being all-encompassing, it provides an interesting approach to substructurality.

According to Sorensen (2003), Greek mathematical practice is proof-centered and deductive. For the Pythagoreans, the proof of a mathematical statement should be public to make it possible for the target audience to check each step of the reasoning in the proof. According to Novaes (2016), persuasion plays a fundamental role in the public character of the demonstration, demanding the prover to justify the deduction. In this sense, deductions internalize a dialogical component because the prover must show to an idealized skeptic that each step of her deduction necessarily follows the previous reasoning (Dutilh Novaes & French, 2018). That is, the prover must ensure that her deduction is not amenable to the counterexamples of the skeptic. The reasoning is invalid if the skeptic provides a counterexample to the prover's deduction. If the skeptic cannot do so, the argument is valid.

On the account put forward by Dutilh Novaes and French (2018), sequents might be understood as expressing incoherent discursive positions. That is, a sequent says that the joint assertion of each member of the premises precludes the denial of each member of the conclusion. In a dialogical setting, this amounts to the idea that one should not accept all the premises and refuse to grant part of the conclusion. So, the prover will ask the skeptic to concede the propositions in the premises, and show them they must accept some proposition that follows from what has been conceded. The prover wins the dialogue if the only sequents that can be challenged by the skeptic are of the form $\varphi \Rightarrow \varphi$. Furthermore, dialogues are, to different degrees, adversarial or cooperative. In an adversarial context, the prover is looking to force the conclusion. In a cooperative one, the aim is for the skeptic to understand why the conclusion is accepted. Deductive proof has a conversational, semi-adversarial, dialogical basis, since it has a built-in skeptic asking for each inferential step to be immune to counterexamples, evident, and convincing.

In this setting, there are different considerations related to the structural rules and their failures. According to the dialogical interpretation, Id means that if one grants φ as a premise, then one must grant φ as a conclusion. Left Contraction says that if one grants the argument having two copies of φ in the premises to the conclusion ψ, one has to accept the argument having one occurrence of φ in the premises. Left Weakening says that, if one accepts an argument from ψ to γ, one cannot reject the conclusion γ in adding an additional premise δ. Finally, Cut refers to the possibility of concatenating arguments in order to generate a new one. Although these structural rules make sense in a dialogical framework, Dutilh Novaes and French (2018) argue that there are reasons to challenge some of them. We now turn to such reasons.

According to the authors, the validity of Identity can be challenged in a multi-agent situation because it can be seen as uninformative. If the prover asks the skeptic to grant φ as a premise, it is to be expected that something different from φ should be accepted as a conclusion. In a dialogical context, the prover wants to convince the skeptic about the consequences of her assumptions. Since Id allows the conclusion to be the same as the premise, it could be rejected according to this interpretation of logical consequence. Moreover, dialogical contexts may be sensitive to the repetition of a proposition φ. If a person repeats φ multiple times in this context, it may be the case that she is more committed to φ than if it were stated once. Thus, in a dialogical interpretation of the logical consequence, it makes sense to restrict Contraction. Among the rules, it seems as if only the restriction of Cut would violate a basic characteristic of argumentation, namely, the possibility of concatenating arguments. So, according to Dutilh Novaes and French (2018), it is difficult to find a reason for restricting Cut in dialogical contexts.

Dutilh Novaes and French approach is proof-theoretical. A semantic approach to this interpretation can also help to make sense of the restriction of these structural rules (Clerbout & McConaughey, 2022). In (Álvarez Lisboa, 2023) we find a dialogical semantics for non-reflexive and Cut-free theories of truth. In this paper, Álvarez Lisboa shows that Cut-free logics are characterized by games where the prover wins if and only if the skeptic does not make the last move. Since we relied on a sequent-style presentation of the semantic theories, unfortunately we will not develop this proposal in full.

Exercises

Exercise 109 Does the failure of Cut make sense within a dialogical framework? If so, explain how Cut may fail.

6.3 Internalization of Valid Metainferences

Being immune to the paradoxes and counting with a philosophical reading is not sufficient. Theories of truth and validity are expected to have other virtues as well. As Barrio et al. (2021) and Rosenblatt (2017) argue, it is particularly important to pose the question on whether substructural logics can derive claims expressing that a given valid metarule is, in fact, validity preserving. Given that naïve validity stands for a more general notion of validity, internalization means that the valid metarules are valid in this general sense. We will call this the 'Internalization Problem', and this section will be devoted to this topic.

Recall the distinction between the two approaches to semantic theories. One view, paradigmatically expressed by Tarski, proposes that theories intending to explain how semantic concepts of a given language work should be given in richer languages. An opposing view holds that a semantic theory should be couched in a language with enough resources to express its own concepts. Hence, there is no need to look for a distinction between language levels. This last line of thought looks for a naïve approach.

As Rosenblatt (2017) explains, a naïve theory of validity would amount to a theory capable of truthfully talking about its own notion of validity. Consequently, a validity predicate for a given **T** is naïve only if, for every **T**-provable argument from Γ to Δ, **T** proves the statement expressing that the argument from the former to the latter is naïvely valid. In other words, **T** also proves that $Val(\ulcorner \bigwedge \Gamma \urcorner, \ulcorner \bigvee \Delta \urcorner)$. However, the problem begins when turning to metainferences.

Substructural theories are generally characterized in terms of their metarules. In light of this, it seems appropriate to call for a system **T** to prove internalized versions of all its correct metarules. The idea is that for a theory of validity to be adequate, it should prove that its metarules are validity preserving.

Definition 6.3.1 A theory **T** *internalizes* a metarule R of the form

$$\text{R}\frac{\Gamma_1 \Rightarrow \Delta_1, \ldots, \Gamma_n \Rightarrow \Delta_n}{\Gamma \Rightarrow \Delta}$$

If **T** proves every instance of

$$\Rightarrow (Val(\ulcorner \bigwedge \Gamma_1 \urcorner, \ulcorner \bigvee \Delta_1 \urcorner) \wedge \ldots \wedge Val(\ulcorner \bigwedge \Gamma_n \urcorner, \ulcorner \bigvee \Delta_n \urcorner))$$
$$\rightarrow Val(\ulcorner \bigwedge \Gamma \urcorner, \ulcorner \bigvee \Delta \urcorner).$$

Now we will analyse how well the substructural solutions presented above deal with the internalization of valid metainferences, starting with Cut-free solutions. Barrio et al. (2021) show that **STTV** fails at internalizing the rule of left negation:

$$\text{L}\neg\frac{\Gamma \Rightarrow \varphi, \Delta}{\Gamma, \neg\varphi \Rightarrow \Delta}$$

The reason for this is that $\Rightarrow Val(\ulcorner \bigwedge \Gamma \urcorner, \ulcorner \varphi \vee \bigvee \Delta \urcorner) \rightarrow Val(\ulcorner \bigwedge \Gamma \wedge \neg\varphi \urcorner, \ulcorner \bigvee \Delta \urcorner)$ is not a theorem of **STTV**. The fact that VP is context-free makes it impossible to prove this latter formula. They show that to obtain an internalization result, one needs to strengthen the rule somehow, for instance by allowing contexts:

$$\text{VP}^+\frac{Val(\ulcorner \bigwedge \Gamma_1 \urcorner, \ulcorner \bigvee \Delta_1 \urcorner), \ldots, Val(\ulcorner \bigwedge \Gamma_n \urcorner, \ulcorner \bigvee \Delta_n \urcorner), \Gamma \Rightarrow \Delta}{Val(\ulcorner \bigwedge \Gamma_1 \urcorner, \ulcorner \bigvee \Delta_1 \urcorner), \ldots, Val(\ulcorner \bigwedge \Gamma_n \urcorner, \ulcorner \bigvee \Delta_n \urcorner) \Rightarrow Val(\ulcorner \bigwedge \Gamma \urcorner, \ulcorner \bigvee \Delta \urcorner)} \quad (6.1)$$

Let **STTV**$^+$ be the theory obtained when putting VP$^+$ in the place of VP. As Barrio et al. (2021) show, **STTV**$^+$ internalizes all of its primitive and derivable metarules, but it internalizes an instance of Cut, as the following derivation illustrates:

$$\text{R}\rightarrow\frac{\text{L}\wedge\frac{\text{VP}^+\frac{\text{LW}\frac{\text{LC}\frac{\text{VD}\frac{}{Val(\ulcorner \pi \urcorner, \ulcorner \bot \urcorner), \pi \Rightarrow \bot}}{Val(\ulcorner \pi \urcorner, \ulcorner \bot \urcorner) \Rightarrow \bot}}{Val(\ulcorner \top \urcorner, \ulcorner \pi \urcorner), Val(\ulcorner \pi \urcorner, \ulcorner \bot \urcorner), \top \Rightarrow \bot}}{Val(\ulcorner \top \urcorner, \ulcorner \pi \urcorner), Val(\ulcorner \pi \urcorner, \ulcorner \bot \urcorner) \Rightarrow Val(\ulcorner \top \urcorner, \ulcorner \bot \urcorner)}}{Val(\ulcorner \top \urcorner, \ulcorner \pi \urcorner) \wedge Val(\ulcorner \pi \urcorner, \ulcorner \bot \urcorner) \Rightarrow Val(\ulcorner \top \urcorner, \ulcorner \bot \urcorner)}}{\Rightarrow (Val(\ulcorner \top \urcorner, \ulcorner \pi \urcorner) \wedge Val(\ulcorner \pi \urcorner, \ulcorner \bot \urcorner)) \rightarrow Val(\ulcorner \top \urcorner, \ulcorner \bot \urcorner)}$$

Since **LK**$^{-c}$ is a Cut-free logical system, **STTV**$^+$ can be accused of saying false things about its base logical system. That is, **STTV**$^+$ is unsound concerning $\mathbb{ST}$. It can be said, in defense of **STTV**$^+$, that such an overinternalization is not a problem at all, because it does not trivialize the theory (Ripley, 2021). However, as Barrio et al. (2021) argue, accepting VP^+ instead of VP not only implies abandoning the idea that the validity predicate should only capture the valid (meta)inferences of the logical theory. Even worse, as it is well known, it is the presence of sentences like λ and π what makes Cut invalid in $\mathbb{ST}$. Thus, it is not clear the sense in which $\mathbf{STTV}^+$ constitutes an adequate response to the internalization problem.[18]

As we can see, $\mathbb{ST}$-based semantic theories face a difficult situation concerning their validity predicate. If one keeps a validity predicate obeying the rules VP and VD, the theory fails at internalizing the valid metarules of $\mathbb{ST}$. On the other hand, the most straightforward strengthening of the rules governing Val overinternalizes metainferences, such as Cut. According to Hlobil (2019), it is possible to overcome this difficult situation if one adopts a 'faithful' validity predicate in the following sense:

Definition 6.3.2 (Hlobil (2019)) A validity predicate, Val, is *faithful* just in case that

$$\Rightarrow Val(\ulcorner \bigwedge \Gamma_1 \urcorner, \ulcorner \bigvee \Delta_1 \urcorner), \ldots, Val(\ulcorner \bigwedge \Gamma_n \urcorner, \ulcorner \bigvee \Delta_n \urcorner)$$
$$\Rightarrow Val(\ulcorner \bigwedge \Gamma \urcorner, \ulcorner \bigvee \Delta \urcorner)$$

Is provable iff $\Gamma \Rightarrow \Delta$ follows from $\Gamma_1 \Rightarrow \Delta_1, \ldots, \Gamma_n \Rightarrow \Delta_n$.

Hlobil shows that an $\mathbf{LK}^{-c}_{+E}$-based semantic theory extended with a faithful validity predicate internalizes only the valid meta-inferences of $\mathbb{ST}$. His proposal, on the other hand, has a cost: a faithful validity predicate cannot obey VD, on pain of proving a stronger version of the validity paradox. In his paper, he presents some reasons to dispense of VD. For example, he argues that validity predicates should not obey such a rule because it contradicts Löb's Theorem, since $\top, Val(\ulcorner \top \urcorner, \ulcorner \varphi \urcorner) \Rightarrow \varphi$ is an instance of it.[19] Thus, VP and faithfulness are all we need to have a validity theory based on $\mathbf{LK}^{-c}_{+E}$ that internalizes all valid meta-inferences.

Let us turn to the non-contractive semantic theories. Since Contraction is not needed to internalize non-structural metarules, non-contractive logics such as Affine Logic (Sect. 5.3.1.1) will internalize all its derivable metarules.

Despite this, there are some problems regarding a kind of incompleteness, namely, regarding admissible metarules. For example, Rosenblatt (2017) shows this for the case of $\circ$:

$$\frac{\Rightarrow \varphi}{\Rightarrow \varphi \circ \varphi}$$

[18] Barrio et al. (2021) explore other formulations of the rule VP. As it happens with VP^+, none of them is satisfactory.

[19] Field (2017) argues that VD is incompatible with the Second Incompleteness Theorem and, for this reason, no validity predicate should obey such a rule. However, Field's interpretation (*a fortiori*, Hlobil's one) was criticized in the literature (e.g., Murzi & Rossi, 2021; Bezerra, 2023; Barrio & Bezerra, 2024). These criticisms point out that the concept of validity expressed by the naive predicate is considerably distinct from the arithmetic provability predicate.

Let us briefly spell out the idea. The rule is admissible since, if there is a derivation of $\Rightarrow \varphi$ the sequent can be used twice, yielding a derivation of $\Rightarrow \varphi \circ \varphi$. However, there will be no derivation of $Val(\ulcorner\top\urcorner, \ulcorner\varphi\urcorner) \Rightarrow Val(\ulcorner\top\urcorner, \ulcorner\varphi \circ \varphi\urcorner)$ since, for deriving such a sequent, an application of Contraction is indispensable.

Finally, we turn our attention to non-reflexive semantic theories. The internalization problem for these theories is largely unexplored. The theory **LPC** has Id restricted to atomic sentences that do not belong to Var^*. Thus, **LPC** does not prove every instance of the sequent previously mentioned, because the derivations in this theory starts with Id^-. Then, this restriction makes **LPC** expressively weak with respect to the metarules that it internalizes.

So, it seems that substructural logics face some problems regarding their expressive adequacy. In the non-transitive case, **STTV** is incomplete, because it fails at internalizing some valid metainferences. In the case of $\mathbf{STTV}^+$, the theory turns out to be unsound because it internalizes Cut. In the non-contractive case, the theory turns out to be incomplete. Finally, in the non-reflexive case, **LPC** does not prove every instance of the sequent required in the Definition 6.3.1.

Exercises

Exercise 110 Show that Contraction is indispensable for deriving $Val(\ulcorner\top\urcorner, \ulcorner\varphi\urcorner) \Rightarrow Val(\ulcorner\top\urcorner, \ulcorner\varphi \circ \varphi\urcorner)$.

6.4 Conclusion

In this chapter, we provided a philosophical perspective on the substructural logics presented in this book. Concretely, we considered non-transitive, non-reflexive, non-contractive and non-monotonic solutions to semantic paradoxes of truth and validity. We showed how these logics can be conceptually motivated, and their philosophical significance. The logics presented in this book proved to be fruitful in many ways. The aim was to provide a gentle introduction to substructural logics and some of their merits.

Bibliography

Álvarez Lisboa, M. (2023). Outline of a dialogical theory of truth. *Crítica, 55*(164), 3–25.

Andrade-Lotero, E., & Novaes, C. (2012). Validity, the squeezing argument and alternative semantic systems: The case of aristotelian syllogistic. *Journal of Philosophical Logic, 41*(2), 387–418.

Barrio, E. A. (2014a). *La lógica de la verdad*. Eudeba.

Barrio, E. A. (2014b). *Paradojas, paradojas y más paradojas*. College Publications.

Barrio, E. A., & Bezerra, E. (2024). The logics of a universal language. *Asian Journal of Philosophy, 3*(1), 12.

Barrio, E. A., & Da Ré, B. (2018) Paraconsistency and its philosophical interpretations. *The Australasian Journal of Logic, 15*(2), 151–170.

Barrio, E. A., Pailos, F., & Szmuc, D. (2017). A paraconsistent route to semantic closure. *Logic Journal of the IGPL, 25*(4), 387–407.

Barrio, E. A., Pailos, F., & Szmuc, D. (2020a). A hierarchy of classical and paraconsistent logics. *Journal of Philosophical Logic, 49*(1), 93–120.

Barrio, E. A., Pailos, F., & Szmuc, D. (2020b). A recovery operator for nontransitive approaches. *The Review of Symbolic Logic, 13*(1), 80–104

Barrio, E. A., Rosenblatt, L., & Tajer, D. (2021). Capturing naive validity in the cut-free approach. *Synthese, 199*(3), 707–723.

Barrio, E. A., Bezerra, E., & Da Ré, B. (2025a). Philosophical Interpretations Matter. *Principia: An International Journal of Epistemology, 29*(2), 7.

Barrio, E. A., Bezerra, E., & Pailos, F. (2025b). A validity theory for the ST-hierarchy. *Logic Journal of the IGPL*, 1–24. Forthcoming.

Beall, J. (2013). LP^+, $K3^+$, FDE^+, and their 'classical collapse'. *The Review of Symbolic Logic, 6*(4), 742–754.

Beall, J., & Murzi, J. (2013). Two flavors of Curry's paradox. *The Journal of Philosophy, 110*(3), 143–165.

Benito-Monsalvo, C. (2022). Local applications of logics via model-theoretic interpretations. *Logic and Logical Philosophy, 31*(4), 535–556.

Bezerra, E. (2023). On validity paradoxes and (some of) their solutions. *Principia: An International Journal of Epistemology, 27*(3), 519–538.

Clerbout, N., & McConaughey, Z. (2022). Dialogical logic. In E. N. Zalta, & U. Nodelman (Eds.), *The Stanford encyclopedia of philosophy* (Fall 2022 ed.). Metaphysics Research Lab, Stanford University.

Cobreros, P., Egré, P., Ripley, E., & van Rooij, R. (2012). Tolerant, classical, strict. *Journal of Philosophical Logic, 41*(2), 347–385.

Cobreros, P., Egré, P., Ripley, E., & van Rooij, R. (2015). Vagueness, truth and permissive consequence. In *Unifying the philosophy of truth* (pp. 409–430). Springer.

Cobreros, P., La Rosa, E., & Tranchini, L. (2022). Higher-level inferences in the Strong-Kleene setting: A proof-theoretic approach. *Journal of Philosophical Logic, 51*(6), 1417–1452.

Da Ré, B. (2021). Structural weakening and paradoxes. *Notre Dame Journal of Formal Logic, 62*(2), 369–398.

Duncombe, M. (2014). Irreflexivity and Aristotle's syllogisms. *The Philosophical Quarterly, 64*(256), 434–452.

Dutilh Novaes, C., & French, R. (2018). Paradoxes and structural rules from a dialogical perspective. *Philosophical Issues, 28*(1), 129–158.

Etchemendy, J. (1999). *The concept of logical consequence*. CSLI Publications.

Field, H. (1972). Tarski's theory of truth. *The Journal of Philosophy, 69*(13), 347–375.

Field, H. (1991). Metalogic and modality. *Philosophical Studies, 62*(1), 1–22.

Field, H. (2008). *Saving truth from paradox*. Oxford University Press.

Field, H. (2017). Disarming a paradox of validity. *Notre Dame Journal of Formal Logic, 58*(1), 1–19.

Fitch, F. B. (1964). Universal metalanguages for philosophy. *The Review of Metaphysics*, 396–402.

Frankowski, S. (2004). Formalization of a plausible inference. *Bulletin of the Section of Logic, 33*, 41–52.

French, R. (2016). Structural reflexivity and the paradoxes of self-reference. *Ergo, 3*(5), 113–131.

Goodship, L. (1996). On dialethism. *Australasian Journal of Philosophy, 74*(1), 153–161.

Halbach, V. (2014). *Axiomatic theories of truth*. Cambridge University Press.

Hlobil, U. (2019). Faithfulness for naive validity. *Synthese, 196*(11), 4759–4774.

Humberstone, L. (1988). Heterogeneous logic. *Erkenntnis, 29*(3), 395–435.

Ketland, J. (2012). Validity as a primitive. *Analysis, 72*(3), 421–430.

Kremer, M. (1988). Kripke and the logic of truth. *Journal of Philosophical Logic, 17*, 225–278.

Kripke, S. (1975). Outline of a theory of truth. *Journal of Philosophy, 72*(19), 690–716.

Mares, E., & Paoli, F. (2014). Logical consequence and the paradoxes. *Journal of Philosophical Logic, 43*(2–3), 439–469.

Meadows, T. (2014). Fixed points for consequence relations. *Logique et Analyse, 227*, 333–357.
Murzi, J. (2014). The inexpressibility of validity. *Analysis, 74*(1), 65–81.
Murzi, J., & Rossi, L. (2021). Naïve validity. *Synthese, 199*(3), 819–841.
Myhill, J. (1960). Some remarks on the notion of proof. *The Journal of Philosophy, 57*(14), 461–471.
Nicolai, C., & Rossi, L. (2023). Systems for non-reflexive consequence. *Studia Logica, 111*(6), 947–977.
Novaes, C. D. (2016). Reductio ad absurdum from a dialogical perspective. *Philosophical Studies, 173*(10), 2605–2628.
Pailos, F. (2020a). A fully classical truth theory characterized by substructural means. *The Review of Symbolic Logic, 13*(2), 249–268.
Pailos, F. (2020b). Validity, dialetheism and self-reference. *Synthese, 197*(2), 773–792.
Pailos, F., & Da Ré, B. (2023). *Metainferential logics*. Springer.
Picollo, L. (2020). Alethic reference. *Journal of Philosophical Logic, 49*(3), 417–438.
Porter, B. (2023). *Three essays on substructural approaches to semantic paradoxes*. Ph.D. Thesis, City University of New York.
Priest, G. (2005). *Doubt truth to be a liar*. Clarendon Press.
Priest, G. (2006). *In contradiction*. Oxford University Press.
Priest, G. (2024). Substructural solutions to the semantic paradoxes: a dialetheic perspective. In *Paradoxes Between Truth and Proof (In print)* (pp. 163–190). Springer.
Quine, W. V. O. (1986). *Philosophy of logic*. Harvard University Press.
Restall, G. (1993). How to be really contraction free. *Studia Logica, 52*, 381–391.
Restall, G. (1994). *On logics without contraction*. Ph.D. Thesis, University of Queensland.
Restall, G. (2005). Multiple conclusions. In *Logic, methodology, and philosophy of science: Proceedings of the twelfth international congress* (pp. 189–205). Kings' College Publications.
Ripley, E. (2012). Conservatively extending classical logic with transparent truth. *Review of Symbolic Logic, 5*(2), 354–378.
Ripley, E. (2013). Paradoxes and failures of cut. *Australasian Journal of Philosophy, 91*(1), 139–164.
Ripley, E. (2015). Comparing substructural theories of truth. *Ergo, 2*(13), 299–328.
Ripley, E. (2021). Précis de uncut. *Análisis Filosófico, 41*(2), 235–260.
Rosenblatt, L. (2017). Naive validity, internalization, and substructural aproaches to paradox. *Ergo, an Open Access Journal of Philosophy, 4*.
Rosenblatt, L. (2019). Noncontractive classical logic. *Notre Dame Journal of Formal Logic, 60*(4).
Rosenblatt, L. (2021a). Expressing consistency consistently. *Thought: A Journal of Philosophy, 10*(1), 33–41.
Rosenblatt, L. (2021b). On structural contraction and why it fails. *Synthese, 198*, 2695–2720.
Scambler, C. (2020). Classical logic and the strict tolerant hierarchy. *Journal of Philosophical Logic, 49*(2), 351–370.
Scharp, K. (2013). *Replacing truth*. OUP Oxford.
Smoryński, C. (1985). *Self-reference and modal logic*. Springer Verlag.
Sorensen, R. (2003). *A brief history of the paradox: philosophy and the labyrinths of the mind*. Oxford University Press.
Spade, Paul V. (1979). Roger Swyneshed's Insolubilia: Edition and Comments. *Archives d'histoire doctrinale et littéraire du moyen age 46*, 177–220.
Tarski, A. (1956). *Logic, semantics, metamathematics: Papers from 1923 to 1938*. Oxford Clarendom Press.
Terzian, G. (2015). Norms of truth and logical revision. *Topoi, 34*(1), 15–23.
van Rooij, R. (2011). Vagueness, tolerance and non-transitive entailment. In *Understanding vagueness* (pp. 205–221). College Publications.
Weber, Z. (2014). Naïve validity. *The Philosophical Quarterly, 64*(254), 99–114.
Williamson, T. (2002). *Vagueness*. Routledge.
Zardini, E. (2014). Naive truth and naive logical properties. *The Review of Symbolic Logic, 7*(2), 351–384.

MIX
Papier aus verantwortungsvollen Quellen
Paper from responsible sources
FSC® C105338

If you have any concerns about our products,
you can contact us on
ProductSafety@springernature.com

In case Publisher is established outside the EU,
the EU authorized representative is:
Springer Nature Customer Service Center GmbH
Europaplatz 3, 69115 Heidelberg, Germany

Printed by Libri Plureos GmbH
in Hamburg, Germany